CORROSION INHIBITORS, PRINCIPLES AND RECENT APPLICATIONS

Edited by **Mahmood Aliofkhazraei**

Corrosion Inhibitors, Principles and Recent Applications
http://dx.doi.org/10.5772/intechopen.70101
Edited by Mahmood Aliofkhazraei

Contributors

Shimaa Ali, Khadijah Emran, Hamedh Al Lehaibi, Ojo Sunday Fayomi, Aa Daniyan, Api Popoola, Saraswathy V, Han Seung Lee, Seung Jun Kwon, Subbiah Karthick, Jorge Morales-Hernandez, Araceli Mandujano Ruiz, Luis-Enrique Corona Almazán, Héctor Herrera Hernández, Daniel Véras Ribeiro, Marianna Rivetti, José Andrade Neto, Nilson Amorim Júnior, Bogumil E. Brycki, Iwona Kowalczyk, Olga Kaczerewska, Adrianna Szulc, Marta Pakiet, Omnia Shehata, Adel Attia, Lobna Korshed, Michael Schorr, Benjamin Valdez, Juan Ricardo Salinas, Nelson Cheng, Yuli Yetri, Gunawarman -, Emriadi -, Novesar J, Amornvadee Veawab, Adisorn Aroonwilas, Sureshkumar Srinivasan, Norinsan Kamil Othman, Zulhusni Dasuki, Emee Marina Salleh, Mohamad Azwani Shah Mat Lazim

Published by InTech
Janeza Trdine 9, 51000 Rijeka, Croatia

Publishing Process Manager Maja Bozicevic
Technical Editor SPi Global
Cover InTech Design team

First published April, 2018
Printed in Croatia
Legal deposit, Croatia: National and University Library in Zagreb

Additional hard copies can be obtained from orders@intechopen.com

Corrosion Inhibitors, Principles and Recent Applications, Edited by Mahmood Aliofkhazraei
p. cm.
Print ISBN 978-953-51-3917-1
Online ISBN 978-953-51-3918-8

For EU product safety concerns:
IN TECH d.o.o., Prolaz Marije Krucifikse Kozulić 3, 51000 Rijeka, Croatia,
info@intechopen.com or visit our website at intechopen.com.

Meet the editor

Dr. Mahmood Aliofkhazraei works in the corrosion and surface engineering group at the Tarbiat Modares University. He is the head of Aliofkhazraei research group (www.aliofkhazraei.com). Dr. Aliofkhazraei has received several honors, including the Khwarizmi award and the best young nanotechnologist award of Iran. He is a member of the International Association of Corrosion Engineers (NACE International) and International Society of Electrochemistry (ISE). His research focuses on materials science, nanotechnology and its use in surface and corrosion science.

Contents

Preface

The most undesirable and noticeable deterioration of different kinds of materials is corrosion of metals, which is defined by ISO (International Organization for Standardization) as "physicochemical interaction between a metal and its environment that results in changes in the properties of the metal, and which may lead to significant impairment of the function of the metal, the environment, or the technical system, of which these form a part." Total costs of corrosion include "the design and construction or manufacturing, the cost of corrosion related maintenance, repair and rehabilitation, and the cost of depreciation or replacement of structures that have become unusable as a result of corrosion" and are estimated to be US $2.5 trillion that corresponds to approximately 4% of the global GDP. To protect metals or alloys from corrosion, some methods can be used such as isolating the structure from the aggressive media (using of coatings or film-forming chemicals) or compensating the loss of electrons (corrosion is oxidation process) from the corroded structure (e.g., cathodic protection by impressed current or by using sacrificial anodes). The use of corrosion inhibitors may include organic and inorganic compounds that adsorb on the metallic structure to isolate it from its surrounding media to decrease oxidation-reduction processes. Organic inhibitors create their inhibition by adsorbing their molecules on the metal or alloy surface to form a protective layer. Alternatively, inorganic inhibitors act as anodic inhibitors and their metallic atoms are enclosed in the film to improve its corrosion resistance.

This book collects new developments about corrosion inhibitors and their recent usages. I would like to appreciate all the contributors of this book and thank them for their high-quality manuscripts. I wish the open access publishing of this book would help all researchers to benefit from this collection.

Dr. Mahmood Aliofkhazraei
Tarbiat Modares University, Iran

Section 1

Introduction

Chapter 1

Organic Corrosion Inhibitors

Bogumił Eugeniusz Brycki, Iwona H. Kowalczyk,
Adrianna Szulc, Olga Kaczerewska and Marta Pakiet

Additional information is available at the end of the chapter

http://dx.doi.org/10.5772/intechopen.72943

Abstract

Organic corrosion inhibitors are one of the five ways, besides material selection, design, cathodic protection and coatings, to protect materials against corrosion. Corrosion is an ubiquitous phenomena that deteriorates all materials, metals, plastics, glass and concrete. The costs of corrosion are tremendous and amounts to 4.0% of gross domestic product (GDP) in USA. The similar losses of GDP are noted in all countries around the world. At this point of time, there is no way to completely stop the corrosion processes. Some new solutions can only slow this process. Organic corrosion inhibitors are widely used in industry because of their effectiveness at wide range of temperatures, compatibility with protected materials, good solubility in water, low costs and relatively low toxicity. Organic corrosion inhibitors adsorb on the surface to form protective film which displace water and protect it against deteriorating. Effective organic corrosion inhibitors contain nitrogen, oxygen, sulfur and phosphorus with lone electron pairs as well can contain structural moieties with π-electrons that interact with metal favoring the adsorption process. This review presents mechanisms and monitoring of corrosion, laboratory methods for corrosion study, relationship between structure and efficacy of corrosion inhibitions, theoretical approach to design new inhibitors and some aspects of biocorrosion.

Keywords: organic corrosion inhibitors, biocorrosion, sulfate-reducing bacteria gemini surfactants, corrosion tests

1. Introduction

The world we live in is chemical. Everything what surrounds us is composed of natural or synthetic chemical compounds. Some of them are very durable, some of them are less stable. But all of them are subjected to interactions with the environment what adversely affects the structural performance, including reliability over time. The deterioration process concerns all materials, not only metals but also plastics, glass, concrete, wood, leather and paper. The cracking, swelling, crazing, discoloration, phase separation or delimitation of plastics is caused by UV-light, heat, moisture or biological activity, that is, by physical, chemical or biological reactions [1, 2]. Corrosion of glass is due to reactions with atmospheric pollutants such as SO_2 or CO_2 as well as hydroxyl ions attack on siloxane bonds what leads to extraction of silica [3, 4]. Corrosion of concrete materials is also a very important economic problem. The maintenance costs of concrete microbial corrosion (CMC) of sewer pipelines in Hamburg (Germany) in 1970s reached up to €25 million whereas in Los Angeles (USA), the sewer pipe of 208 km in a total length of 1900 km had been damaged by CMC, and the rehabilitation costs were as high as $400 million [5]. Wood may also cleave or decompose what is related to its chemical structure based on cellulose, lignin and hemicelluloses. In natural environment, wood is rapidly colonized by microorganism and insects and the process of decomposition begins [6–8]. Deterioration of wood materials is a serious problem especially for historic wood pieces and monuments [9–11].

The deterioration of different kind of materials is defined as corrosion by American Society for Testing and Materials (ASTM International), that is, "*the chemical or electrochemical reaction between a material, usually a metal, and its environment that produces a deterioration of the material and its properties*" [12]. However, the most undesirable and noticeable is corrosion of metals which is defined by ISO (International Organization for Standardization) as "*physicochemical interaction between a metal and its environment that results in changes in the properties of the metal, and which may lead to significant impairment of the function of the metal, the environment, or the technical system, of which these form a part*" [13]. Total costs of corrosion include "*the design and construction or manufacturing, the cost of corrosion-related maintenance, repair and rehabilitation, and the cost of depreciation or replacement of structures that have become unusable as a result of corrosion*" [14] and is estimated to be US$2.5 trillion what corresponds to 3.4% of the global GDP (2013). The use of current corrosion control practice and procedures, including organic corrosion inhibitors, would allow to save 15–35% costs of corrosion [14]. The cost of corrosion varies depending on branch of industry and usually it is the highest in transportation and chemical industry. In China in 2014, direct corrosion cost in transportation was 268.72 billion RMB that constitutes 23.97% of total corrosion costs [15].

Corrosion is an ubiquitous phenomena and there is no way to completely stop it. Some new solutions can only slow this process. However, the environmental pollution, global warming and climate change are the direct cause for increasing corrosion costs. Only the increase of global temperature by 2°C causes an increase of corrosion rates by up to 15% [16].

2. Mechanism of corrosion

Corrosion is mainly induced by chemical and electrochemical processes. Chemical corrosion takes place in dry gases and nonconductive liquids where there is no current/electron flow. The

http://dx.doi.org/10.5772/intechopen.72943

main effect of chemical corrosion is an oxide layer as a result of oxidation in the air [17]. Electrochemical corrosion takes place in solution, between metallic materials and electrolytes due to different potentials on the surface of the corroded metal and due to redox reactions. One part of the metal is anode where metal oxidizes and becomes ion.

$$M \rightarrow M^{n+} + n\,e \tag{1}$$

The another part is cathode, where depolarization takes place, mainly reduction of oxygen and hydrogen cation [15, 16].

$$O_2 + 4\,H^+ + 4\,e \rightarrow 2\,H_2O \tag{2}$$

$$2\,H^+ + 2\,e \rightarrow H_2\uparrow \tag{3}$$

A scheme of electrochemical corrosion is shown in **Figure 1**.

Another type of corrosion is biocorrosion, commonly called microbial-induced corrosion (MIC) and is defined as deterioration of metallic material caused by consortia of microorganisms [20]. Microbes form a biofilm on the metal surface which is very good environment for their growth [21]. Biofilms are multicellular communities of bacteria encased in an extracellular matrix of exopolysaccharide, protein and sometime extracellular nucleic acids [22]. Biofilms are very hard to eradicated from any environment. They are resistant to most antimicrobial agents [23]. Sulfate-reducing bacteria (SRB) and sulfur-oxidizing bacteria which form biofilm to large extend are the main reason of the biocorrosion [21].

Another term associated with material's degradation is erosion which means "*a progressive loss of original material from a solid surface due to mechanical interaction between the surface and a fluid, a multicomponent fluid or impinging liquid or solid particles*" [24]. A process involving conjoint corrosion and erosion is named corrosion-erosion [24].

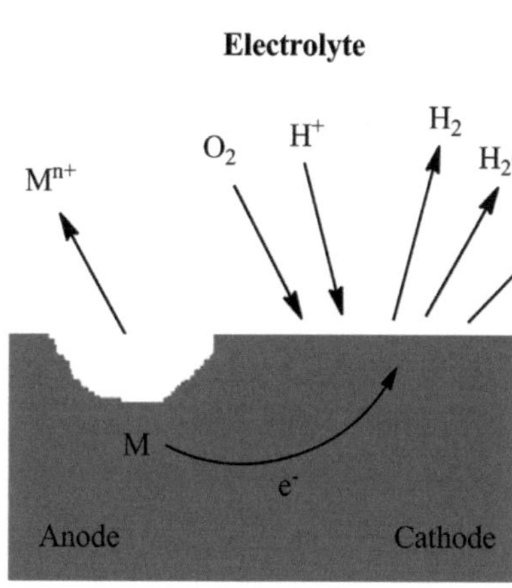

Figure 1. Scheme of electrochemical corrosion.

3. Monitoring of corrosion

To estimate real corrosion damage, many direct and indirect physicochemical analytical methods are used [25]. The most basic method is visual inspection [15, 23]. However, many non-intrusive methods are commonly applied, like ultrasonic [27], potential measurements, radiography, eddy current, magnetic particle inspection [28] and acoustic emission [29]. Ultrasonic allows to control thickness by using sonic waves with high frequency (1–6 MHz). The method can be used for liquids and solids but not for gases [28]. The equipment is calibrated on two thicknesses and by collecting data of material sound velocity, the thickness is calculated. The method can be used in-service and also allows to measure pipe with high temperature and over coatings [27]. The advantage of this method is a possibility to access the material only from one side. The surface of material has to be prepared carefully to have a good contact with the equipment [18, 19]. Acoustic emission measures acoustic sound waves which are emitted due to deformation in the monitored material. The sensors are very sensitive microphones. The sound waves are produced from mechanical stress (pressure or temperature changes) so the technique does not need excitation or human intervention [30]. This technique is a source of large amount of data which requires elaborate filtering and analysis [29]. Potential measurements to control corrosion can be done with a voltmeter with high internal resistance. To get results, the equipment has to be connected to a reference electrode (usually Ag/AgCl electrode). Another method, radiography, uses ability of gamma radiation to penetrate the investigated material. When the beam passes through the material, some energy is absorbed in the material. The thicker the material, the larger amount of the absorbed energy. As a result, a photographic film is obtained where dark color is associated with high intensity of the transmitted beam and light is associated to low intensity. The gamma radiation can penetrate up to 5 cm of the metal with acceptable signal of attenuation [29]. The method is commonly used to control of the permanence of the weld [28]. Radiography is also used for corrosion pits monitoring. Limitation of the methods is a fact that the material has to be available from both sides [18]. Eddy current is a technique used for monitoring of cracking and pitting corrosion of metallic materials. The method depends on the eddy current produced in the surface of the metallic material. The method is restricted to a small layer in the surface of the metal [28]. Magnetic particle inspection method is applied only to ferromagnetic materials and is based on the fact that the surface distortion imparts to a magnetic field. The tested material can be magnetized by several methods like flow an electrical current through material or using electromagnets [28]. Another possibility to control corrosion is to measure indirect changes such as hydrogen evolution which is a product of cathodic reaction. Hydrogen monitoring is applicable to oil and petrochemical industries. The monitoring is based on techniques such as hydrogen pressure or vacuum probe, electrochemical hydrogen patch probe and hydrogen fuel cell probe [29]. Water chemistry analyses can provide interesting information to corrosion monitoring program such as measurement of pH, conductivity or dissolved oxygen [29].

Corrosion monitoring methods like weight loss measurements, linear polarization measurements or electrical sensor utilizing electrical resistance are also used in the field. However, the measurements are indirect which make it difficult to estimate real corrosion damage [25]. All used methods have to be standardized. Examples of extensive collections are the NACE (*National Association of Corrosion Engineers*) standards and the ASTM (*American Society for Testing and Materials*) standards [18].

4. Laboratory methods for corrosion study

The most widespread methods for laboratory corrosion study as well to calculate a corrosion rate (CR) are gravimetric and electrochemical (potentiometry and impedance spectroscopy) techniques, whereas scanning electron microscope (SEM), atomic force microscope (AFM) and transmission electron microscope (TEM) are used to show morphology of surface. The gravimetric method, that is, the weight loss measurements, is one of the simplest methods to carry out. A sample of tested metal, previously carefully degreased and polished is weighed and immersed in an electrolyte solution for specified time (t). After that time, the sample is taken out, sluiced, degreased and weighted. The average weight loss (ΔW) is calculated by following equation:

$$\Delta W = W_1 - W_2 \tag{4}$$

where W_1 and W_2 is samples weight before and after immersion in electrolyte solution for t time, respectively [31]. ΔW is represented in grams. Based on obtained results, corrosion rate can be estimated as

$$CR = (K^* \Delta W)/(A^* t^* d) \tag{5}$$

where K is a constant (8.76×10^4) which allows to represent CR in mm/year; A is the surface of the metal sample (cm^2); t is the immersion time (hours); d is the density of the metal (g/cm^3) [31].

On the other hand, open circuit potential (OCP), named also corrosion potential (E_{corr}), linear polarization resistance (LPR) and potentiodynamic measurements (Tafel slopes) are included in potentiometric methods [32]. The measurements are conducted in standard, three electrode system such as reference (saturated calomel electrode), auxiliary (platinum electrode) and working (sample of the tested metal).

Electrodes are immersed in the electrolyte solution. After reaching an equilibrium, OCP (E_{corr}), which is a difference in potential of microcells of the metal, is registered. At that potential, oxidation and reduction reactions occur which allows to estimate if the metal is resistant to corrosion in tested environment. The higher value of E_{corr}, the higher corrosion resistance of the metal [33]. Graphical representation of results of the potentiodynamic measurements are Tafel slopes: graphs of applied potential (E, V) versus registered current density (i, A/cm^2) (**Figure 2**) [34].

The graph of applied potential versus registered current density is a straight line but that behavior is not observed for measured current (black lines). Due to that, corrosion current density (i_{corr}) is estimated by extrapolation of straight part of the measured current (green lines). The value of corrosion current density is equal for anodic and cathodic reactions, that is, the parameter has a direct influence on the corrosion rate [34].

The value of i_{corr} can also be estimated by measuring linear polarization resistance which is a quick testing technique for this method the material is polarized. The material's resistance (R_p) is found by taking the slope of the potential versus current and the corrosion current density is calculated by using Stern-Geary Equation [35]:

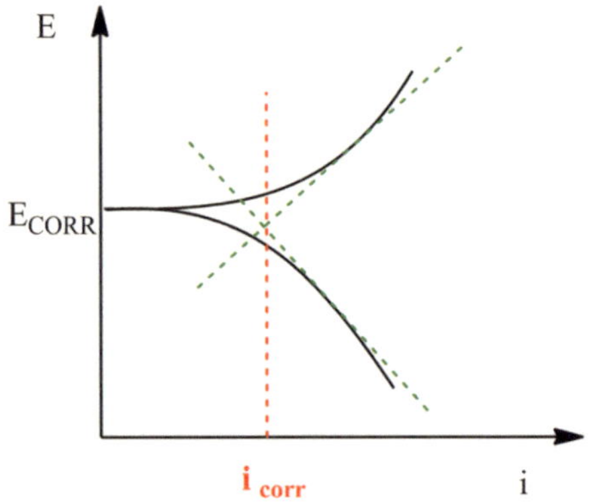

Figure 2. Tafel slopes.

$$i_{corr} = (\beta_a \beta_c / 2.303(\beta_a + \beta_c))/R_p \tag{6}$$

where β_a, β_c are slopes of anodic and cathodic of Tafel slopes, respectively [34].

Knowing the i_{corr} value, corrosion rate can be estimated:

$$CR = i_{corr}{}^{*}\ E_q{}^{*}\ 10^{*}\ 3.15^{*}10^{7}/F^{*}\ d \tag{7}$$

where E_q is the equivalent mass of metal exposed to corrosion (g); F is the Faraday constant (96,500 C); d is the density of metal (g/cm^3) and $10\times3.15\times10^7$ is the conversion factor used to obtain the result in mm/year [36].

Electrochemical impedance spectroscopy is also carried out in the standard three electrode system. After reaching equilibrium, impulse with known potential and frequency is applied. It disturbs balance of the electrochemical system. A measured quantity is impedance of the working electrode, Z (Ω), which is described by two types of impedance: real, Re(Z) and imaginary, Im(Z). The results are presented as Nyquist diagrams: a curve in Re(Z)-Im(Z) system [34]. Analysis of obtained data consists in describing the studied system with an equivalent electrical circuit. When the processes are related to charge transfer between metal and electrolyte, then the equivalent circuit consists in: solution resistance (R_s), charge transfer resistance (R_{ct}) and double layer capacitance (C_{dl}) formed at the metal:solution interface. An example of the Nyquist diagram and the equivalent circuit is presented in **Figure 3**.

In practice, instead of double layer capacitance, constant phase element (CPE) is used. Thanks to that, a fact that the system is not ideal capacitor to be considered. CPE is converted to C_{dl} by following equation [37]:

$$C_{dl} = \left[(CPE^{*}R_{ct})^{1/n}\right]/R_{ct} \tag{8}$$

Where n is a chase shift which represents degree of imperfection [38].

http://dx.doi.org/10.5772/intechopen.72943

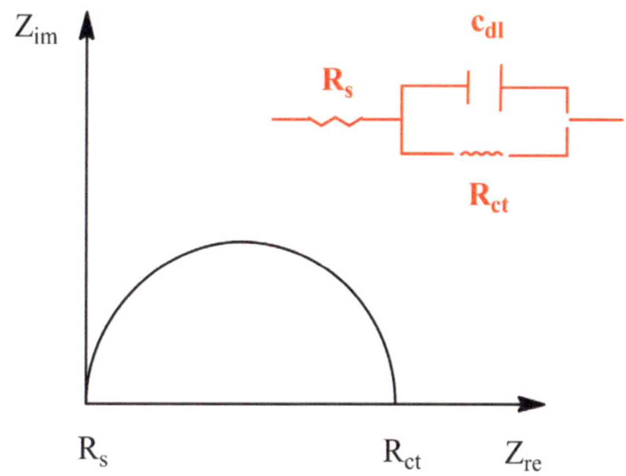

Figure 3. Nyquist diagram and equivalent circuit.

Corrosion rate is calculated from the equation:

$$CR = i_{corr}{}^{*}\ E_q{}^{*}\ 10^{*}\ 3.15^{*}10^{7}/F^{*}\ d \tag{9}$$

Corrosion current density is calculated from Stern-Geary equation [36].

Surface morphology of the metal is examined by using microscope techniques, SEM, AFM or TEM. It allows to observe all surface defects/damage done by corrosion [39]. Very often confocal laser scanning microscope (CLSM) is used to choose the right area of the surface for further SEM/AFM/TEM examinations [37]. The mentioned methods are the most popular but there are also other possibilities to estimate corrosive damage such as analytical methods, volumetric, radiography or magnetic-powder testing. Analytical methods are based on the determination of the metal ions content in environment and qualitative and quantitative analysis of corrosion products by using mass spectrometry, absorption atomic spectrometry or spectrophotometry [31, 37]. Volumetric method measures volume of evolving gas in the corrosive reaction. Radiography is based on phenomenon of radiation absorption by materials. Radiographic images are compared with the image of undamaged sample [41]. Magnetic-powder testing consists in irregularity of distribution of the magnetic field in material's defects. Tested sample has to be magnetized and then magnetic particles are applied and assemble in damage. Magnetic-powder testing is applied only for ferromagnetic materials [34].

5. Corrosion protection methods

There are five primary methods of corrosion control: (I) material selection, (II) design, (III) cathodic protection, (IV) coatings and (V) inhibitors. The simplest method for controlling the corrosion is the selection of the structural materials that change composition, change micro-structure stress and eliminate tensile stress [39, 40]. Another method is an application of rational design principles which can eliminate many corrosion problems and reduce the time and costs associated with corrosion maintenance and repair [39, 40]. Cathodic protection is an electrical method to reduce corrosion rate of metallic structures in electrolytes such as soil or water [42]. To achieve the protection, the impressed current cathodic protection (ICCP) system and the sacrificial anodes cathodic protection (SACP) system are used [44]. Corrosion control by anodic protection is known from the literature [43] but currently it is rarely used due to high restrictions.

Coatings, that is, the isolation of the metal from the corrosive environment is one of the most significant method of protection against corrosion. Coatings could be metallic (tin-plated steel and galvanized steel) or non-metallic (organic or inorganic). The most important are organic coatings, polymers made from epoxides, polyurethanes, polyesters, melamine formaldehyde resins, polyacrylates and phenolic polymers [42, 43].

Recently, many new protective coatings have been developed; they could be based on graphene [47], hybrid based on graphene oxide (GO) and reduced graphene oxide [22, 45, 46], polypyrrole [50], polyaniline/polyvinyl chloride blended [51], nano-hybrid epoxy coatings [52], bio-based polymers [53] or inorganic: siloxane based sol-gel coatings [54], silica-titaniahybrid [55] and iron aluminide coating [56]. Some of them can be used to form the superhydrophobicity surface [48, 49, 54–59, 61]. The coatings can also contain anticorrosive pigments, for example, lanthanum molybdate [62] or sodium phosphomolybdate [63], or anticorrosive coating additives like multiwalled carbon nanotubes [64]. To get the best protection against corrosion as well the optimal economy coatings and cathodic protection can be complementary used [44]. Corrosion inhibitors are special group of substances or their mixtures that prevent or minimize the corrosion. Inhibitors are adsorbed on the surface of the metal and form a protective thin film [39, 62].

The consumption of corrosion inhibitors reached nearly \$1.1 billion in USA in 1998 and is forecasted to increase to \$2.5 billion in 2017 [66].

Inorganic corrosion inhibitors, besides the oldest one, that is, molybdate anion, belong to calcium nitrite, rare earth metals salts, zinc phosphate, chromates and lanthanide compounds. However, the most numerous class of corrosion inhibitors are organic once [67]. Many of them are surfactants with hydrophilic and hydrophobic molecular moieties [68]. The corrosion inhibitors can be introduced as protective coating [69], bio-based lubricants [67, 68] and smart coatings that are released by the action of specific stimulus (e.g. change of pH, ionic strength or the change of electrode potential) [42, 60, 69–72].

6. Organic corrosion inhibitors

Organic corrosion inhibitors are widely used in industry because of their effectiveness at wide range of temperatures, compatibility with protected materials, good solubility and relatively low toxicity [78, 79] This is a very important issue for researchers which is confirmed by an increasing number of papers **(Figure 4)**. These compounds act as cathodic and anodic inhibitors. Cathodic corrosion inhibitors move the corrosion potential toward lower values and inhibit or delay the reactions occurring at the cathode (oxygen reduction and hydrogen evolution).

http://dx.doi.org/10.5772/intechopen.72943

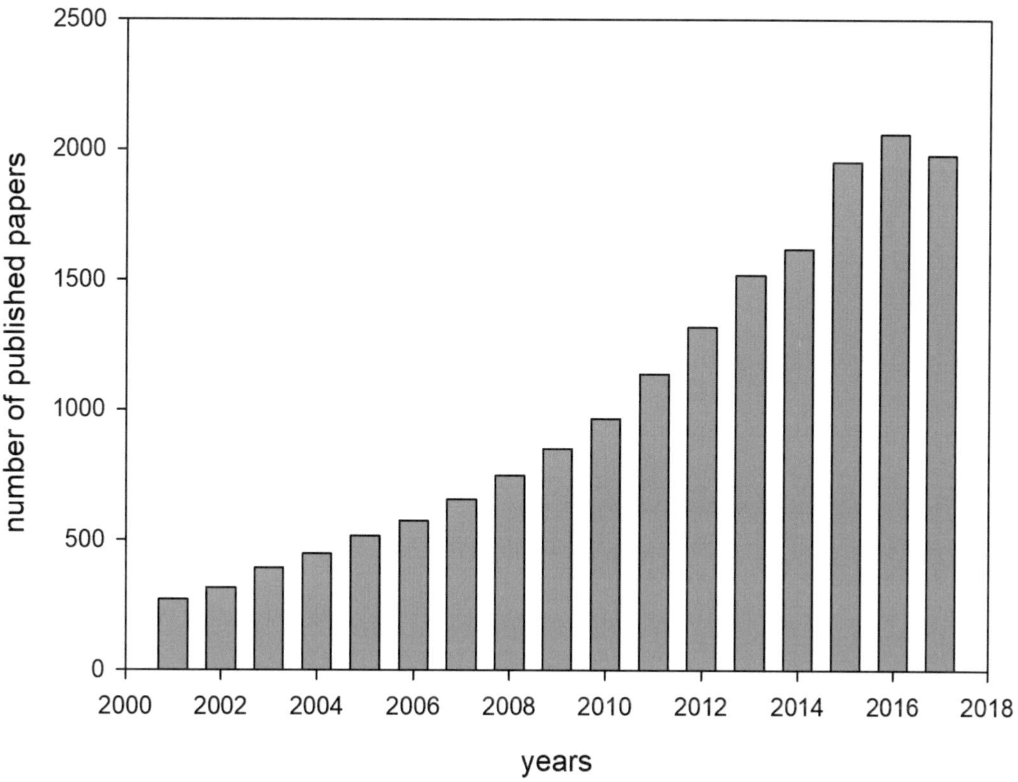

Figure 4. Number of published papers about organic corrosion inhibitors versus year of publication.

In contrast, anode corrosion inhibitors react with the metal cation to form an insoluble hydroxide, block the active sites on the metal surface and move the corrosion potential in the direction of positive values, which prevents further oxidation (dissolution) of the metal thus reducing the rate of corrosion. It is very important to use the right amount of an anode inhibitor, because insufficient concentration to cover all the active sites can lead to localized corrosion which is difficult to detect. Mixed inhibitors provide the highest protection because they affect both cathodic and anodic reactions.

The mechanism of action of organic corrosion inhibitors is based on the adsorption on the surface to form protective film which displace water from the metal surface and protect it against deteriorating. This process is not either purely physical or purely chemical adsorption. Adsorption is influenced by the chemical structure of organic inhibitors, nature and surface charge, the distribution of charge in the molecule and type of aggressive media (pH and/or electrode potential). The physical adsorption is based on electrostatic interaction between the charged metal surface and charged inhibitor molecule. Chemical adsorption is connected with the donor-acceptor interactions between free electron pairs and vacant, low energy d-orbital of metal **(Figure 5)**.

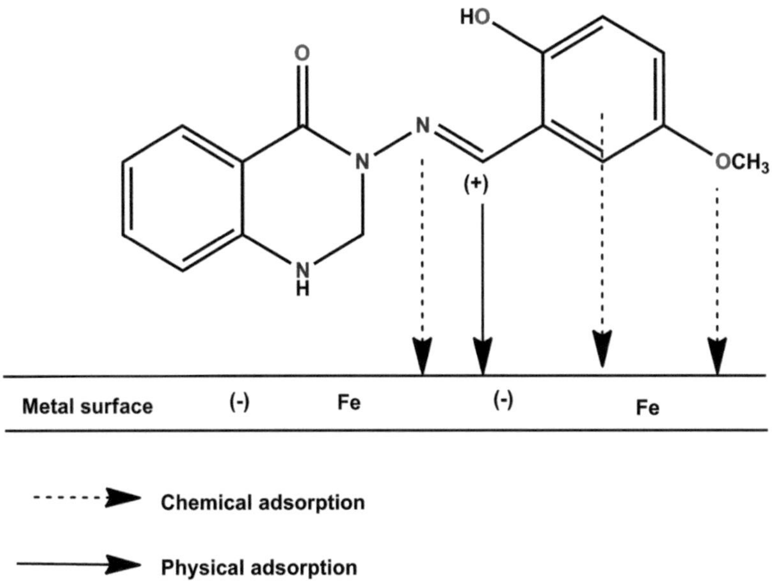

Figure 5. Schematic diagram representing the adsorption mechanism of Shiff bases on mild steel surface [82].

Effective organic corrosion inhibitors should contain heteroatoms (nitrogen, oxygen, sulfur and phosphorus) with lone electron pairs and moiety with π-electrons (aromatic rings and multiple bonds) that can interact with free orbital d metal, favoring the adsorption process [73].

The standard adsorption free energy (ΔG°_{ads}) gives information about type of adsorption. Values up to −20 kJ/mol are connected with the electrostatic interaction (physical adsorption). More negative values, below −40 kJ/mol, correspond to chemisorption process. Negative values mean that both processes are spontaneous. Also the standard enthalpy of adsorption provides valuable information about the mechanism of corrosion inhibition. An endothermic adsorption process ($\Delta H^{\circ}_{ads} > 0$) is attributed to chemisorptions, whereas an exothermic adsorption ($\Delta H^{\circ}_{ads} < 0$) is connected to physical or physical/chemical adsorption process [26]. Presence of the heteroatoms with lone pair of electrons like nitrogen, oxygen, sulfur or phosphor as well π-electrons of multiple bonds or aromatic rings enhance adsorption phenomena [74]. Chemisorption involves transfer or sharing of unbounded electrons between the inhibitor molecule and the metal surface [75]. The order of corrosion inhibition is the reverse order of the electronegativity of these atoms

$$P > S > N > O \tag{10}$$

In acid environment, heteroatoms are protonated that additionally promotes the interactions between the inhibitor and the surface. The adsorption of organic corrosion inhibitors onto the surface of a corroding metal may be regarded as a substitution process between the organic compound, especially aliphatic chain in aqueous phase and water molecules adsorbed on the metal surface:

$$Org_{(sol)} + xH_2O_{(ads)} \leftrightarrow Org_{(ads)} + xH_2O_{(sol)} \quad (11)$$

where Org (sol) and Org (ads) are, respectively, the organic species dissolved in the aqueous solution and adsorbed onto the metallic surface; H_2O (ads) and H_2O (sol) is the water molecule adsorbed onto the metallic surface and that in the bulk solution; x is the size ratio representing the number of water molecules replaced by one organic adsorbate.

The aliphatic chain has an influence on the corrosion protection due to the repulsion of nonpolar hydrophobic part of inhibitor and polar medium. The hydrophobic chains form a protective layer at the metal/water interface. The size and molecular weight of organic inhibitor also have an impact on the efficiency of inhibition [76]. Larger the molecule, greater is the inhibition efficiency:

$$R_3N > R_2NH > RNH_2 \quad (12)$$

where R is a hydrocarbon chain.

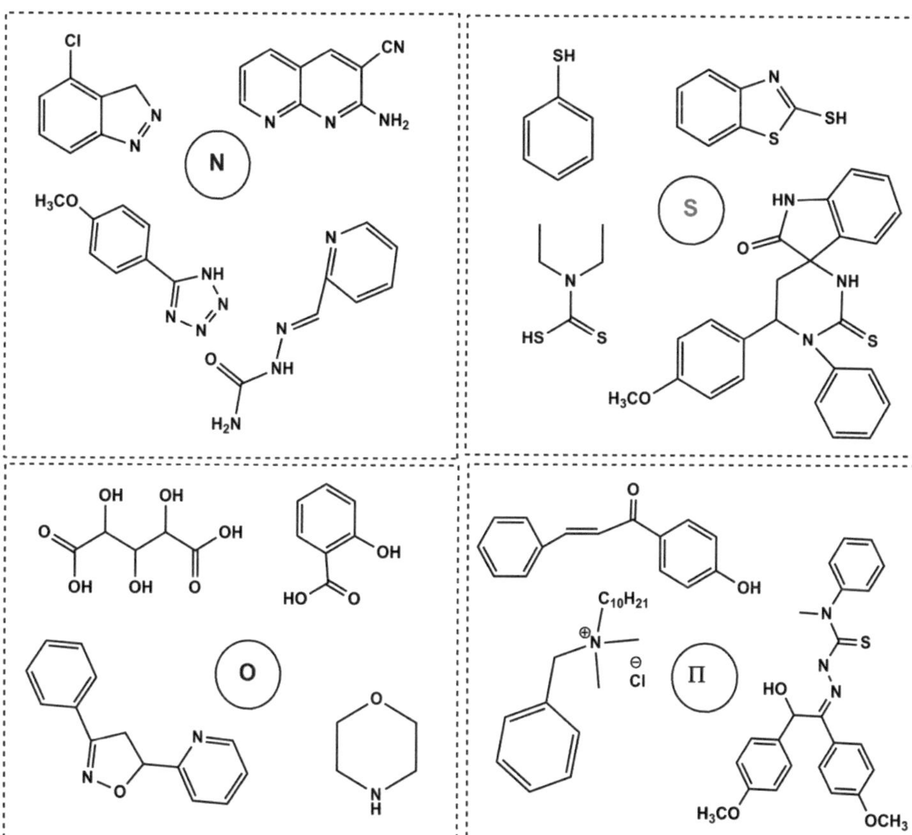

Figure 6. Nitrogen, sulfur, oxygen and π clusters of organic corrosion inhibitors.

Concentration of corrosion inhibitors has an important impact on the inhibitor efficiency. The corrosion rate decreases with increasing concentration of inhibitors because the adsorption of the inhibitor also increases.

6.1. Structures of organic corrosion inhibitors

The large number of organic corrosion inhibitors can be divided for some clusters with specific elements, like **(Figure 6)**:

- *compounds containing nitrogen:* amines [77], pyridine derivatives [74, 75], quaternary ammonium salts [80], triazole derivatives [81], Schiff base [82], amino acids [64] and indazole [83],
- *compounds containing nitrogen and sulfur:* imidazole derivatives [84], thiadiazole derivatives and thiazole derivatives [85],
- *compounds containing sulfur:* thiourea derivatives [86] ans sulfonates,
- *compounds containing nitrogen and oxygen:* oxazol derivatives [87], phtalimides [88] and plant extracts/natural [89].

Data about material, kind of inhibitor and inhibition efficiency are summarized in **Table 1**.

Polymers have also a high anti-corrosion efficacy. This involves the ability to interact with many, so that surface adsorption is stronger in comparison with monomers. Polymers can be a protective coating, but they can also be used as corrosion inhibitors. For example, deoxyribonucleic acid is a biopolymer with high inhibition efficiency against steel reinforcement [94]. Similarly, natural polymer, chitosan, is used as corrosion inhibitor of copper in hydrochloric acid [95].

Material	Medium	Corrosion inhibitors	Conc. of inhibitor	Inhbition efficiency (%)	Reference
Mild steel	1 M HCl	2-(4 -nitrophenyl) benzimidazole	1 mM	93.7	[90]
Mild steel	1 M HCl	(6-methyl-3-oxopyridazine-2-yl) acetate	10^{-1} mM	66.7	[91]
Mild steel	1 M HCl	Shiff base	1 mM	92.0	[82]
Low-alloy steel specimen	1% NaCl	L-cysteine	400 ppm	65.7	[64]
Carbon steel (API N80)	1.65 wt%NaCl con. 1 g/L acetic acid	Imidazoline	0.1 g/L	99.3	[84]
Copper	3% NaCl	N-decyl-3-amino-1,2,4-triazole	10^{-3} M	99.2	[92]
Tin	0.5 M HCl	Adenosine	10^{-3} M	73.0	[93]
Indium	0.5 M HCl	Adenosine	10^{-3} M	72.0	[93]
Tin-Indium Alloy (5%Indium)	0.5 M HCl	Adenosine	10^{-3} M	73.0	[93]

Table 1. Data about material, kind of inhibitor and inhibition efficiency.

http://dx.doi.org/10.5772/intechopen.72943

7. Quaternary ammonium salts as corrosion inhibitors

Due to the presence of positively charged nitrogen atom and the amphiphilic structure, quaternary ammonium salts are the center of interest for using them as highly effective corrosion inhibitors [96–100]. A special attention is devoted to new generation of quaternary ammonium salts—gemini surfactants. These compounds contain two hydrophilic head groups and two hydrophobic tails connected by a spacer at the head groups or closed to them. The spacer can have different structure; it can be rigid or flexible. It can also be hydrophobic or hydrophilic. The neutral charge of the molecule is retained by the presence of organic or inorganic anions. The gemini alkylammonium salts possess a very low critical micelle concentrations (cmc), which is up to hundred times lower than cmc's of corresponding monomeric surfactants. Gemini surfactants also have a larger molecular area in comparison to monomeric analogs which cause them to act more efficiently as corrosion inhibitors [80, 101–110]. Dimeric surfactant 1,4-tetramethylene-bis(*N*-dodecyl-*N*,*N*-dimethylammonium bromide) (12-4-12) is more efficient in decreasing corrosion rate of carbon steel in 1 M HCl than its monomeric analogue *N*-dodecyl-*N*,*N*,*N*-trimethylammonium bromide (DTAB). The relationships of concentration and corrosion rate for both surfactants are presented in **Figure 7.**

Resistance of the carbon steel in a system-containing dimeric surfactant is much higher in comparison to blank solution (1 M HCl) or to solution-containing monomeric surfactant (**Table 2**). It means that metal is less susceptible to corrosion [111].

Inhibition efficiency (IE%) is affected not only by number of positively charged nitrogen atoms but also depends on the length of alkyl chain. Longer the alkyl chain higher is the inhibition efficiency [97, 105] (**Table 3**).

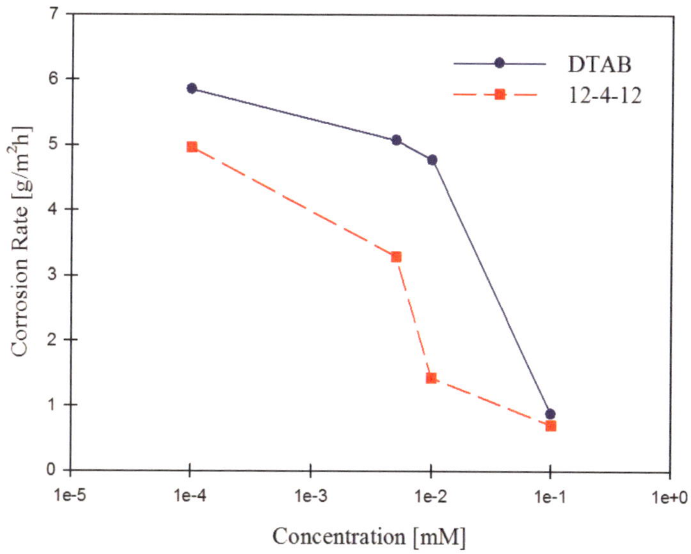

Figure 7. Relationship between corrosion rate of carbon steel and concentrations of surfactants.

	Concentration of surfactant (mM)	R (Ωcm^2)
Blank	—	43
DTAB	0.05	159
12-4-12	0.001	364

Table 2. Resistance of the carbon steel immersed in solution without and with cationic inhibitors.

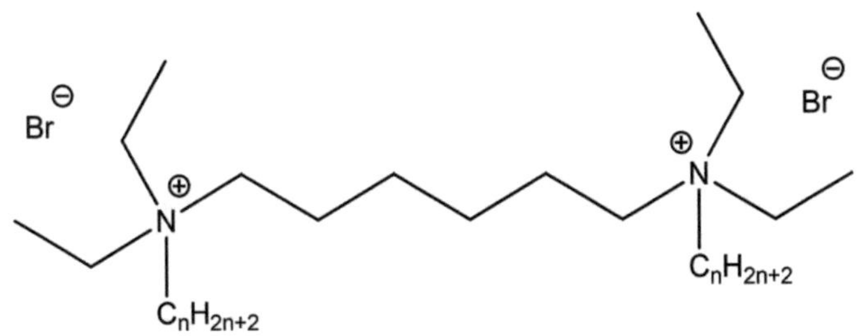

n	IE (%)
10	93.7
12	95.3
14	96.5
16	98.2

Table 3. Inhibition efficiency of dimeric surfactants for aluminum in 1 M HCl; concentration of surfactants: 1 mM [112].

The same correlation is observed in the spacer length. Longer the spacer, higher is the inhibition efficacy. For gemini surfactant $((C_{12}H_{25})_3N^+$-s-$N^+(C_{12}H_{25})_3)$ with two methylene groups as spacer (s) inhibition efficiency is 89% whereas IE is 93% for six methylene units at concentration 5 mM [109].

Additional heteroatoms or π electrons also favor adsorption onto the metal surface [113]. The order of the effective action increases with decreasing the electronegativity of the heteroatom: O < N < S < P [114]. Replacement of isopropyl group to hydroxyethyl or benzyl group leads to the increase of the inhibition efficiency from 95 to 96 or 97%, respectively (**Table 4**) [115]. Oxygen is a source of two unbonded electron pairs whereas benzene ring is a source of three pairs of π electrons, which can interact with free d orbitals of the metal.

Increasing the number of heteroatoms also affects the inhibition efficiency (**Figure 8**) where 12–6-12 is 1,6-hexamethylene-bis(*N*-dodecyl-*N*,*N*-dimethylammonium) dibromide; G6-MOH-12 is (1,6-hexamethylene-bis(*N*-dodecyl-*N*-hydroxyethyl-*N*-methylammonium) dibromide; 12-MOH-O-MOH-12 is 3-oxa-1,5-pentamethylene-bis(*N*-dodecyl-*N*-hydroxy-ethyl-*N*-methyl-ammonium) dichloride [116].

The plots in **Figure 8** present a semicircular shape with a diameter corresponding to the corrosion resistance (large diameter and higher resistance). The stainless steel impedance response in

http://dx.doi.org/10.5772/intechopen.72943

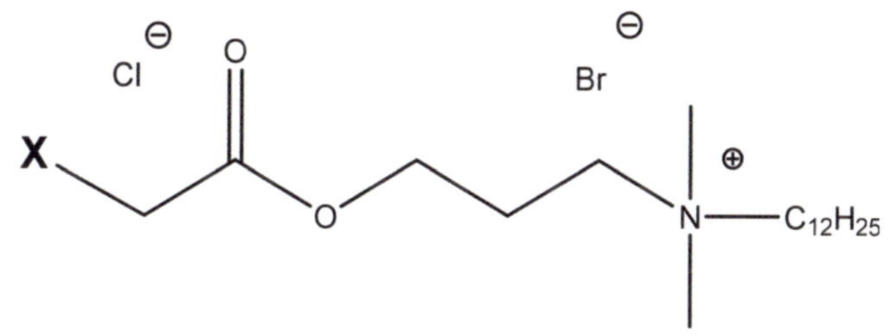

X	Inhibition efficiency (%)
$(CH_3)_2CHN^+(CH_3)_2$	95.43
$OHCH_2CH_2N^+(CH_3)_2$	96.24
$PhN^+(CH_2CH_3)_2$	97.19

Table 4. Influence of the presence of heteroatoms and π electrons on inhibition efficiency of steel in 1 M HCl.

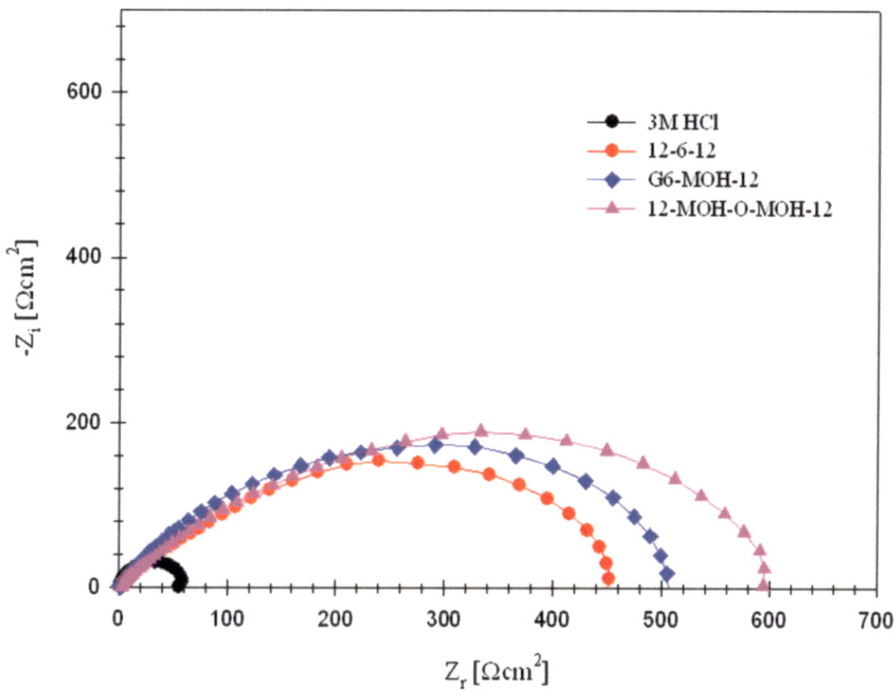

Figure 8. Structure of the gemini surfactants and Nyquist plots for stainless steel in 3 M HCl in the presence and absence of the synthesized inhibitors (naturally aerated, 7 days of immersion).

the blank solution (black) shows a smaller diameter compared to the diameters of the plots for both gemini surfactants, the latter indicating higher resistance, that is, lower corrosion rate. The diameter is the biggest for surfactants functionalized with three oxygen atoms (12-MOH-O-MOH-12) [116]. Some of gemini surfactants, which are already commercially used, are a part of

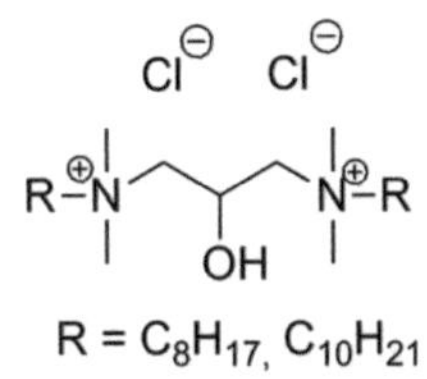

Figure 9. Structure of one commercially used inhibitor.

special compositions which are based on synergistic effect in order to lower the concentrations used [30, 110–112]. Some gemini surfactants are a part of patent about multifunctional corrosion inhibitors for iron alloys (tanks transporting oil and liquid fuel) which are subjected to acidic pollution, sulfur compounds, water, oxygen as well as calcium and magnesium cations [120]. Another patent is about using gemini surfactants as corrosion inhibitors of metallic materials which are used for gas and oil extraction [121]. The inhibitors contain heteroatoms and π electrons, their efficiency is higher than 90%. One inhibitor's structure is presented in **Figure 9** [120].

One of the interesting organic corrosion inhibitors are ionic liquids (ILs) as potentially green chemicals. Amidation of chitosan subsequently quaternized with oleic acid and p-toluene sulfonic acid gave new ionic liquid. The corrosion inhibition of the prepared polymeric ionic liquid on steel in acidic medium was investigated by using different electrochemical techniques [122].

8. Biocorrosion

Microbially induced corrosion (MIC) is one of the fundamental problems in marine industry, pulp and paper industry, natural gas transmission, industrial water transmission, metalworking and chemical process industries. This kind of corrosion is also called as biocorrosion, bacterial corrosion, microbial corrosion is deterioration of metal or non-metal materials as a result of the metabolic activity of microorganisms [116, 117]. Biocorosion is mostly a result of the interaction of mechanical, physical, chemical and/or biological factors. Wind, water, particles of dust, pollution atmosphere and water, and, in the case of stone materials, action light and temperature changes cause damage to materials, allowing penetration moisture and colonization of microorganisms on their surface. This leads to the uprising biofilm, a biologically active layer of various types of microorganisms as well as mucus being the product of their metabolic activity. Biofilm creates very good environment for growing microorganism, and increase rate of corrosion even to 10,000 times. The key to avoid this problem is understanding the dynamics of microbes [118, 119].

8.1. Microorganism and environment

Prerequisites for microbial-induced corrosion are the presence of microorganisms. If the corrosion is influenced by their activity, further requirements are (a) an energy source, (b) a carbon source, (c) an electron donator and (d) water. The kinetics of biocorrosion is strongly influenced by the concentration of oxygen, presence of salts, pH value, redox potential and conductivity. The living bacteria produce extracellular polymeric substances (EPS) and form biofilms on the metal surface. Biofilms are characterized by strong heterogeneity [120, 121]. Biofilm consists of bacterial cells and extracellular polymeric substances (mixture of polysaccharides, proteins, nucleic acids and fats) which facilitate the attachment of bacterial cells to the surface. The biofilm also includes inorganic sludge from water and/or corrosion products [122, 123]. The most common methods to observe biocorrosion effects are scanning electron microscope (SEM) and confocal laser scanning microscope (CLSM) [128] as well as X-ray photoelectron spectroscopy [124–126] and time of flight-secondary ion mass spectroscopy (ToF-SIMS) [129].

Bacteria involved in corrosion can be divided into following groups:

Sulfate-reducing bacteria (SRB) are anaerobic microorganisms that can obtain energy by oxidizing organic compounds or molecular hydrogen (H_2) while reducing sulfate (SO_4^{2-}) to hydrogen sulfide (H_2S). In a sense, these organisms "breathe" sulfate rather than oxygen in a form of anaerobic respiration [20, 21]. Sea water is a primary source of sulfate-reducing bacteria (SRB) (**Figure 10**).

Metal-reducing bacteria (MRB) affects the corrosion of iron and its alloys by dissolving the passive film on the surface of the metal or by transformation of the sediment to a less stable reduced form that does not inhibit corrosion process. Included in this group are the bacteria of *Pseudomonas* and *Shewanella* have the ability to reduce iron oxide and manganese oxides whereby the speed reduction depends on the type of sediment [130, 133].

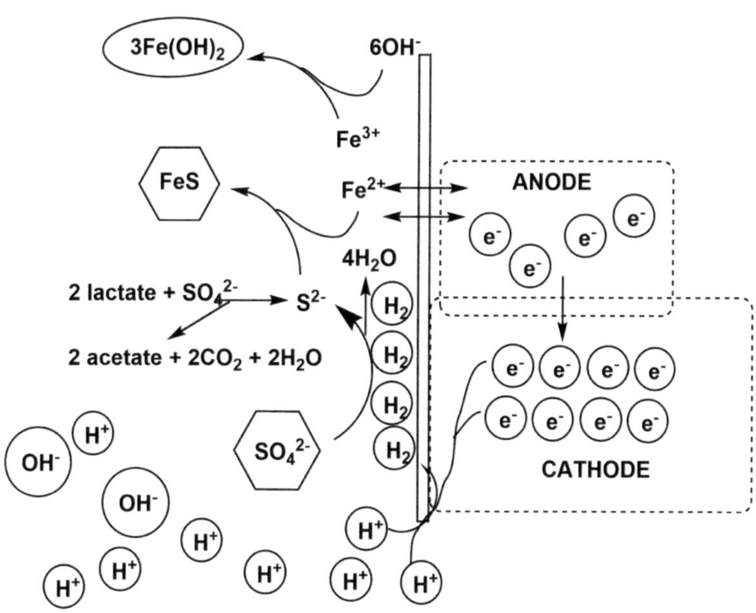

Figure 10. Mechanism of action SRB.

Metal-depositing bacteria (MDB)—Siderocapsa, Gallionella, Leptothrix, Sphaerotilus, Crenothrix and *Clonothrix* are involved in the biotransformation of iron oxide and manganese. Iron-depositing bacteria (e.g. *Gallionella* and *Leptothrix*) gain energy by oxidizing Fe (II) ions (dissolved or bound in sediments) to Fe (III). All these types of bacteria have the ability to oxidize Mn (II) to Mn (IV) with the precipitation of manganese dioxide that occurs in the rapid filter beds. *Filamentous bacteria* are associated with the formation of pitting corrosion [21].

Hydrogen sulfide-producing bacteria (SPB) are bacteria producing a large amount of extracellular polymeric substances (EPS) during the development of biofilm (e.g. *Clostridium, Flavobacterium* and *Desulfovibrio*) their role in the corrosion process consists in covering the metal surface with a EPS layer facilitating the attachment and multiplication of others bacteria [21].

Acid-producing bacteria (APB) are bacteria that secrete inorganic and organic acids as products by-pass metabolism while simple organic acids (acetic, formic and lactic) are metabolites of bacteria heterotrophic [21].

8.2. Biocorrosion inhibitors

Methods to significantly slow have concern on inhibition of the growth of microorganisms and modification of the environment in which the corrosion process takes place. The basic steps to prevent and control biocorrosion are (i) cleaning procedures; (ii) microbiocides; (iii) coatings and (iv) cathodic protection [65].

Obviously it is not easy to stop process of growing bacteria with one compounds. Because these kind of species should have antimicrobial activity and also should be corrosion inhibitors. Example of that kind of multifunctional compounds are gemini surfactants [127, 128]. Biocidal activities of the synthesized surfactants were achieved by dropping the redox potential and confirmed preventing sulfide production in the reactor's bulk phase. This means that all sulfidogenic bacteria are active in the reactor's bulk phase and on the metal surface. Labena et al. describe potential biocorrosion inhibitors that contain quaternary ammonium atom (**Figure 11**) [135].

Quite new approach is using surface modification technologies such as short anti-biofilm peptides applying by immobilization method [136] also using small lipopeptides [137]. Using of coatings based on on silicone and epoxy resins are also method to protection against biocorrosion.

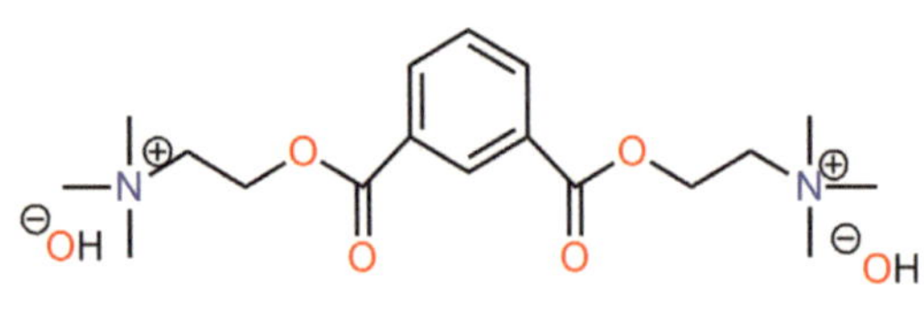

Figure 11. Structure of potential biocorrosion inhibitors.

http://dx.doi.org/10.5772/intechopen.72943

9. Designing organic corrosion inhibitors

Quantum chemical methods are useful for designing new, effective organic corrosion inhibitors as they relate electron structure of the compounds to their reactivity. Every year density functional theory (DFT) is more often applied for predicting a theoretical ability to inhibit corrosion process according to some quantum chemical parameters: energy of the highest occupied molecular orbital (E_{HOMO}), energy of the lowest unoccupied molecular orbital (E_{LUMO}), gap energy (ΔE), chemical hardness (η), softness (σ), ionization potential (I_P) and electron affinity (E_A), electronegativity (χ) and fraction of electron transferred (ΔN) [94, 99, 131, 138]. The energy of the highest occupied molecular orbital (HOMO) is associated with the ability of a molecule to donate electrons to the free *d* orbital of a metal. Compounds with higher E_{HOMO} are more capable of donating electrons. The energy of the lowest unoccupied molecular orbital (LUMO) is related to the ability to accept electrons from the metal. Lower values indicate higher tendency of accepting electrons. Moreover, positive values are connected with chemisorption, whereas negative values with physisorption [94, 132]. Another important parameter is energy gap:

$$\Delta E = E_{LUMO} - E_{HOMO} \tag{13}$$

The lower ΔE, the more reactive molecule, is related to better adsorption of inhibitor's molecules onto the metal surface [133, 134]. The dipole moment (μ) is also an important parameter which gives information about polarity in a bond. Corrosion inhibition efficiency increases with increasing the value of μ, due to the stronger dipole-dipole interactions with the metal surface which results in stronger adsorption and efficient corrosion inhibition [140]. Chemical hardness (η) and softness (σ) provide information about the resistance of a molecule to charge transfer and about the capacity of a molecule to receive electrons. They are calculated according to the equations:

$$\eta = -1/2\,(E_{HOMO} - E_{LUMO}) \tag{14}$$

$$\sigma = 1/\eta \tag{15}$$

The higher σ value suggests softer nature of the molecule and greater tendency to donate electrons to the metal [101]. Energy of HOMO and LUMO orbitals can be used for calculating ionization potential (I_P) and electron affinity (E_A) by the following equations:

$$I_P = -E_{HOMO} \tag{16}$$

$$E_A = -E_{LUMO} \tag{17}$$

The calculated values are used for estimating the electronegativity (χ) [141]:

$$\chi = (I_P + E_A)/2 \tag{18}$$

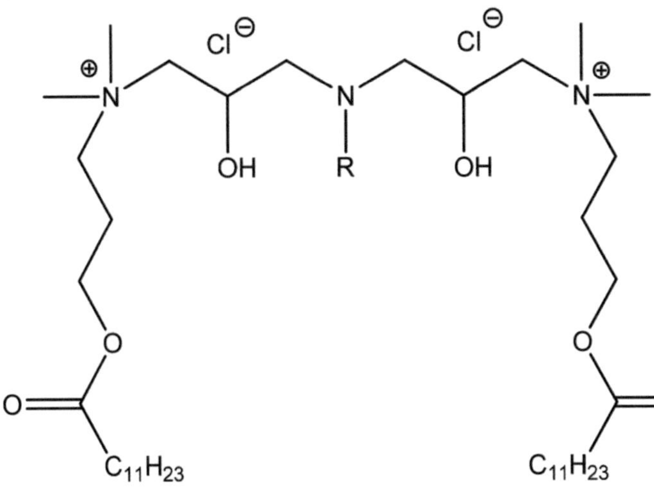

R	IE (%)	E_{HOMO} (eV)	E_{LUMO} (eV)	ΔE (eV)	μ (De)	ΔN
C_4H_9	79.86	−4.03	0.86	4.89	2.16	1.11
C_6H_{13}	79.49	−2.10	0.83	2.93	5.45	2.17
C_8H_{17}	81.32	−1.43	0.73	2.16	6.42	3.08

Table 5. Electron parameters and IE% for the gemini surfactants.

High value of χ suggests strong ability to attract electrons from the metal which leads to greater interactions and higher corrosion protection. The last parameter which is calculated is the fraction of electron transferred (ΔN):

$$\Delta N = \chi_{Fe} - \chi_{inh}/2(\eta_{Fe} - \eta_{inh}) \quad (19)$$

where χ_{Fe} equals 7 eV and $\eta_{Fe} = 0$ [101]. If $\Delta N > 0$, electrons are transferred from the molecule to the metal and if $\Delta N < 0$, from the metal to the molecule. For all tested gemini surfactants, the values of the fraction of electron transferred are negative indicating transfer from the metal to the molecules. Some quantum parameters for gemini surfactants and corrosion inhibition efficiency (IE%) of carbon steel are presented in **Table 5** [139].

According to the presented results, elongating the alkyl chain (R) leads to increasing corrosion inhibition efficiency from 79.86 to 81.32%. Based on the electron parameters, all values of E_{LUMO} are negative, which are associated with physisorption. The lowest value of E_{LUMO} and the highest value of E_{HOMO} were noticed for a surfactant with octyl group, as well as the lowest ΔE and the highest dipole moment. The electron parameters suggest the compound with octyl group should be the most efficient inhibitor of the tested group. This observation is in agreement with experimental results (IE%). The values of ΔN which are higher than 0 indicate that electrons are transferred from inhibitor to free d orbital of the metal.

http://dx.doi.org/10.5772/intechopen.72943

10. Summary

Organic corrosion inhibitors with heteroatoms and π-electron moieties are very efficient compounds to fight corrosion. The reviewed literature data clearly indicate that a new way to inhibit deterioration processes can be multifunctional gemini surfactants. Gemini alkylammonium surfactants with the highest corrosion inhibition efficacy can be synthesized according to prediction by theoretical calculations structures. These organic corrosion inhibitors can be also immobilized and used as biocorrosion inhibitors.

Acknowledgements

The work has been supported by the National Centre of Research and Development (Poland TANGO1/266340/NCBR/2015).

Author details

Bogumił Eugeniusz Brycki*, Iwona H. Kowalczyk, Adrianna Szulc, Olga Kaczerewska and Marta Pakiet

*Address all correspondence to: brycki@amu.edu.pl

Faculty of Chemistry, Laboratory of Microbiocides Chemistry, Adam Mickiewicz University, Poznan, Poland

References

[1] Raja PB, Ismail M, Ghoreishiamiri S, Mirza J, Ismail MC, Kakooei S, et al. Reviews on corrosion inhibitors: A short view. Chemical Engineering Communications. 2016 Sep; **203**(9):1145-1156

[2] Shah AA, Hasan F, Hameed A, Ahmed S. Biological degradation of plastics: A comprehensive review. Biotechnology Advances. 2008 May;**26**(3):246-265

[3] Franz H. Durability and corrosion of silicate glass surfaces. Journal of Non-Crystalline Solids. 1980;**42**(1–3):529-534

[4] Melcher M, Wiesinger R, Schreiner M. Degradation of glass Artifacts: Application of modern surface analytical techniques. Accounts of Chemical Research. 2010 Jun 15;**43**(6): 916-926

[5] Kong L, Zhang B, Fang J. Study on the applicability of bactericides to prevent concrete microbial corrosion. Construction and Building Materials. 2017 Sep;**149**:1-8

[6] Eriksson K-EL, Blanchette RA, Ander P. In: Timell TE, editor. Microbial and Enzymatic Degradation of Wood and Wood Components. Berlin, Heidelberg: Springer Berlin Heidelberg; 1990. Springer Series in Wood Science

[7] Weslien J, Djupström LB, Schroeder M, Widenfalk O. Long-term priority effects among insects and fungi colonizing decaying wood: Species interactions during wood decay. The Journal of Animal Ecology. 2011 Nov;**80**(6):1155-1162

[8] Ulyshen MD. Interacting effects of insects and flooding on wood decomposition. Shaw P, editor. PLoS One. 2014 Jul 10;**9**(7):e101867

[9] Blanchette RA, Nilsson T, Daniel G, Abad A. Biological degradation of wood. In: Rowell RM, Barbour RJ, editors. Archeological Wood: Properties, Chemistry and Preservation. Washington, USA: American Chemical Society; 1989. pp. 141-174

[10] Forrest W. Building Materials Evaluation Handbook. New York: Springer Verlag; 2012

[11] Sterflinger K, Piñar G. Microbial deterioration of cultural heritage and works of art — Tilting at windmills? Applied Microbiology and Biotechnology. 2013 Nov;**97**(22):9637-9646

[12] ASTM International. ASTM G15-02. Standard Terminology Relating to Corrosion and Corrosion Testing. West Conshohocken: ASTM International; 2002

[13] ISO 8044:2015. Corrosion of metals and alloys – Basic terms and definitions. 8044:2015; 2015

[14] Koch G. Cost of corrosion. In: Trends in Oil and Gas Corrosion Research and Technologies. Kidlington: Elsevier; 2017. pp. 3-30

[15] Hou B, Li X, Ma X, Du C, Zhang D, Zheng M, et al. The cost of corrosion in China. npj Materials Degradation. 2017 Dec;**1**(1)

[16] Stewart MG, Wang X, Nguyen MN. Climate change adaptation for corrosion control of concrete infrastructure. Structural Safety. 2012 Mar;**35**:29-39

[17] McCafferty E. Introduction to Corrosion Science. London: Springer; 2010. p. 575

[18] Bardal E. Corrosion and Protection. London: Springer; 2004. p. 315

[19] Sastri VS. Challenges in Corrosion; Costs, Causes, Consequences, and Control. New Jersey: John Wiley & Sons Inc.; 2015. p. 408

[20] Xu D, Li Y, Gu T. Mechanistic modeling of biocorrosion caused by biofilms of sulfate reducing bacteria and acid producing bacteria. Bioelectrochemistry. 2016 Aug;**110**:52-58

[21] Liengen T. Understanding Biocorrosion: Fundamentals and Applications. Boston: Elsevier; 2014

[22] Kostakioti M, Hadjifrangiskou M, Hultgren SJ. Bacterial biofilms: Development, dispersal, and therapeutic strategies in the Dawn of the Postantibiotic era. Cold Spring Harbor Perspectives in Medicine. 2013 Apr 1;**3**(4):a010306-a010306

[23] Costertion JW, Stewart P, Greenberg E. Bacterial biofilms: A common cause of persistnet infections. Microbes, Immunity and Disease. 1999;**284AD**:1318-1322

[24] ISO/TR 16203: 2016. Corrosion of metals and alloys — Guidelines for the selection of methods for particle-free erosion corrosion testing in flowing liquids. 16203; 2016

[25] Dong B, Fang G, Liu Y, Dong P, Zhang J, Xing F, et al. Monitoring reinforcement corrosion and corrosion-induced cracking by X-ray microcomputed tomography method. Cement and Concrete Research. 2017 Oct;**100**:311-321

[26] Sastri VS. Green Corrosion Inhibitors: Theory and Practice. Hoboken, N.J: Wiley; 2011. p. 310. (Wiley series in corrosion)

[27] Salujaa A, Costainb J, Van der Ledenc E. Non-intrusive online corrosion monitoring. In: NDE2009. Tiruchirappalli, India: National Seminar & Exhibition on Non-Destructive Evaluation; 2009

[28] Bardal E, Drugli JM. Corrosion detection and diagnosis. In: Materials Science and Engineering. Oxford: EOLSS Publications; 2009. pp. 144-164

[29] Roberge PR. Corrosion Inspection and Monitoring. Hoboken, N.J: Wiley-Interscience; 2007. p. 383. (Wiley series in corrosion)

[30] Yang L, Yang TT, Zhou YC, Wei YG, RT W, Wang NG. Acoustic emission monitoring and damage mode discrimination of APS thermal barrier coatings under high temperature CMAS corrosion. Surface and Coating Technology. 2016 Oct;**304**:272-282

[31] ASTM International. ASTM G31-72. Standard Practice for Labolatory Immersion Corrosion Testing of Metals. West Conshohocken: ASTM International; 1999

[32] ASTM International. ASTM G59. Standard Test Method for Conducting Potentiodynamic Polarization Resistance Measurements. West Conshohocken: ASTM International; 2002

[33] Migahed MA, Hegazy MA, Al-Sabagh AM. Synergistic inhibition effect between Cu2+ and cationic gemini surfactant on the corrosion of downhole tubing steel during secondary oil recovery of old wells. Corrosion Science. 2012 Aug;**61**:10-18

[34] Sastri VS, Ghali E, Elboujdaini M. Corrosion Prevention and Protection; Practical Solutions. England: John Wiley & Sons Ltd.; 2007

[35] Perez N. Electrochemistry and Corrosion Science. Cham: Springer International Publishing; 2016

[36] Dehdab M, Yavari Z, Darijani M, Bargahi A. The inhibition of carbon-steel corrosion in seawater by streptomycin and tetracycline antibiotics: An experimental and theoretical study. Desalination. 2016 Dec;**400**:7-17

[37] Leiva-García R, Fernandes JCS, Muñoz-Portero MJ, García-Antón J. Study of the sensitisation process of a duplex stainless steel (UNS 1.4462) by means of confocal microscopy and localised electrochemical techniques. Corrosion Science. 2015 May;**94**:327-341

[38] Hegazy MA, Rashwan SM, Kamel MM, El Kotb MS. Synthesis, surface properties and inhibition behavior of novel cationic gemini surfactant for corrosion of carbon steel tubes in acidic solution. Journal of Molecular Liquids. 2015 Nov;**211**:126-134

[39] Yadav DK, Maiti B, Quraishi MA. Electrochemical and quantum chemical studies of 3,4-dihydropyrimidin-2(1H)-ones as corrosion inhibitors for mild steel in hydrochloric acid solution. Corrosion Science. 2010 Nov;**52**(11):3586-3598

[40] Joseph LA, Israel OK, Edet EJ, Ekwumemgbo PA. Determination of metal ions released by stainless steel arch bar into bio-fluids. Bulletin of the Chemical Society of Ethiopia. 2009;**23**(1)

[41] Michel A, Pease BJ, Geiker MR, Stang H, Olesen JF. Monitoring reinforcement corrosion and corrosion-induced cracking using non-destructive x-ray attenuation measurements. Cement and Concrete Research. 2011 Nov;**41**(11):1085-1094

[42] Davis JR. Corrosion: Understanding the Basics. Material Park, Ohio: ASTM International; 2000

[43] Salas BV, Wiener MS, Stoytcheva M, Zlatev R, Beltran MC. Corrosion in the food industry and its control. In: Food Industrial Processes-Methods and Equipment. Rijeka: InTech; 2012

[44] Tezdogan T, Demirel YK. An overview of marine corrosion protection with a focus on cathodic protection and coatings. Brodogradnja: Teorija i praksa brodogradnje i pomorske tehnike. 2014;**65**(2):49-59

[45] Fernández-Solis CD, Vimalanandan A, Altin A, Mondragón-Ochoa JS, Kreth K, Keil P, et al. Fundamentals of electrochemistry, corrosion and corrosion protection. In: Lang P, Liu Y, editors. Soft Matter at Aqueous Interfaces [Internet]. Cham: Springer International Publishing; 2016 ; [cited 2017 Oct 16]. pp. 29-70 Available from: http://link.springer.com/10.1007/978-3-319-24502-7_2

[46] Grundmeier G, Schmidt W, Stratmann M. Corrosion protection by organic coatings: Electrochemical mechanism and novel methods of investigation. Electrochimica Acta. 2000;**45**(15):2515-2533

[47] Prasai D, Tuberquia JC, Harl RR, Jennings GK, Bolotin KI. Graphene: Corrosion-inhibiting coating. ACS Nano. 2012 Feb 28;**6**(2):1102-1108

[48] Quezada-Rentería JA, Cházaro-Ruiz LF, Rangel-Mendez JR. Synthesis of reduced graphene oxide (rGO) films onto carbon steel by cathodic electrophoretic deposition: Anticorrosive coating. Carbon. 2017 Oct;**122**:266-275

[49] Promphet N, Rattanawaleedirojn P, Rodthongkum N. Electroless NiP-TiO_2 sol-RGO: A smart coating for enhanced corrosion resistance and conductivity of steel. Surface and Coating Technology. 2017 Sep;**325**:604-610

[50] Nautiyal A, Qiao M, Cook JE, Zhang X, Huang T-S. High performance polypyrrole coating for corrosion protection and biocidal applications. Applied Surface Science. 2018 Jan;**427**:922-930

[51] Bandeira RM, van Drunen J, Tremiliosi-Filho G, dos Santos JR, de Matos JME. Polyaniline/polyvinyl chloride blended coatings for the corrosion protection of carbon steel. Progress in Organic Coating. 2017 May;**106**:50-59

[52] Saravanan P, Jayamoorthy K, Ananda Kumar S. Design and characterization of non-toxic nano-hybrid coatings for corrosion and fouling resistance. Journal of Science: Advanced Materials and Devices - Elsevier. 2016 Sep;**1**(3):367-378

[53] Finkenstadt VL, Bucur CB, Côté GL, Evans KO. Bacterial exopolysaccharides for corrosion resistance on low carbon steel. Journal of Applied Polymer Science. 2017 Aug 5; **134**(29):45032

[54] El-Meligi AA. Nanostructure of materials and corrosion resistance. In: Aliofkhazraei M, editor. Developments in Corrosion Protection. Rijeka: InTech; 2014

[55] Lakshmi RV, Aruna ST. Corrosion protection behaviour of silica–titania hybrid coatings embedded with silica nanoparticles. Surface Engineering. 2017 Jun 3;**33**(6):467-473

[56] Majumdar S, Paul B, Chakraborty P, Kishor J, Kain V, Dey GK. Formation of Al_2O_3/FeAl coatings on a 9Cr-1Mo steel, and corrosion evaluation in flowing Pb-17Li loop. Journal of Nuclear Materials. 2017 Apr;**486**:60-65

[57] Mohamed AMA, Abdullah AM, Younan NA. Corrosion behavior of superhydrophobic surfaces: A review. Arabian Journal of Chemistry. 2015 Nov;**8**(6):749-765

[58] Liu L, Xu F, Ma L. Facile fabrication of a superhydrophobic cu surface via a selective etching of high-energy facets. Journal of Physical Chemistry C. 2012 Sep 6;**116**(35): 18722-18727

[59] Liu L, Xu F, Yu Z, Dong P. Facile fabrication of non-sticking superhydrophobic boehmite film on al foil. Applied Surface Science. 2012 Sep;**258**(22):8928-8933

[60] Zhou X, Kong J, Sun J, Li H, He C. Stable superhydrophobic porous coatings from hybrid ABC Triblock copolymers and their anticorrosive performance. ACS Applied Materials & Interfaces. 2017 Sep 6;**9**(35):30056-30063

[61] Alpysbaeva DA, Vershok DB, Emel'yanenko AM, Batishchev OV, Kuznetsov YI, Boinovich LB. Superhydrophobization of low-carbon steel with conversion coatings. Protection of Metals and Physical Chemistry of Surfaces. 2014 Dec;**50**(7):898-902

[62] Roselli SN, Lendvay-Györik G, Mészáros G, Deyá C, Romagnoli R. Anticorrosive water borne paints free from zinc and with reduced phosphate content. Progress in Organic Coating. 2017 Nov;**112**:27-36

[63] Zea C, Barranco-García R, Alcántara J, Chico B, Morcillo M, de la Fuente D. Hollow mesoporous silica nanoparticles loaded with phosphomolybdate as smart anticorrosive pigment. Journal of Coating Technology and Research. 2017 Jul;**14**(4):869-878

[64] Qiu P, Yang HF, Yang LJ, Chen ZS, Lv LJ, Song Y, et al. Enhanced inhibition of steel corrosion by L-cysteine under visible-light illumination. Materials and Corrosion. 2017 Sep;**68**(9):1004-1012

[65] Dariva GC, Galio FA. Corrosion inhibitors – Principles, mechanisms and applications. In: Aliofkhazraei M, editor. Developments in Corrosion Protection [Internet]. Rijeka: InTech; 2014

[66] Bassey I. Recent advances in computational design of organic materials for corrosion protection of steel in aqueous media. In: Aliofkhazraei M, editor. Developments in Corrosion Protection. Rijeka: InTech; 2014

[67] Martinez R, Olivares-Xomelt O, Likhanov VN. Environmentally friendly corrosion inhibitors. In: Aliofkhazraei M, editor. Developments in Corrosion Protection. Rijeka: InTech; 2014

[68] Zhu Y, Free ML, Woollam R, Durnie W. A review of surfactants as corrosion inhibitors and associated modeling. Progress in Materials Science. 2017 Oct;**90**:159-223

[69] Caldas CM, Calheiros LF, Soares GB. Silica-polyaniline hybrid materials prepared by inverse emulsion polymerization for epoxy-based anticorrosive coating. Journal of Applied Polymer Science. 2017 Dec 15;**134**(47):45505

[70] Maia F, Silva AP, Fernandes S, Cunha A, Almeida A, Tedim J, et al. Incorporation of biocides in nanocapsules for protective coatings used in maritime applications. Chemical Engineering Journal. 2015 Jun;**270**:150-157

[71] Mousaa IM, Radi H. Photosynthesis of anticorrosive protective coatings for steel substrate based on acrylated oil containing unsaturated amino acid compounds. Progress in Organic Coating. 2017 Jun;**107**:18-28

[72] Keyvani A, Yeganeh M, Rezaeyan H. Application of mesoporous silica nanocontainers as an intelligent host of molybdate corrosion inhibitor embedded in the epoxy coated steel. Progress in Natural Science: Materials International. 2017 Apr;**27**(2):261-267

[73] Oguzie EE, Li Y, Wang SG, Wang F. Understanding corrosion inhibition mechanisms—Experimental and theoretical approach. RSC Advances. 2011;**1**(5):866

[74] Guo L, Dong W, Zhang S. Theoretical challenges in understanding the inhibition mechanism of copper corrosion in acid media in the presence of three triazole derivatives. RSC Advances. 2014;**4**(79):41956-41967

[75] Guo L, Qi C, Zheng X, Zhang R, Shen X, Kaya S. Toward understanding the adsorption mechanism of large size organic corrosion inhibitors on an Fe(110) surface using the DFTB method. RSC Advances. 2017;**7**(46):29042-29050

[76] Malik MA, Hashim MA, Nabi F, Al-Thabaiti SA, Khan Z. Anti-corrosion ability of surfactants: A review. International Journal of Electrochemical Science. 2011;**6**(6):1927-1948

[77] Sığırcık G, Yildirim D, Tüken T. Synthesis and inhibitory effect of N,N′-bis(1-phenylethanol)ethylenediamine against steel corrosion in HCl media. Corrosion Science. 2017 May;**120**:184-193

[78] Meng Y, Ning W, Xu B, Yang W, Zhang K, Chen Y, et al. Inhibition of mild steel corrosion in hydrochloric acid using two novel pyridine Schiff base derivatives: A comparative study of experimental and theoretical results. RSC Advances. 2017;**7**(68):43014-43029

[79] Krim O, Elidrissi A, Hammouti B, Ouslim A, Benkaddour M. Synthesis, characterization, and comparative study of pyridine derivatives as corrosion inhibitors of mild steel in HCl medium. Chemical Engineering Communications. 2009 Aug 31;**196**(12):1536-1546

[80] Heakal FE-T, Elkholy AE. Gemini surfactants as corrosion inhibitors for carbon steel. Journal of Molecular Liquids. 2017 Mar;**230**:395-407

[81] Agafonkina MO, Andreeva NP, Kuznetsov YI, Timashev SF. Substituted benzotriazoles as inhibitors of copper corrosion in borate buffer solutions. Russian Journal of Physical Chemistry A. 2017 Aug;**91**(8):1414-1421

[82] Khan G, Basirun WJ, Kazi SN, Ahmed P, Magaji L, Ahmed SM, et al. Electrochemical investigation on the corrosion inhibition of mild steel by Quinazoline Schiff base compounds in hydrochloric acid solution. Journal of Colloid and Interface Science. 2017 Sep; **502**:134-145

[83] Qiang Y, Zhang S, Yan S, Zou X, Chen S. Three indazole derivatives as corrosion inhibitors of copper in a neutral chloride solution. Corrosion Science. 2017 Sep;**126**:295-304

[84] Li YZ, Xu N, Guo XP, Zhang GA. Inhibition effect of imidazoline inhibitor on the crevice corrosion of N80 carbon steel in the CO_2 -saturated NaCl solution containing acetic acid. Corrosion Science. 2017 Sep;**126**:127-141

[85] Zheludkevich ML, Yasakau KA, Poznyak SK, Ferreira MGS. Triazole and thiazole derivatives as corrosion inhibitors for AA2024 aluminium alloy. Corrosion Science. 2005 Dec; **47**(12):3368-3383

[86] Guo L, Kaya S, Obot IB, Zheng X, Qiang Y. Toward understanding the anticorrosive mechanism of some thiourea derivatives for carbon steel corrosion: A combined DFT and molecular dynamics investigation. Journal of Colloid and Interface Science. 2017 Nov;**506**:478-485

[87] Fouda AS, Elmorsi MA, Fayed T, El Said IA. Oxazole derivatives as corrosion inhibitors for 316L stainless steel in sulfamic acid solutions. Desalination and Water Treatment. 2016;**57**(10):4371-4385

[88] Fouda AS, El-Aal AA, Kandil AB. The effect of some phthalimide derivatives on corrosion behavior of copper in nitric acid. Desalination. 2006 Nov;**201**(1–3):216-223

[89] Oguzie EE, Adindu CB, Enenebeaku CK, Ogukwe CE, Chidiebere MA, Oguzie KL. Natural products for materials protection: Mechanism of corrosion inhibition of mild steel by acid extracts of Piper Guineense. Journal of Physical Chemistry C. 2012 Jun 28; **116**(25):13603-13615

[90] Dutta A, Saha SK, Adhikari U, Banerjee P, Sukul D. Effect of substitution on corrosion inhibition properties of 2-(substituted phenyl) benzimidazole derivatives on mild steel in 1 M HCl solution: A combined experimental and theoretical approach. Corrosion Science. 2017 Jul;**123**:256-266

[91] Ghazoui A, Benchat N, El-Hajjaji F, Taleb M, Rais Z, Saddik R, et al. The study of the effect of ethyl (6-methyl-3-oxopyridazin-2-yl) acetate on mild steel corrosion in 1M HCl. Journal of Alloys and Compounds. 2017 Feb;**693**:510-517

[92] Rhattas K, Benmessaoud M, Doubi M, Hajjaji N, Srhiri A. Corrosion inhibition of copper in 3% NaCl solution by derivative of Aminotriazole. Materials Sciences and Applications. 2011;**02**(04):220-225

[93] El-Sayed A-R, Shaker AM, Abd El-Lateef HM. Corrosion inhibition of tin, indium and tin–indium alloys by adenine or adenosine in hydrochloric acid solution. Corrosion Science. 2010 Jan;**52**(1):72-81

[94] Jiang S, Jiang L, Wang Z, Jin M, Bai S, Song S, et al. Deoxyribonucleic acid as an inhibitor for chloride-induced corrosion of reinforcing steel in simulated concrete pore solutions. Construction and Building Materials. 2017 Sep;**150**:238-247

[95] Jmiai A, El Ibrahimi B, Tara A, Oukhrib R, El Issami S, Jbara O, et al. Chitosan as an eco-friendly inhibitor for copper corrosion in acidic medium: Protocol and characterization. Cellulose. 2017 Sep;**24**(9):3843-3867

[96] Aiad I, El-Sukkary MM, Soliman EA, El-Awady MY, Shaban SM. Inhibition of mild steel corrosion in acidic medium by some cationic surfactants. Journal of Industrial and Engineering Chemistry. 2014 Sep;**20**(5):3524-3535

[97] Badr EA. Inhibition effect of synthesized cationic surfactant on the corrosion of carbon steel in 1M HCl. Journal of Industrial and Engineering Chemistry. 2014 Sep;**20**(5):3361-3366

[98] Palomar ME, Olivares-Xometl CO, Likhanova NV, Pérez-Navarrete J-B. Imidazolium, pyridinium and dimethyl-ethylbenzyl ammonium derived compounds as mixed corrosion inhibitors in acidic medium. Journal of Surfactants and Detergents. 2011 Apr;**14**(2):211-220

[99] Sangeetha Y, Meenakshi S, SairamSundaram C. Corrosion mitigation of N-(2-hydroxy-3-trimethyl ammonium)propyl chitosan chloride as inhibitor on mild steel. International Journal of Biological Macromolecules. 2015 Jan;**72**:1244-1249

[100] Shihab MS, Mahmood AF. Experimental and theoretical study of some N-pyridinium salt derivatives as corrosion inhibitors for mild-steel in 1 M H_2SO_4: Port. Electrochimica Acta. 2017;**35**(6):39-51

[101] Abd El-Lateef HM, Abo-Riya MA, Tantawy AH. Empirical and quantum chemical studies on the corrosion inhibition performance of some novel synthesized cationic gemini surfactants on carbon steel pipelines in acid pickling processes. Corrosion Science. 2016 Jul;**108**:94-110

[102] Hegazy MA, Abd El-Rehim SS, Badr EA, Kamel WM, Youssif AH. Mono-, di- and tetra-cationic surfactants as carbon steel corrosion inhibitors. Journal of Surfactants and Detergents. 2015 Nov;**18**(6):1033-1042

[103] Kaczerewska O, Leiva-Garcia R, Akid R, Brycki B. Efficiency of cationic gemini surfactants with 3-azamethylpentamethylene spacer as corrosion inhibitors for stainless steel in hydrochloric acid. Journal of Molecular Liquids. 2017;**247**:6-13

[104] Labena A, Hegazy MA, Horn H, Muller E. Cationic gemini surfactant as a corrosion inhibitor and a biocide for high salinity sulfidogenic bacteria originating from an oil-field water tank. Journal of Surfactants and Detergents. 2014;**17**:419-431

[105] Mobin M, Aslam R, Aslam J. Non toxic biodegradable cationic gemini surfactants as novel corrosion inhibitor for mild steel in hydrochloric acid medium and synergistic effect of sodium salicylate: Experimental and theoretical approach. Materials Chemistry and Physics. 2017 Apr;**191**:151-167

[106] Mobin M, Aslam R, Zehra S, Ahmad M. Bio−/environment-friendly cationic gemini surfactant as novel corrosion inhibitor for mild steel in 1 M HCl solution. Journal of Surfactants and Detergents. 2017 Jan;**20**(1):57-74

[107] Mobin M, Masroor S. Cationic gemini surfactants as novel corrosion inhibitor for mild steel in 1M HCl. International Journal of Electrochemical Science. 2012;**7**:6920-6940

[108] Sharma V, Borse M, Jauhari S, Pai KB, Devi S. New hydroxylated cationic gemini surfactants as effective corrosion inhibitors for mild steel in hydrochloric acid medium. Tenside, Surfactants, Detergents. 2005;**42**(3):163-167

[109] Tawfik SM. Ionic liquids based gemini cationic surfactants as corrosion inhibitors for carbon steel in hydrochloric acid solution. Journal of Molecular Liquids. 2016 Apr;**216**: 624-635

[110] Tawfik SM, Abd-Elaal AA, Aiad I. Three gemini cationic surfactants as biodegradable corrosion inhibitors for carbon steel in HCl solution. Research on Chemical Intermediates. 2016 Feb;**42**(2):1101-1123

[111] Asefi D, Arami M, Mahmoodi NM. Comparing chain length effect of single chain and gemini surfactants on corrosion inhibition of steel in acid. ECS Transactions. 2011;**35**(17): 89-101

[112] Zhang Q, Gao Z, Xu F, Zou X. Adsorption and corrosion inhibitive properties of gemini surfactants in the series of hexanediyl-1,6-bis-(diethyl alkyl ammonium bromide) on aluminium in hydrochloric acid solution. Colloids and Surfaces A: Physicochemical and Engineering Aspects. 2011 May;**380**(1–3):191-200

[113] Hegazy MA, Atlam FM. Three novel bolaamphiphiles as corrosion inhibitors for carbon steel in hydrochloric acid: Experimental and computational studies. Journal of Molecular Liquids. 2016 Jun;**218**:649-662

[114] Abdallah M, Eltass HM, Hegazy MA, Ahmed H. Adsorption and inhibition effect of novel cationic surfactant for pipelines carbon steel in acidic solution. Protection of Metals and Physical Chemistry of Surfaces. 2016 Jul;**52**(4):721-730

[115] Hegazy MA, Abdallah M, Awad MK, Rezk M. Three novel di-quaternary ammonium salts as corrosion inhibitors for API X65 steel pipeline in acidic solution. Part I: Experimental results. Corrosion Science. 2014 Apr;**81**:54-64

[116] Kaczerewska O, Brycki B, Leiva-García R, Akid R. Cationic Gemini Surfactants as Corrosion Inhibitors of Stainless Steel (AISI 304) in 3M Hydrochloric Acid; Praga; 2017

[117] Qiu L-G, Wu Y, Wang Y-M, Jiang X. Synergistic effect between cationic gemini surfactant and chloride ion for the corrosion inhibition of steel in sulphuric acid. Corrosion Science. 2008 Feb;**50**(2):576-582

[118] Z-Y W, Fang Z, Qiu L-G, Wu Y, Li Z-Q, Xu T, et al. Synergistic inhibition between the gemini surfactant and bromide ion for steel corrosion in sulphuric acid. Journal of Applied Electrochemistry. 2009 Jun;**39**(6):779-784

[119] Zhao J, Duan H, Jiang R. Synergistic corrosion inhibition effect of quinoline quaternary ammonium salt and Gemini surfactant in H_2S and CO_2 saturated brine solution. Corrosion Science. 2015 Feb;**91**:108-119

[120] Altamirano RH, Cervantes VYM, Rivera LSZ, Conde HIB, Ramirez SL. Gemini surfactants, process of manufacture and use as multifunctional corrosion inhibitors. US 9023 785 B2; 2015

[121] Henry KM, Hicks KD. Bis-quaternary ammonium salt corrosion inhibitors. US 8999315 B2; 2015

[122] El-Mahdy GA, Atta AM, Al-Lohedan HA, Ezzat AO. Influence of green corrosion inhibitor based on chitosan ionic liquid on the steel corrodibility in chloride solution. International Journal of Electrochemical Science. 2015;**10**:5812-5826

[123] Muyzer G, Stams AJM. The ecology and biotechnology of sulphate-reducing bacteria. Nature Reviews. Microbiology [Internet]. 2008 May;**7**; [cited 2017 Sep 19] Available from: http://www.nature.com/doifinder/10.1038/nrmicro1892

[124] Maluckov BS. Corrosion of steels induced by microorganisms. Metallurgical and Materials Engineering. 2012;**18**(3):223-232

[125] Kip N, van Veen JA. The dual role of microbes in corrosion. The ISME Journal. 2015 Mar;**9**(3):542-551

[126] Videla HA. Prevention and control of biocorrosion. International Biodeterioration and Biodegradation. 2002;**49**(4):259-270

[127] Miranda E, Bethencourt M, Botana FJ, Cano MJ, Sánchez-Amaya JM, Corzo A, et al. Biocorrosion of carbon steel alloys by an hydrogenotrophic sulfate-reducing bacterium *Desulfovibrio capillatus* isolated from a Mexican oil field separator. Corrosion Science. 2006 Sep;**48**(9):2417-2431

[128] Wagner M, Ivleva NP, Haisch C, Niessner R, Horn H. Combined use of confocal laser scanning microscopy (CLSM) and Raman microscopy (RM): Investigations on EPS – Matrix. Water Research. 2009 Jan;**43**(1):63-76

[129] Pradier CM, Rubio C, Poleunis C, Bertrand P, Marcus P, Compère C. Surface characterization of three marine bacterial strains by Fourier transform IR, X-ray photoelectron

spectroscopy, and time-of-flight secondary-ion mass spectrometry, correlation with adhesion on stainless steel surfaces. The Journal of Physical Chemistry. B. 2005 May;**109**(19): 9540-9549

[130] Lee M, Kim H, Seo J, Kang M, Kang S, Jang J, et al. Surface zwitterionization: Effective method for preventing oral bacterial biofilm formation on hydroxyapatite surfaces. Applied Surface Science. 2018 Jan;**427**:517-524

[131] Enning D, Garrelfs J. Corrosion of iron by sulfate-reducing bacteria: New views of an old problem. Applied and Environmental Microbiology. 2014 Feb 15;**80**(4):1226-1236

[132] Michalska J, Sowa M, Socha RP, Simka W, Cwalina B. The influence of *Desulfovibrio desulfuricans* bacteria on a Ni-Ti alloy: Electrochemical behavior and surface analysis. Electrochimica Acta. 2017 Sep;**249**:135-144

[133] Seo H, Roh Y. Biotransformation and its application: Biogenic Nano-catalyst and metal-reducing-bacteria for remediation of Cr(VI)-contaminated water. Journal of Nanoscience and Nanotechnology. 2015 Aug 1;**15**(8):5649-5652

[134] Labena A, Hegazy MA, Horn H, Müller E. The biocidal effect of a novel synthesized gemini surfactant on environmental sulfidogenic bacteria: Planktonic cells and biofilms. Materials Science and Engineering: C. 2015 Feb;**47**:367-375

[135] Labena A, Hegazy MA, Horn H, Müller E. Cationic gemini surfactant as a corrosion inhibitor and a biocide for high salinity sulfidogenic bacteria originating from an oil-field water tank. Journal of Surfactants and Detergents. 2014 May;**17**(3):419-431

[136] Mishra B, Lushnikova T, Golla RM, Wang X, Wang G. Design and surface immobilization of short anti-biofilm peptides. Acta Biomaterialia. 2017 Feb;**49**:316-328

[137] Mishra B, Lushnikova T, Wang G. Small lipopeptides possess anti-biofilm capability comparable to daptomycin and vancomycin. RSC Advances. 2015;**5**(73):59758-59769

[138] Danaee I, Ghasemi O, Rashed GR, Rashvand Avei M, Maddahy MH. Effect of hydroxyl group position on adsorption behavior and corrosion inhibition of hydroxybenzaldehyde Schiff bases: Electrochemical and quantum calculations. Journal of Molecular Structure. 2013 Mar;**1035**:247-259

[139] Migahed MA, Shaban MM, Fadda AA, Ali TA, Negm NA. Synthesis of some quaternary ammonium gemini surfactants and evaluation of their performance as corrosion inhibitors for carbon steel in oil well formation water containing sulfide ions. RSC Advances. 2015;**5**(126):104480-104492

[140] Obot IB, Macdonald DD, Gasem ZM. Density functional theory (DFT) as a powerful tool for designing new organic corrosion inhibitors. Part 1: An overview. Corrosion Science. 2015 Oct;**99**:1-30

[141] Zarrok H, Oudda H, Zarrouk A, Salghi R, Hammouti B, Bouachrine M. Weight loss measurement and theoretical study of new pyridazine compound as corrosion inhibitor for C38 steel in hydrochloric acid solution. Pharma Chemica. 2011;**3**(6):576-590

Chapter 2

Corrosion Inhibitors for Reinforced Concrete

Marianna Luna Sousa Rivetti,
José da Silva Andrade Neto,
Nilson Santana de Amorim Júnior and
Daniel Véras Ribeiro

Additional information is available at the end of the chapter

http://dx.doi.org/10.5772/intechopen.72772

Abstract

Reinforced concrete has been widely used in the last century, however, due to aggressive agents such as carbonic gas and chloride ions, it suffers premature deterioration. The concrete is a physical barrier that protects steel from corrosion, and the alkalinity of concrete leads to the formation of a passive layer around the reinforcement, which increases protection against corrosive processes. However, concrete is a porous material and has cracks that allow the entrance of aggressive agents, destabilizing the passive layer and corroding steel. The corrosion is the major cause of deterioration of concrete structures and several methods of protection and repair have been developed to increase the durability of such structures. Corrosion inhibitors, chemical substances that reduce the corrosion rate, have been widely used, both for prevention and correction. Inhibitors are classified according to their method of application, their mechanism of protection and chemical composition. In this chapter, through a literature review the main inhibitors used in reinforced concrete structures and their acting mechanisms are presented, as well as their efficiency and some side effects on concrete. It is very important to know the different types of inhibitors for correct use, thus increasing the life span of reinforced concrete structures.

Keywords: reinforced concrete, deterioration, corrosion inhibitors

1. Introduction

Reinforced concrete is a successful steel and concrete combination that has been widely used for over 100 years [1]. Steel provides tensile strength to the set, while concrete, as a high alkalinity material and in the role of a physical barrier, protects the steel from corrosion. The alkalinity of the concrete, which, according to Abdulrahman et al. [2], with pH values of about

12 to 14, provides the conditions for the formation of a passive film protecting the steel reinforcement from corrosion [3].

However, the protection provided by the concrete is not sufficient because the material is porous and has fissures, thus allowing the penetration of aggressive agents, such as chloride ions that lead to the corrosion of the reinforcement. Also of note is the action of carbon dioxide (CO_2) that reacts with the cement hydration products, a process known as carbonation, resulting in the formation of calcium carbonate and consequently, a pH decrease, thus destabilizing the reinforcement passivating film and making it susceptible to corrosion [4].

The corrosion of steel reinforcement is one of the main causes of premature deterioration of reinforced concrete, leading to significant economic losses [5, 6]. Rapid deterioration can be caused by chloride ions in a marine zone or due to the use of thaw salts [7] or by carbonation in urban zones [8].

In order to provide additional protection and increase the life span of reinforced concrete structures, several products identified as corrosion inhibitors have been used in recent years [2]. Due the advantages that they offer, corrosion inhibitors have been receiving more attention in recent years. The use of inhibitors is a promising method to increase the durability of concrete structures exposed to chloride ions [9] and carbonation.

Inhibitors are chemicals which, when present at in certain concentrations, reduce the corrosion rate without significantly altering the concentration of any other corrosion agent [10]. Corrosion inhibitors may be a good alternative compared to other conventional methods of protection and repair, due to their lower cost and ease of application [11]. In addition, inhibitors can be applied preventively or as a corrective action [2].

Corrosion inhibitors can be classified according to their application methods, protection mechanism, or composition [12], which can be organic and inorganic, inhibiting the corrosive process through the formation of a passivating film (anodic inhibitors) or by increasing the polarity and decreasing the corrosion potential (cathodic inhibitors). There are also inhibitors that act both ways. These products can be applied directly to the reinforcement, by a pretreatment through the immersion of the reinforcement in an inhibitor solution, added in the kneading water during the concrete mixing, or applied to the surface of the reinforced concrete structure, with the inhibiting solution penetrating the concrete by capillarity.

Known corrosion inhibitors used in reinforced concrete structures are sodium nitrite ($NaNO_2$), calcium nitrite [$Ca(NO_2)_2$], sodium monofluorophosphate (Na_2PO_3F), amine-based inhibitors, sodium and potassium chromate, sodium phosphate, and bauxite residue (red mud) [13]. In parallel, surface protection systems are adopted, such as painting, applying varnishes, epoxy resins, chlorinated rubber resins, and water-repellant silicones. **Figure 1** shows the main inhibitors used for reinforced concrete, and **Table 1** shows some recent studies that used these inhibitors.

In this chapter, the main inhibitors used in reinforced concrete structures and their acting mechanisms are presented through a literature review, as well as their efficiency and some side effects on the concrete. Knowing the different types of inhibitors is essential for the correct use of them, thus increasing the life span of reinforced concrete structures.

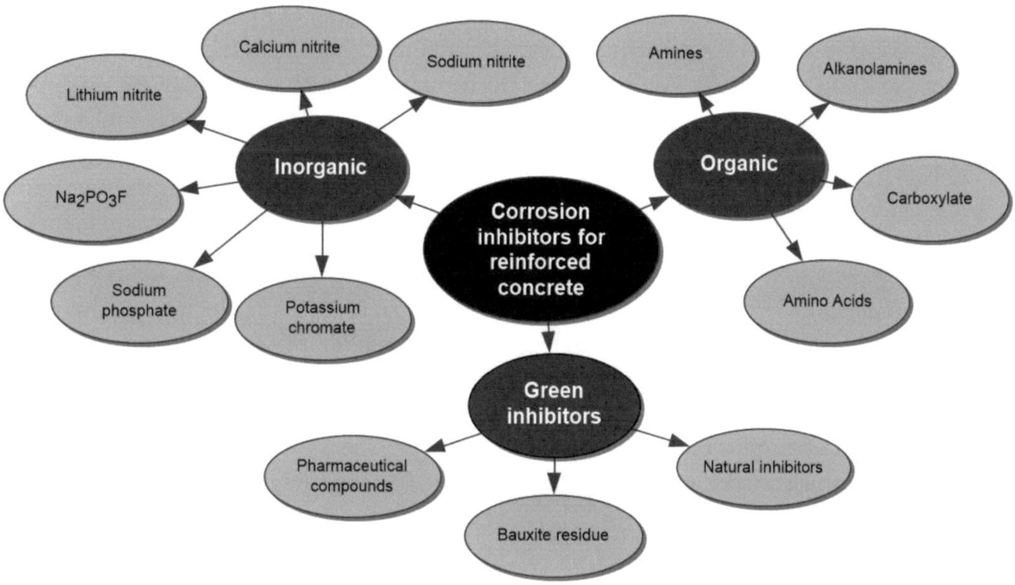

Figure 1. Corrosion inhibitors usually used for reinforced concrete.

Inhibitor	Type	Ref.	Inhibitor	Type	Ref.
Sodium nitrite	Inorganic	[14–16]	Amine	Organic	[15, 27–29]
Calcium nitrite	Inorganic	[15, 17–20]	Carboxylates	Organic	[27, 28]
Lithium nitrate	Inorganic	[7, 21, 22]	Amino acids	Organic	[27, 28]
Sodium monofluorophosphate	Inorganic	[23–25]	Pharmaceutical compounds	Green inhibitors	[30, 31]
Sodium phosphate	Inorganic	[25]	Red mud	Inorganic	[32]
Potassium chromate	Inorganic	[16, 26]	Natural inhibitors	Green inhibitors	[12, 33–35]

Table 1. Recent studies about the main corrosion inhibitors used for reinforced concrete structures.

2. Corrosion mechanisms in reinforced concretes

2.1. Corrosion

For the correct understanding of the mechanisms of action of the inhibitors, it is first necessary to understand the phenomena involved in the corrosion process in reinforced concrete because the mode of action of the inhibitors is closely related to the corrosion cell [36]. Corrosion can be defined as a process of deterioration of the material due to the chemical or electrochemical action of the environment, resulting in loss of material mass. The corrosion of the steel reinforcement of a reinforced concrete structure is an electrochemical process and occurs when there is difference in the concentration of dissolved ions inside the concrete, creating electrochemical potential cells or corrosion cells, characterized by a flow of electrons and ions between the cathodic and anodic regions [37]. The cathode and the anode are different regions of the same reinforcement, the region with section loss being the anodic region where corrosion of the steel occurs. The iron turns into ferrous ions and electrons, which migrate from the anode to the cathode. The reinforcement areas that have a more positive electrochemical potential act like cathodes, reducing oxygen and consuming the electrons coming from the anodic areas, in the presence of water, to form hydroxyl ion. In the electrochemical cell formed, the steel bars act as electrical conductors, and the interstitial fluid of the concrete is the electrolytic environment in which the ions move [38]. The schematic representation of the electrochemical corrosion process is shown in **Figure 2**.

According to Souza [37], the presence of different electrochemical potentials that form the corrosion cells is a consequence of physical and/or chemical heterogeneities on the metal surface. In steel reinforcement, these heterogeneities are mainly caused by differences in the concentration of ions in the surroundings of steel, such as alkalis, chlorides, and oxygen. Causes of potential difference between the points of the rebar include the difference in humidity, aeration, and stress in concrete and steel. As a result, in the presence of the electrolyte, part of the metal becomes anodic and another part cathodic, thus forming an electrochemical cell.

Eqs (1)–(3) represent the reactions that occur at the anode and at the cathode.

Anodic reaction: $$Fe \rightarrow Fe^{2+} + 2e^- \quad (oxidation\ reaction) \tag{1}$$

Cathodic reaction: $$O_2 + 2H_2O + 4e^- \rightarrow 4OH^- (redox\ reaction) \tag{2}$$

$$2Fe + O_2 + 2H_2O + 4e^- \rightarrow 2Fe(OH)_2 \tag{3}$$

According to Baroghel-Bouny et al. [38], ferrous hydroxide [$2Fe(OH)_2$] may undergo new oxidation reactions, forming various hydroxides and oxides, depending on the pH, temperature, amount of oxygen, and water availability. Souza [37] states that generally $Fe(OH)_2$ is oxidized to insoluble ferric hydroxide [$2Fe(OH)_3$], also represented by $Fe_2O_3{\cdot}H_2O$, commonly known as rust. The formed oxides and hydroxides have a higher volume than steel, as shown in **Figure 3**. The volume of the corrosion products is compared to the initial volume of iron (Fe).

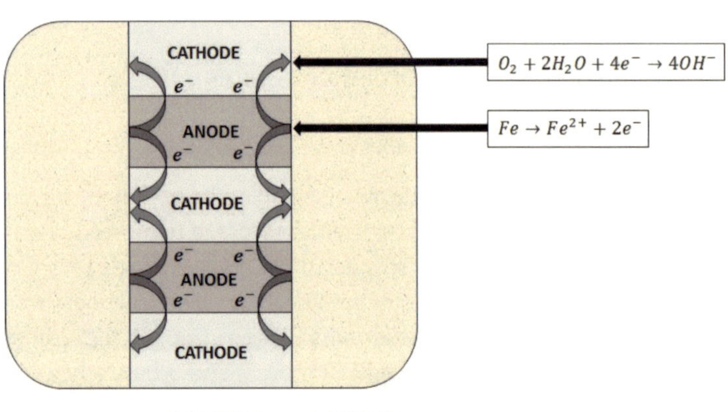

Figure 2. Representation of the corrosion process [37].

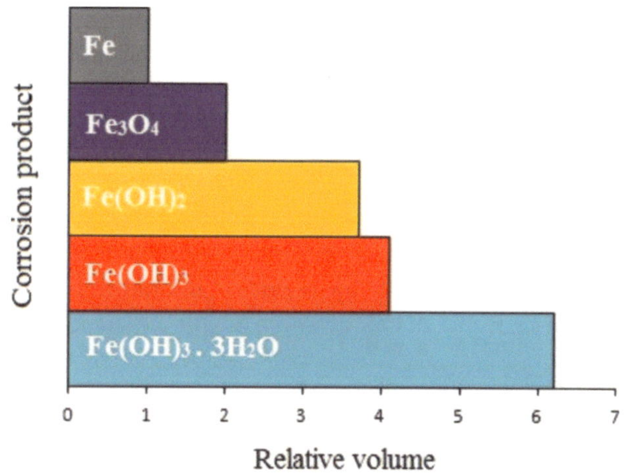

Figure 3. Corrosion product relative volume [38].

These higher volume corrosion products lead to pressure, generating cracks and thus reducing the adhesion between the steel bars and the concrete and also the chipping and detachment of concrete [38]. The formation of these expansive products causes the deterioration of concrete structures and, therefore, reduces their life span.

However, not all corrosion products are harmful to concrete structures. The formation of Fe_3O_4, one of the products of reinforcement corrosion, generates a compact and adherent passive film (passivating film) on rebar, acting as a physical barrier between the corrosive environment and the steel surface, protecting it from the corrosion process in the absence of chloride ions [37]. The formation of this compound occurs only at high pH values when the concrete is alkaline. This is the passive state, while the concrete is alkaline and free from aggressive ions such as chloride. However, the passive protective film can be destroyed in a phenomenon called depassivation, by decreasing the pH of the concrete to values between 8 and 9 (carbonation) and/or by the existence of chloride ions dissolved in the pore solution [36]. After the reinforcement has been depassivated, it will become susceptible to corrosion.

The main causes of corrosion in reinforced concrete structures are the penetration of chloride ions in marine regions and the carbonation process in urban environments. Thus, to better understand the mechanisms of inhibition of corrosion, these two processes must be understood.

2.1.1. Carbonation

In the absence of chloride ions, the steel passivating film is considered stable because the pH of the solution in the concrete pores is greater than 11.5. As Portland cement has high alkali contents and solid calcium hydroxide (about 20%) in the solution, the pH of the environment under normal conditions exceeds 12 [39]. However, reinforced concrete structures over time can undergo a decline in pH due to a process known as carbonation, which destabilizes the passivating layer and therefore leads to the beginning of the corrosion process. Carbonation occurs mainly in urban areas where there are large amounts of CO_2 emitted by vehicles and industrial parks [40]. **Figure 4** shows an example of corrosion on a viaduct caused by carbonation [41].

Carbonation is a process in which carbon dioxide (CO_2) penetrates the concrete by diffusion, reacting with the water present in the pores and forming carbonic acid. Carbonic acid reacts with the portlandite [$Ca(OH)_2$] formed during the cement hydration, resulting in the formation of calcium carbonate ($CaCO_3$) (Eq. (4)). This reaction causes a reduction in the environment pH to around 8 as the formed product ($CaCO_3$) is less alkaline than portlandite [42]. From the Pourbaix diagram (**Figure 5**), which shows the regions of immunity, passivation, and corrosion in an iron-water system at 25°C, the pH reduction to 8 with the same corrosion potential leads

Figure 4. Corrosion on the reinforced concrete of a viaduct [41].

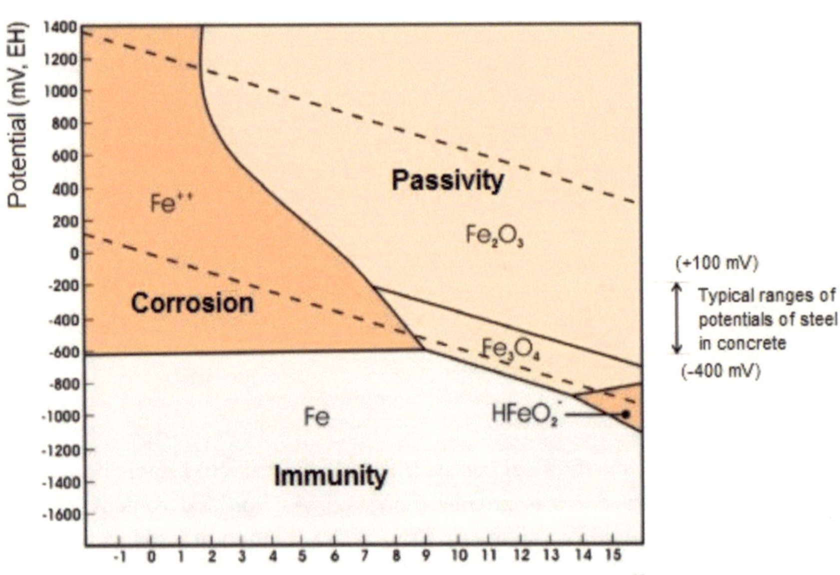

Figure 5. Iron-water system, without chloride, at 25°C Pourbaix diagram [43].

http://dx.doi.org/10.5772/intechopen.72772

the system from the passivation region to the region of corrosion. This pH reduction is responsible for the destabilization of the passivating film, thus initiating the corrosive process. The summarized carbonation reaction is presented in Eq. (4):

$$Ca(OH)_2 + CO_2 + H_2O \rightarrow CaCO_3 + 2H_2O \tag{4}$$

2.1.2. Chloride attack

The presence of chlorides in concrete, arising from seawater, marine atmosphere, setting time accelerator additives containing $CaCl_2$, water or contaminated aggregates from industrial pollutants, is one of the main causes of corrosion of reinforced concrete reinforcements [44]. **Figure 6** shows an example of corrosion caused by the action of chloride ions on Qeshm Port, Iran [45].

In the presence of chlorides, the passivating film can be destroyed even at pH values greater than 11.5. Small amounts can do this as this phenomenon is related to the amount of hydroxides available in the solutions present in the concrete pores. The maintenance of the passivating film depends on the balance between the alkalinity (verified by the amount of OH^- ions) and the acidity (verified by the quantity of Cl^- ions). Thus, the relationship between chloride ions and hydroxyl ions controls the rate of pitting corrosion [39, 46]. According to studies carried out by Hausmann [46], for Cl^-/OH^- molar ratios higher than 0.6, the passivating film becomes permeable or unstable, thus leaving the steel reinforcement unprotected against corrosion.

The corrosion process due to chloride ions is different from the corrosion process caused by carbonation because corrosion does not occur in a generalized way, but rather by pitting. In corrosion due to the action of chlorides, there is the formation of micro piles, in which the pitches constitute the anodic zones and the passivating film the cathodic surface [38].

The complete mechanism of corrosion due to the action of chloride ions is not yet fully understood. The most accepted theory suggests that the presence of ions causes a destabilization of the passivating layer, reacting with iron to form iron hydroxide (Eqs. (5) and (6)). In this process, the chloride ions are not consumed and are thus available to continue to react.

Figure 6. Corrosion of reinforced concrete in Qeshm port, Iran [45].

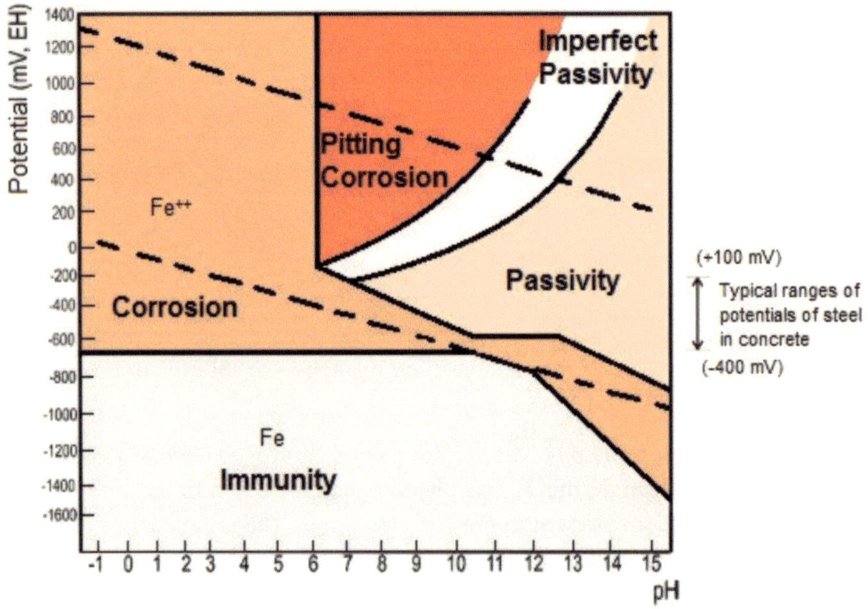

Figure 7. Iron-water system with the presence of chloride, at 25°C Pourbaix diagram [43].

$$Fe + 3Cl^{-} \rightarrow FeCl_3^{-} + 2e^{-} \tag{5}$$

$$FeCl_3^{-} + 2OH^{-} \rightarrow Fe(OH)_2 + 3Cl^{-} \tag{6}$$

Several aspects influence the probability of occurrence of corrosion due to chlorine ions. The decrease in the concrete water/cement ratio and concrete relative humidity, resistivity and increasing concrete cover thickness, the use of mineral additions, and the use of inhibitors tend to hinder the occurrence of corrosion.

The influence of the chloride ions on the depassivation of reinforcement can be observed by the alteration in the Pourbaix diagram for the iron-water system in the presence of chlorides (**Figure 7**). In this diagram, a decrease in the passivity region is observed in comparison to the Pourbaix diagram without the presence of chlorides (**Figure 5**). This reduction is due to the presence of pitting corrosion and imperfect passivity regions occurring due to the presence of these ions. From this diagram, it is possible to observe that corrosion can occur for all pH values, including values between 8.5 and 14, the zone of passivity or immunity for the iron-water system without chloride ions.

3. Corrosion inhibitors

Corrosion inhibitors are chemicals that when present in suitable quantities are capable of promoting a reduction in corrosion rate without significantly altering the concentration of any corrosive agent [10]. Thus, this concept excludes other corrosion protection methods that alter the concentration of aggressive agents, such as some mineral additives and other surface protection mechanisms such as coatings and pore blockers.

Corrosion inhibitors can act in the corrosive process in two ways: delaying the reinforcement depassivation, strengthening the passivating film, or reducing the corrosion rate after depassivation [36]. Thus, the use of corrosion inhibitors, when in adequate concentrations, delays the corrosive process and therefore prolongs the life span of reinforced concrete structures [13]. Nowadays, with increasing concern about the durability of reinforced concrete structures, the use of corrosion inhibitor additives has been gradually increasing so as to prolong the life span of the structures [15].

For reinforced concrete structures, different products are used to protect new and settled structures. For new structures, the inhibitors can be added to the fresh concrete mix, in a preventive way, to avoid or delay the onset of the corrosive process. In the preexisting structures, inhibitors can be added to repair mortars or repair concrete, applied to the concrete surface, or introduced through holes or grooves on the surface in order to accelerate the diffusion through the covering [13].

Among the inhibitors applied in existing concrete structures are the liquid products, usually phosphate substances such as monofluorophosphate [13]. For this category of inhibitors, it is essential that these substances have absorptive capacity so that they can penetrate the concrete structure through the interstitial pores to reach the reinforcement.

Inhibitors, especially those mixed with concrete, can alter the concrete properties both in the fresh and hardened state and may influence the setting time, the strength gain rate, and other properties. Thus, it is fundamental to understand its mechanism of action and its possible side effects in order to avoid unforeseen damage.

3.1. Classification of inhibitors

Corrosion inhibitors used in reinforced concrete structures can be classified according to their chemical composition (organic and inorganic), their protection mechanism (anodic, cathodic and both), or their method of application (mixed into the concrete and applied to the surface) and may also be grouped into another class known as green inhibitors.

3.1.1. Chemical composition

The action mechanisms and use of inhibitors are derived from their base (chemical composition). According to Ryu et al. [22], different types of inhibitors are used, traditionally divided into inorganic chemical compounds (basically nitrites) and organic (alkanolamine and carboxylates).

3.1.1.1. Inorganic inhibitors

Regarding inorganic inhibitors, nitrites stand out as the most used. These inhibitors, mainly added to the blend of fresh concrete as a preventive action, began to be tested in the 1950s. Initially sodium nitrite was studied. However, due to side effects, such as poorer of concrete strength and increased occurrence of the probability of alkali-silica reaction (RAS), which have negative effects on the concrete, it was replaced by calcium nitrite, the first to be widely commercialized on a large scale [13].

According to Ryu et al. [20], greater care is needed in estimating the quantity of calcium nitrite-based inhibitors to be used, because if the amount is less than necessary, it will have a negative impact on both the concrete and steel. Thus, it is of utmost importance to define the optimal amount to be added for the inhibitor work properly.

According to Abdulrahman et al. [2], the main advantage of using calcium nitrite is the possibility of using rational methods, based on the quality of the concrete, the exposure conditions, and the amount of calcium nitrite, to design it for a long life span with a certain chloride/nitrite expected ratio. In addition, nitrites are effective in inhibiting the corrosion process caused by both chloride ions by carbonation and the action of sulfate ions [47]. Among the side effects in concrete, nitrites accelerate the setting of fresh concrete and increase its strength gain rate [36].

Despite their efficiency in reducing the corrosion rate of concrete structure reinforcement, nitrites are carcinogenic and toxic [48] and banned in several European countries, such as Germany and Switzerland [5]. Because of this, alternative inhibitors are being studied. Among other inorganic inhibitors are sodium monofluorophosphate (Na_2PO_3F) and red mud.

Sodium monofluorophosphate (Na_2PO_3F) has the ability of both delaying the corrosion onset and reducing the corrosion rate, even in carbonate concretes. The Na_2PO_3F in aqueous and neutral environment undergoes hydrolysis, forming orthophosphate and fluoride, which reacts with the corrosion products, forming Fe_3O_4, γ-Fe_2O_3, and $FePO_4.H_2O$ passive films.

Red mud is a residue from the bauxite processing in the manufacture of aluminum. It has the capacity to increase reinforced concrete structures resistance to corrosion. Its performance as corrosion inhibitor is discussed in Section 3.1.4. of this chapter.

3.1.1.2. Organic inhibitors

Like inorganic, organic inhibitors have been widely used in recent years, with several studies on their mechanisms of action and efficiency [27, 49–51]. Organic compounds can donate electrons to or receive electrons from the surface of the metal, forming covalent bonds and are therefore excellent corrosion inhibitors [51]. According to Zhi-Yong et al. [49], organic inhibitors applied on the concrete surface are considered one of the simplest and most effective methods to control the corrosion in existing reinforced concrete structures. Ryu et al. [52] argue that organic inhibitors are popular and frequently used to control corrosion in reinforcement concrete, including in the presence of chloride ions.

Those inhibitors are developed based on organic compounds which have nitrogen (N), oxygen (O), and sulfur (S) atoms and multiple bonds within molecules which aid the adsorption of these on the metallic surface because its efficiency is related to the adsorption properties. Additionally, adsorption depends on the nature and state of the metallic surface, the corrosive environment type, and the chemical structure of the inhibitor [2]. According to Hmamou et al. [53] and Singh et al. [54], the adsorption on the steel surface depends on various factors. These are mainly on the physical-chemical properties of the inhibitor group, as with the functional group, the donated atom's electronic density and the character p of the orbital.

Thus, the inhibition efficiency depends on the molecular electronic structure containing active centers of adsorption (S, N, O), the molecular size, the method of absorption, the formation of metallic compounds, and the rate of inhibitor cover on the metallic surface [51].

Within the corrosion inhibitors, the amines and alkanolamines are widely used due to their high solubility in water and small influence on the properties of concrete in its fresh and hardened state [27]. These inhibitors have a high capacity to spread themselves through the concrete. They are called volatile inhibitors or migration inhibitors as they can be mixed on the concrete, repair mortar, or be applied on the concrete surface during service [13].

The amines are organic compounds which possess a nitrogen atom with a lone pair that provides the material with the capacity to adhere to the metal surface, protecting it from corrosion [27]. According to Lourenço and Souza [13], the amine can be adsorbed on the metal surface, preventing contact between the metal and the corrosive environment, forming a protective steel barrier. Additionally, the amine is capable of removing corrosion products from the metallic surface, granting direct contact of the metallic surface with the protection layer [13].

Another group used as organic inhibitors is substances such as carboxylate, characterized by the presence of a carboxylic acid group (-COOH) [27]. As amines, these substances have the ability of adhere to the surface of the metal, thus forming an organic layer, protecting the steel from corrosion [36]. Among the effects generated in the concrete, carboxylate-based inhibitors tend to decrease the cement hydration rate, increasing the setting time and slowing the increase in resistance rate [36].

Amino acids, molecules composed by functional groups of amine and carboxylate (**Figure 8**), are also used as organic inhibitors [27].

In a study comparing 80 organic inhibitors, Ormellese et al. [27] conclude that amines have a low effect of inhibition, amino acids presented an inhibiting effect, but not enough for industrial

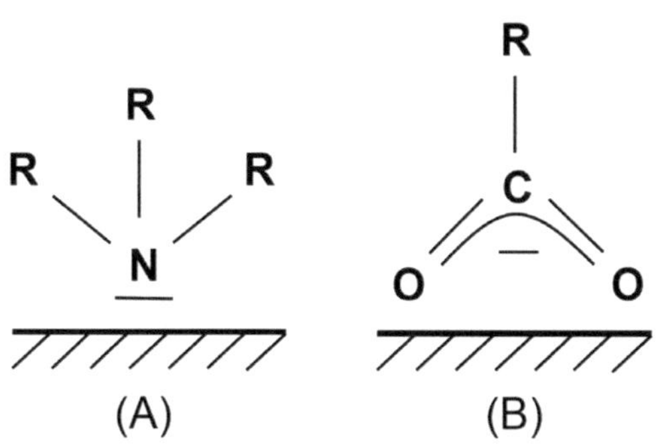

Figure 8. Functional groups of amines (A) and carboxylates (B) [27].

application. The authors also concluded that the substances with carboxylate, particularly polycarboxylates, present high efficiency in corrosion rate reduction. These are the most promising among the tested inhibitors.

3.1.2. Protection mechanisms

The inhibitors are classified as anodic, cathodic, or mixed [13] according to the electrochemical protection mechanism.

3.1.2.1. Anodic mechanism of protection

Anodic inhibitors, also known as passivation inhibitors, act as oxidizing agents, gaining electrons [36]. Such inhibitors act by promoting the oxidation of the ferrous ions on ferritic ions, which precipitate, forming a passive dense film on the steel surface and ceasing the corrosion process anodic reaction [13, 36]. The cohesive and insoluble film formed on the metallic surface is a result of the reaction between the inhibitors and the product of corrosion [55]. According to Myrdal [36], the effects of these inhibitors on the cathodic reaction are negligible.

According to Myrdal [36], the action of the anodic inhibitors increases the electrochemical potential of the anodic region, and therefore the anodic and cathodic curves intercept with a lower corrosion current and higher corrosion potential, as illustrated in **Figure 9**.

The anodic inhibitors are effective if present in high concentrations, which is defined according to the level of chloride existing in the environment to which the steel will be exposed [56].

According to Elsener [57], reinforced concrete counts on the use of calcium nitrite [$Ca(NO_2)_2$] and calcium nitrate [$Ca(NO_3)_2$] as anodic inhibitor, as they present similar inhibiting properties [36]. Nitrite alkali salts, benzoate, and chromate are also used as anodic inhibitors [11].

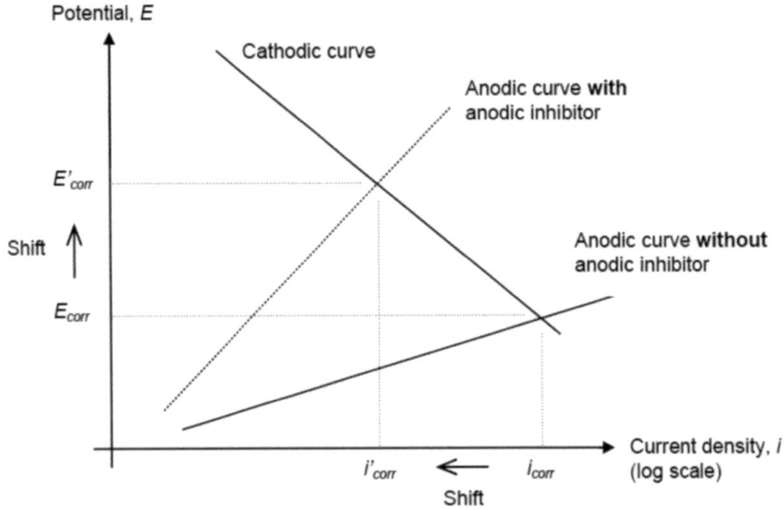

Figure 9. Corrosion potential with anodic inhibit effect (Ecorr = corrosion potential and icorr = corrosion rate) [36].

3.1.2.2. Cathodic mechanism of protection

The cathodic inhibitors act during the corrosion process on the cathodic region and prevent a cathodic reaction on the steel [55]. These inhibitors retard the oxygen reduction reaction that occurs on the steel surface, as they produce insoluble compounds that precipitate on the cathodic regions, forming a superficial barrier upon deposition on the metal, thus forming a compact and adherent film [36, 55]. This barrier restricts oxygen diffusion and the conduction electrons in this region [55].

The reduction in the oxygen supply in the corrosion process reduces the corrosion rate and the corrosion potential for lower negative values [36, 55]. By the time the corrosion potential is reduced, it is usual to consider that there is a smaller corrosion risk [36]. It can be observed in the Pourbaix diagrams (**Figures 5** and **7**), in which the lower the potential, the bigger the probability for the steel to be in an immunity zone. **Figure 10** illustrates the cathodic inhibitor behavior on the corrosion process.

As observed in **Figure 8**, the cathodic inhibitor promotes a change in the corrosion potential and in the corrosion rate for the values E'corr and i'corr, respectively. As the corrosion potential is reduced to lower values, the corrosion rate diminishes [36], thus cathodic polarization occurs.

Examples of cathodic inhibitors include carbonates, phosphates, polyphosphates, and silicates [36]; zinc oxide was also reported by Rincón et al. [58] as a cathodic inhibitor. According to Söylev and Richardson [11], sodium hydroxide and sodium carbonate are among the most used cathodic inhibitors, which tend to increase the pH around encompassing the steel, in addition to reducing the oxygen transport and creating a film on the steel surface.

The contents of cathodic inhibitors tend to be higher than anodic inhibitors because the suppression effect of the corrosive process is inferior to the anodic inhibitors [59].

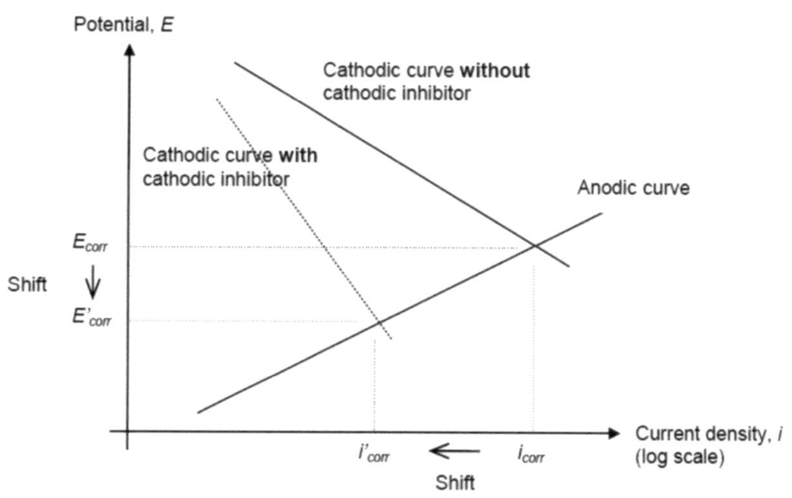

Figure 10. Corrosion potential with cathodic inhibitor effect (Ecorr = corrosion potential and icorr = corrosion rate) [36].

3.1.2.3. Mixed mechanism of protection

The mixed inhibitors act both on the cathodic and the anodic regions, usually by the adsorption on the steel surface, forming a protecting layer [11], therefore being denominated adsorption inhibitors [60] or pellicle forming inhibitors [61]. The inhibitor reduces the corrosion rates, causing no significant changes to the corrosion potential. The corrosion potential is slightly changed for the anodic or cathodic side, so regardless of the most affected reaction, the total effect upon the corrosion potential is negligible [36]. This type of inhibitor is therefore just effective on the reduction of the corrosion rate (icorr) as shown in **Figure 11**.

Several mixed inhibitors are organic compounds such as amines and alkanolamines [12], capable of application both as inhibitors mixed on the fresh concrete, as much as inhibitors applied on the concrete surface which migrate to its interior [36].

3.1.3. Application methods

Inhibitors can also be classified according to the application method. They can be either mixed with the concrete or applied to the surface of the structures.

3.1.3.1. Mixed into the concrete

Additives are mixed to concrete usually to improve its properties in the fresh or hardened states; however, corrosion inhibitors act mainly on the steel surface. Inhibitors are mixed into the concrete in the fresh state, at the moment the mixing water is in contact with the steel surface at the time of concreting. They are considered corrosion inhibitors because they interact chemically or electrochemically with the surface of the steel [61].

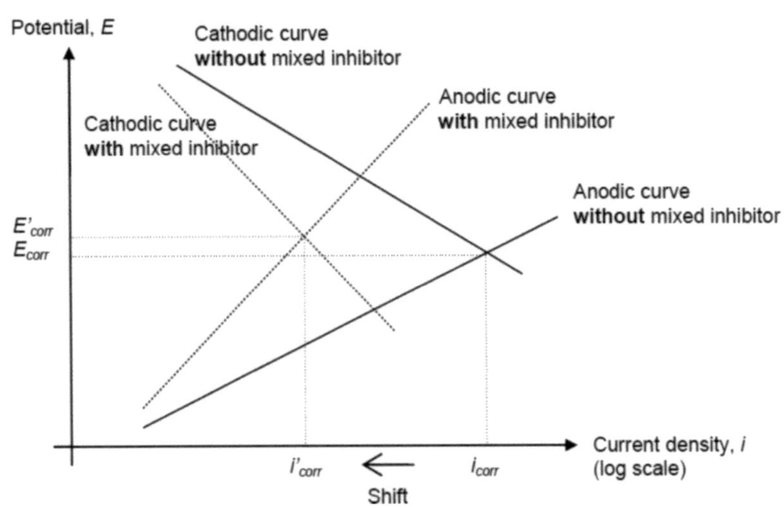

Figure 11. Corrosion potential with mixed inhibitor effect (Ecorr = corrosion potential and icorr = corrosion rate) [36].

These inhibitors should not impair the properties of the concrete, but as a result of being mixed in the fresh state, they usually give other characteristics to the concrete as well as inhibiting corrosion. Given this, some studies have been developed in order to verify the effects of corrosion inhibitors on the properties of the concrete. According to Gaidis [62], inhibitors can influence the concrete hydration process, the gain in mechanical resistance, and other properties.

According to Elsener and Angst [61], monofluorophosphate (MFP) cannot be used as a corrosion inhibitor mixed into the concrete due to its reaction with the fresh concrete, which removes the active substance from the concrete pore solution and retards its hydration; it is therefore recommended only for surface application. The sodium nitrite or potassium nitrites cause a fall in the concrete compressive strength and raise the risk of alkali-aggregate reaction (AAR). The calcium nitrite, on the other hand, acts by accelerating the cement hydration, which requires the addition of a water reducer and a setting retarder in the concrete mix.

3.1.3.2. Applied to the concrete surface (absorption and migration)

The corrosion inhibitors applied to the surface should be capable of migrating to the concrete interior until they reach the metal reinforcement through absorption by capillary pores [36] and become adsorbed on the metallic surface. These are also called migration corrosion inhibitors (MCI) and are considered an effective technique, able to inhibit and retard the reinforcement corrosive process and the concrete structure corrosive process [63]. Several studies have been developed [63, 64] in the use of this technique, and it is becoming one of the main methods used, mainly due to the ease of application.

These inhibitors act by forming an adsorbed film on the steel surface, preventing the depolarizing agents (O_2 and Cl^-) from getting into direct contact with the reinforcement surface, avoiding therefore the onset of the corrosive process [5]. Fei et al. [5] claim that this protective layer formed by the inhibitors and blocks the electrons flow on the concrete/metal interface, also creating a barrier to the aggressive ions, making the reinforcement less prone to corrosion. The efficiency of the inhibitor in protecting the steel from corrosion depends mainly on the interaction between the inhibitor and the metallic surface [51].

Among the corrosion inhibitors applied to concrete, there are the alkanolamines and amines, which are able to penetrate the capillary pores and micro pores of the concrete [63]. Lan et al. [64] found that inhibitors of amino alcohol and carboxylates applied on the high resistance concrete contaminated with chlorides are efficient at inhibiting the corrosive process in medium and low concentrations of chloride ions.

3.1.3.3. Commercial inhibitors

Several corrosion inhibitors for reinforced concrete structures are marketed worldwide. **Table 2** shows some of these inhibitors and their respective manufacturers, chemical basis, and method of

Manufacturer	Product	Chemical basis	Method of use
Sika	CNI	Calcium nitrite	Mixed into concrete
	FerroGard 901	Amino alcohols	Mixed into concrete
	FerroGard 903	Amino alcohols	Applied on concrete
BASF	MasterLife 222	Amines and esters	Mixed into concrete
	MasterLife CI30	Calcium nitrite	Mixed into concrete
W. R. Grace	DCI	Calcium nitrite	Mixed into concrete
Cortec Co.	MCI 2000	Amino alcohols	Mixed into concrete
	MCI 2005	Amine carboxylate	Mixed into concrete
	MCI 2007	Superplasticizer + amine carboxylate	Mixed into concrete
	MCI 2020	Amino alcohols + carboxylic acid salt	Applied on concrete
Euclid chemical	EUCON BCN	Calcium nitrite	Mixed into concrete
	EUCON CIA	Calcium nitrite	Mixed into concrete

Table 2. Examples of commercial inhibitors.

use. As can be seen in **Table 2**, most of the commercially available corrosion inhibitors are organic or calcium nitrite based.

3.1.4. Green inhibitors

Inhibitors may be natural or synthetic [55]. Synthetic inhibitors are artificially produced and can lead to damage to both man and the environment, synthetic organic inhibitors in particular. Due to these problems, several studies have examined the versatility and feasibility of the use of environmentally sustainable inhibitors, known as green or natural inhibitors [33].

These natural and low-cost inhibitors that are capable to feed the need for synthetics [33–35] have been shown to be more effective and highly beneficial to the environment compared to organic and inorganic inhibitors produced by chemical and petrochemical industries [34].

Examples of green inhibitors are natural organic compounds obtained from several types of plants such as herbs, spices, and others [12, 34]. Such plants are an abundant source of naturally synthesized chemical compounds, which can be extracted through simple and low-cost practices, besides being biodegradable. Much research has been carried out into the extraction of natural stems, leaves, or plant seeds [12, 33–35].

Within the classification of green inhibitors is red mud, a compound based on ferric oxide and alumina and a residue generated in bauxite processing for aluminium production. Ribeiro et al. [43] evaluated the incorporation of red mud in the reinforced concrete, in contents of 10, 20, and 30% of the cement mass. The authors experimentally observed a significant decrease in the reinforcement corrosion rate as the amount of red mud increased. The corrosion reduction is probably due to the increase in the alkalinity near the region of the steel reinforcement, which favors the reestablishment of the passivating film, in addition to the high concentrations of aluminates, which hinder the diffusion of chlorides.

http://dx.doi.org/10.5772/intechopen.72772

3.2. Factors influencing inhibitor choice

Given the variety of available inhibitors, choosing the appropriate inhibitor should be based on the physical-chemical properties of the inhibitor, as well as the nature and conditions of the metal surface and the aggressive environment in which the concrete structure is situated [51].

The efficiency of the corrosion inhibitors can be calculated by Eq. (7), which relates the decreasing corrosion rate with the use of the inhibitor:

$$E_f = \frac{R_o - R_i}{R_o} x100 \tag{7}$$

where E_f—efficiency factor (%)

R_o—corrosion rate on concrete without inhibitors ($\mu A/cm^2$)

R_i—corrosion rate on concrete with inhibitors ($\mu A/cm^2$)

Table 3 summarizes the results obtained in several studies with varied inhibitors.

Ref.	Inhibitor	Type	Efficiency (%)	Aggressive conditions
[15]	Calcium nitrite, 25 (g/L)	Inorganic	92.30	Carbonation
[15]	Calcium nitrite, 10 (g/L)	Inorganic	64.90	3.5% NaCl
[22]	Calcium nitrite, 8.94 (g/L)	Inorganic	85.75	0.99 NaCl (g/L)
[22]	Calcium nitrite, 71.55 (g/L)	Inorganic	83.26	7.91 NaCl (g/L)
[15]	Sodium nitrite, 25 (g/L)	Inorganic	95.00	Carbonation
[15]	Sodium nitrite, 10 (g/L)	Inorganic	82.60	3.5% NaCl
[14]	Sodium nitrite, 40 ppm	Inorganic	61.40	–
[20]	Lithium nitrate, 2.15 (g/L)	Inorganic	63.67	0.99 de NaCl (g/L)
[20]	Lithium nitrate, 17.21 (g/L)	Inorganic	31.71	7.91 de NaCl (g/L)
[15]	Sodium molybdate, 10 (g/L)	Inorganic	95.40	Carbonation
[15]	Sodium molybdate, 40 (g/L)	Inorganic	82.60	3.5% NaCl
[14]	Sodium molybdate, 1000 ppm	Inorganic	98.60	–
[15]	Tannin, 10 (g/L)	Organic	64.00	Carbonation
[15]	Tannin, 10 (g/L)	Organic	80.60	3.5% NaCl
[15]	Amine, 10 (g/L)	Organic	84.80	Carbonation
[15]	Amine, 10 (g/L)	Organic	61.80	3.5% NaCl
[14]	Sodium nitroethane, 20 ppm	Organic	87.40	–
[65]	DAESPN, 5×10^{-3} M	Organic	39.40	3.0 M NaCl
[65]	Bis-HSPN, 10^{-3} M	Organic	59.00	3.0 M NaCl
[51]	PT, 10^{-3} M	Organic	93.68	HCl 1 M
[66]	Amine-ether, 10 (g/L)	Organic	41.30	Carbonation

Table 3. Efficiency factors of various inhibitors studied by different authors.

Among the inhibitors shown in **Table 3**, sodium molybdate shows the highest efficiency (82.60–98.60%); however, despite its good performance, it is expensive which makes the use of this inhibitor unviable for some applications.

4. Future trends

The use of inhibitors of low environmental impact is becoming more popular nowadays mostly due to the major concerns about the use of these chemicals. Studies with green inhibitors have been widely discussed, and we should their use increase in the next few years.

There remain some gaps in our knowledge, and further work is needed, for example, the effect of inhibitors upon the properties of fresh concrete and the development of multifunctional blends, besides the use of inhibitor without undesired secondary effects starting from a combination of different types, such as a mixture of an inhibitor that has a setting retarder effect with an inhibitor that has a setting accelerator effect [36].

Lastly, Quraishi et al. [67] point out the importance of studies about association of the effects of chemical admixture (corrosion inhibitors) and minerals (pozzolans) so as to increase the life span of reinforced concrete structures.

5. Final considerations

Among the commercially available inhibitors, there is a wide variety classified according to their chemical composition, method of protection, and method of use. The use of inhibitors generally leads to side effects in the concrete; therefore an understanding of these substances is of fundamental importance in order to avoid damage to structures.

The environmental issue related to toxicity and carcinogenic inhibitors, such as calcium nitrite, which is widely used, has led to several studies reviewing the use of current inhibitors and developing natural inhibitors that do not harm the properties of concrete while being eco-friendly [33]. It is important to investigate any side effects the inhibitor may have on the properties of concrete, both in the fresh and hardened states [36].

The concrete composition and microstructure must be considered in the evaluation of the performance of corrosion inhibitors, as they may influence several aspects of the concrete in the corrosive process [5]. Therefore, when evaluating inhibitors, it is necessary to verify their composition, mechanism of action, and application, as well as the aspects related to concrete and also the influence of the inhibitors on the concrete (changes in the hydration process, porosity, resistance, others).

http://dx.doi.org/10.5772/intechopen.72772

Author details

Marianna Luna Sousa Rivetti, José da Silva Andrade Neto, Nilson Santana de Amorim Júnior and Daniel Véras Ribeiro*

*Address all correspondence to: verasribeiro@hotmail.com

Federal University of Bahia, Brazil

References

[1] Shan H, Xu J, Wanga Z, Jiang L, Xu N. Electrochemical chloride removal in reinforced concrete structures: Improvement of effectiveness by simultaneous migration of silicate ion. Construction and Building Materials. 2016;**127**:344-352. DOI: 10.1016/j.conbuildmat. 2016.09.137

[2] Abdulrahman AS, Ismail M, Hussain MS. Corrosion inhibitors for steel reinforcement in concrete: A review. Scientific Research and Essays. 2011;**6**:4152-4162. DOI: 10.5897/SRE11.1051

[3] Elfmarkova V, Spiesz P, Brouwers HJH. Determination of the chloride diffusion coefficient in blended cement mortars. Cement and Concrete Research. 2015;**78**:190-199. DOI: 10.1016/2015.06.014

[4] Zhang D, Ghouleh Z, Shao Y. Review on carbonation curing of cement-based materials. Journal of CO2 Utilization. 2017;**21**:119-131. DOI: 10.1016/j.cemconres.2015.06.014

[5] Fei F, Jie H, Wei J, Yu Q, Chen Z. Corrosion performance of steel reinforcement in simulated concrete pore solutions in the presence of imidazoline quaternary ammonium salt corrosion inhibitor. Construction and Building Materials. 2014;**70**(15). DOI: 10.1016/j. conbuildmat.2014.07.082

[6] Song N, Wang Z, Liu Z, Zhou J, Zheng D. Study on improvement of durability for reinforced concrete by surface-painting migrating corrosion inhibitor and engineering application. In: 1st International Conference on New Material and Chemical Industry (NMCI2016). IOP Conference Series: Materials Science and Engineering; 2017. p. 167. DOI: 10.1088/1757-899X/167/1/012072

[7] Lee HS, Ryu HS, Park WJ, Ismail MA. Comparative study on corrosion protection of reinforcing steel by using amino alcohol and lithium nitrite inhibitors. Materials (Basel). 2015;**8**:251-269. DOI: 10.3390/ma8010251

[8] Redaelli E, Bertolini L. Electrochemical repair techniques in carbonated concrete. Part I: Electrochemical realkalisation. Journal of Applied Electrochemistry. 2011;**41**:817-827. DOI: 10.1007/s10800-011-0301-4

[9] Xu J, Jiang L, Wang W, Tang L, Cui L. Effectiveness of inhibitors in increasing chloride threshold value for steel corrosion. Water Science and Engineering. 2013;**6**:354-363. DOI: 10.3882/j.issn.1674-2370.2013.03.011

[10] The International Organization for Standardization, ISO 8044. Corrosion of metals and alloys – Basic terms and definitions; 2015

[11] Söylev TA, Richardson MG. Corrosion inhibitors for steel in concrete: State-of-the-art report. Construction and Building Materials. 2008;**22**:609-622. DOI: 10.1016/j.conbuildmat.2006.10.013

[12] Raja PB, Ghoreishiamiri S, Ismail M. Natural corrosion inhibitors for steel reinforcement in concrete – A review. World Scientific Publishing Company. 2015;**22**:68-75. DOI: 10.1142/S0218625X15500407

[13] Lourenço MZ, Souza CAC. Métodos de proteção e aumento da durabilidade do concreto armado. In: Ribeiro DV, editor. Corrosão em Estruturas de Concreto Armado: Teoria, Controle e Métodos de Análise. 1st ed. Rio de Janeiro: Elsevier; 2014. pp. 51-73

[14] Mohammadi S, Ravari FB, Dadgarinezhad A. Improvement in corrosion inhibition efficiency of Molybdate-based inhibitors via addition of Nitroethane and zinc in stimulated cooling water. Corrosion. 2012;**9**. DOI: 10.5402/2012/515326

[15] Bolina CC, Cascudo O. Inibidores de corrosão: análise da eficiência frente à corrosão do aco induzida por cloretos e carbonatação em meio aquaoso. ENGEVISTA. 2013;**15**(1415–7374):81-94

[16] Okeniyi JO, Omotosho OA, Ajayi OO, Loto CA. Effect of potassium-chromate and sodium-nitrite on concrete steel-rebar degradation in sulphate and saline media. Construction and Building Materials. 2014;**50**:448-456. DOI: 10.1016/j.conbuildmat.2013.09.063

[17] Królikowski A, Kuziak J. Impedance study on calcium nitrite as a penetrating corrosion inhibitor for steel in concrete. Electrochimica Acta. 2011;**56**:7845-7853. DOI: 10.1016/j.electacta.2011.01.069

[18] Balonis M, Glasser FP. Calcium nitrite corrosion inhibitor in Portland cement: Influence of nitrite on chloride binding and mineralogy. Journal of the American Ceramic Society. 2011;**94**(7):2230-2241. DOI: 10.1111/j.1551-2916.2010.04362.x

[19] Montes-Garcia P, Jiménez-Quero V, López-Calvo H. Assessment of high performance concrete containing fly ash and calcium nitrite based corrosion inhibitor as a mean to prevent the corrosion of reinforcing steel. VII International Congress of Engineering Physics. Journal of Physics: Conference Series. 2015;582. DOI:10.1088/1742-6596/582/1/012028

[20] Ryu HS, Singh JK, Lee H, Park W. An electrochemical study to evaluate the effect of calcium nitrite inhibitor to mitigate the corrosion of reinforcement in sodium chloride contaminated ca$(OH)_2$ solution. Advances in Materials Science and Engineering. 2017. DOI: 10.1155/2017/626518

[21] Lee H, Shin S. Evaluation on the effect of lithium nitrite corrosion inhibitor by the corrosion sensors embedded in mortar. Construction and Building Materials. 2007;**21**:1-6. DOI: 10.1016/j.conbuildmat.2006.01.004

[22] Ryu HS, Singh JK, Lee H, Ismail MA, Park W. Effect of LiNO2 inhibitor on corrosion characteristics of steel rebar in saturated ca$(OH)_2$ solution containing NaCl: An electrochemical study. Construction and Building Materials. 2017;**133**(15):387-396. DOI: 10.1016/j.conbuildmat.2016.12.086

[23] Alonso C, Andrade C, Argiz C, Malric B. Na_2PO_3F as inhibitor of corroding reinforcement in carbonated concrete. Cement and Concrete Research. 1996;**26**(3):405-415. DOI: 10.1016/S0008-8846(96)85028-9

[24] Douche-Portanguen A, Prince W, Lutz T, Arliguie G. Detection or quantitative analysis of a corrosion inhibitor, the sodium monofluorophosphate, in concrete. Cement and Concrete Composites. 2005;**27**(6):679-687. DOI: 10.1016/j.cemconcomp.2004.11.002

[25] Bastidas DM, Criado M, La Iglesia VM, Fajardo S, La Iglesia A, Bastidas JM. Comparative study of three sodium phosphates as corrosion inhibitors for steel reinforcements. Cement and Concrete Composites. 2013;**43**:31-48. DOI: 10.1016/j.cemconcomp.2013.06.005

[26] Omotosho OA, Okeniyi JO, Ajayi OO. Performance evaluation of potassium dichromate and potassium chromate inhibitors on concrete steel rebar corrosion. Journal of Failure Analysis and Prevetion. 2010;**10**:408-415. DOI: 10.1007/s11668-010-9375-2

[27] Ormellese M, Lazzari L, Goidanich S, Fumagalli G, Brenna A. A study of organic substances as inhibitors for chloride-induced corrosion in concrete. Corrosion Science. 2009;**51**(12):2959-2968. DOI: 10.1016/j.corsci.2009.08.018

[28] Diamanti MV, Pérez Rosales EA, Raffaini G, Ganazzoli F, Brenna A, Pedeferri M, Ormellese M. Molecular modelling and electrochemical evaluation of organic inhibitors in concrete. Corrosion Science. 2015;**100**:231-241. DOI: 10.1016/j.corsci.2015.07.034

[29] Mechmeche LB, Dhouibi L, Ben-Ouezdou M, Triki E, Zucchi F. Investigation of the early effectiveness of an amino-alcohol based corrosion inhibitor using simulated pore solutions and mortar specimens. Cement and Concrete Composites. 2008;**30**(3):167-173. DOI: 10.1016/j.cemconcomp.2007.05.007

[30] Pathak RK, Mishra P. Drugs as corrosion inhibitors: A review. International Journal of Science and Research. 2016;**5**:671-677

[31] Mahdi AS. Amoxicillin as green corrosion inhibitor for concrete reinforced steel in simulated concrete pore solution containing chloride. International Journal of Advanced Research in Engineering and Technology. 2014;**5**:99-107

[32] Ribeiro DV, Labrincha JA, Morelli MR. Effect of the addition of red mud on the corrosion parameters of reinforced concrete. Cement and Concrete Research. 2012;**42**:124-133. DOI: 10.1590/S1983-41952012000400004

[33] Abdulrahman AS, Ismail M. Green plant extract as a passivation-promoting inhibitor for reinforced concrete. International Journal of Engineering Science and Technology. 2011;**3**(8): 6484-6490

[34] Asipita SA, Ismail M, Majid MZA, Majid ZA, Abdullah C, Mirza J. Green bambusa arundinacea leaves extract as a sustainable corrosion inhibitor in steel reinforced concrete. Journal of Cleaner Production. 2014;**67**(15):139-146. DOI: 10.1016/j.jclepro.2013.12.033

[35] Okeniyi JO, Loto CA, Popoola API. Electrochemical performance of anthocleista djalonensis on steel-reinforcement corrosion in concrete immersed in saline/marine simulating-environment. Transactions of the Indian Institute of Metals. 2014;**67**(6):959-969. DOI: 10.1007/s12666-014-0424-5

[36] Myrdal R. Corrosion Inhibitors – State of the art. In: COIN Project Report; 2010; Oslo. Oslo: SINTEF Building and Infrastructure; 2010

[37] Sousa CAA. Princípios da corrosão eletroquímica. In: Ribeiro DV, editor. Corrosão em Estruturas de Concreto Armado: Teoria, Controle e Métodos de Análise. 1st ed. Rio de Janeiro: Elsevier; 2014. pp. 13-50

[38] Baroghel-Bouny V, Capra B, Laurens S. A durabilidade das armaduras e do concreto de cobrimento. In: Ollivier J, Vichot A, eds. Durabilidade do Concreto: Bases científicas para a formulação de concretos duráveis de acordo com o ambiente. 1st ed. São Paulo: IBRACON; 2014

[39] Mehta PK, Monteiro PJM, editors. Concrete: Microstructure, Properties and Materials. 4th ed. New York: MC GRAW Hill Education; 2014

[40] González F, Fajardo G, Arliguie G, Juárez CA, Escadeillas G. Electrochemical realkalisation of carbonated concrete: An alternative approach to prevention of reinforcing steel corrosion. International Journal of Electrochemical Science. 2011;**6**:6332-6349

[41] Pereira JLMB, Miranda MLB, Fainstein A, Aurélio LC, Carvalho CL, Rosa CLM. Manual de recuperação de pontes e viadutos rodoviários. DNIT, Rio de Janeiro; 2010. 159 p

[42] Shi Z, Lothenbach B, Geiker MR, Kaufmann J, Leemann AS, Ferreiro S, Skibsteda J. Experimental studies and thermodynamic modeling of the carbonation of Portland cement, metakaolin and limestone mortars. Cement and Concrete Research. 2016;**88**:60-72. DOI: 10.1016/2016.06.006

[43] Ribeiro DV, Cunha MPT. Deterioração das estruturas de concreto armado. In: Ribeiro DV, editor. Corrosão em estruturas de concreto armado: Teoria, Controle e Métodos de Análise. 1st ed. Rio de Janeiro: Elsevier; 2014. pp. 87-118

[44] Almeida FCR, Sales A. Efeitos da ação do meio ambiente sobre as estruturas de concreto. In: Ribeiro DV, editor. Corrosão em Estruturas de Concreto Armado: Teoria, Controle e Métodos de Análise. 1st ed. Rio de Janeiro: Elsevier; 2014. pp. 51-73

[45] Ramesht MH, Tavasani MAMA. Case study on corrosion in concrete floating docks in Qeshm port. Procedia Engineering. 2013;**54**:109-116. DOI: 10.1016/j.proeng.2013.03.010

[46] Hausmann DA. Steel corrosion in concrete: How does it occur? Materials Protection. 1967;**6**:16-28

[47] Cao Z, Hibino M, Goda H. Effect of nitrite concentration and pH on steel corrosion induced by sulfate in simulated concrete pore environment. Applied Mechanics and Materials. 2013;**368-370**:911-918. DOI: 10.4028/www.scientific.net/AMM.368-370.911

[48] Bryan SN, Alexander DD, Coughlin JR, Milkowski AL, Boffetta P. Ingested nitrate and nitrite and stomach cancer risk: An updated review. Food and Chemical Toxicology. 2012;**50**(10):3646-3665. DOI: 10.1016/j.fct.2012.07.062

[49] Zhi-Yong L, Xiu-lin L, Wen-Jian L, Baohong W. Electrochemical behavior of carbon steel in neutral chloride solution adding amine-carboxylate based corrosion inhibitor. Applied Mechanics and Materials. 1603-1606;**2012**:166-169. DOI: 10.4028/www.scientific.net/AMM.166-169.1603

[50] Li ZY, Tu LQ, Qin MQ. Effectiveness of amino alcohol-based corrosion inhibitor and its influence on concrete properties. Applied Mechanics and Materials. 2014;**578-579**:1454-1458. DOI: 10.4028/www.scientific.net/AMM.578-579.1454

[51] Larouj M, Lgaza AH, Serrard H, Zarroka H, Bourazmib H, Zarroukc A, Elmidaouia A, Guenbourb A, Boukhrisd S, Ouddaa H. Adsorption properties and inhibition of carbon steel corrosion in hydrochloric acid solution by ethyl 3-hydroxy-8-methyl-4-oxo-6-phenyl-2-(p-toly)-4,6- dihydropyrimido[2,1-b] [1,3]thiazine-7-carboxylate. Journal of Materials and Environmental Science. 2015;**6**(11):3251-3267

[52] Ryu HS, Singh JK, Yang HM, Ismail MA. Evaluation of corrosion resistance properties of N, NO -dimethyl ethanolamine corrosion inhibitor in saturated Ca(OH)2 solution with different concentrations of chloride ions by electrochemical experiments. Construction and Building Materials. 2016;**114**:223-231. DOI: 10.1016/j.conbuildmat.2016.03.174

[53] Hmamou DB, Aouad MR, Salghi R, Zarrouk A, Assouag M, Benali O, Messali M, Zarrok H, Hammouti B. Inhibition of C38 steel corrosion in hydrochloric acid solution by 4,5-diphenyl-1H-Imidazole-2-thiol: Gravimetric and temperature effects treatments. Journal of Chemical and Pharmaceutical Research. 2012;**4**:3498-3504

[54] Singh AK, Quraishi MA, Piroxicam A. Novel corrosion inhibitor for mild steel corrosion in HCl acid solution. Journal of Materials and Environmental Science. 2010;**1**(2):101-110

[55] Dariva CG, Galio AF. Corrosion inhibitors – Principles, mechanisms and applications. In: Aliofkhazraei M, editor. Developments in Corrosion Protection. 1st Ed. InTech. 2014. pp. 365-379. DOI: 10.5772/57255

[56] Trépanier SM, Hope BB, Hansson CM. Corrosion inhibitors in concrete. Part III: Effect on time to chloride-induced corrosion initiation and subsequent corrosion rates of steel in

mortar. Cement and Concrete Research. 2001;**31**(5):713-718. DOI: 10.1016/S0008-8846(01)00470-7

[57] Elsener B. Mixed-in inhibitors. In: Weydert R, editor. COST 521: Corrosion of Steel in Reinforced Concrete Structures. Prevention – Monitoring – Maintenance. 1st ed. Luxembourg: Final Workshop ed. 2002. pp. 43-55

[58] Rincón OT, De Pérez O, Paredes E, Caldera Y, Urdaneta C, Sandoval I. Long-term performance of ZnO as a rebar corrosion inhibitor. Cement and Concrete Composites. 2002;**24**(1):79-87. DOI: 10.1016/S0958-9465(01)00029-4

[59] Soeda K, Ichimura T. Present state of corrosion inhibitors in Japan. Cement and Concrete Composites. 2003;**25**(1):117-122. DOI: 10.1016/S0958-9465(01)00058-0

[60] Kondratova IL, Montes P, Bremner TW. Natural marine exposure results for reinforced concrete slabs with corrosion inhibitors. Cement and Concrete Composites. 2003;**25**(4–5):483-490. DOI: 10.1016/S0958-9465(02)00088-4

[61] Elsener B, Angst U. Corrosion inhibitors for reinforced concrete. Science and Technology of Concrete. 2016. DOI: 10.1016/B978-0-08-100693-1.00014-X

[62] Gaidis JM. Chemistry of corrosion inhibitors. Cement and Concrete Composites. 2004;**26**(3):181-189. DOI: 10.1016/S0958-9465(03)00037-4

[63] Liu Z, Miao C, Zhou W, Liu J. Maintenance and improvement of durability of reinforced concrete using migrating corrosion inhibitors. Journal of the Chinese Ceramic Society. 2008;**36**:1494-1500

[64] Lan WJ, Wu B, Liu ZU. Effect of surface applied inhibitor on corrosion rate of steel bar in high strength concrete with various admixed chlorides. Applied Mechanics and Materials. 2012;**174–177**:1184-1187. DOI: 10.4028/www.scientific.net/AMM.174-177.1184

[65] Sezer E, Ustamehmetoglu B, Bayir ZA, Çoban K, Kalkan A. Corrosion inhibition effect of 4-(2-Diethylamino-Ethylsulfonyl)-Phthalonitrile and 4,5-Bis(Hexylsulfonyl)-Phthalonitrile. Intenatonal Journal of Electrochemistry. 2011;**5**. DOI: 10.4061/2011/235360

[66] Kaur K, Goyal S, Bhattacharjee B, Kumar M. Efficiency of migratory-type organic corrosion inhibitors in carbonated environment. Journal of Advanced Concrete Technology. 2016;**14**:548-558. DOI: 10.3151/jact.14.548

[67] Quraishi MA, Nayak DK, Kumar R, Kumar V. Corrosion of reinforced steel in concrete and its control: An overview. Journal of Steel Structures and Construction. 2017;**3**. DOI: 10.4172/2472-0437.1000124

Section 2

Principles of Corrosion Inhibitors

Chapter 3

Green Methods for Corrosion Control

Khadija M. Emran, Shimaa M. Ali and
Hamedh A. Al Lehaibi

Additional information is available at the end of the chapter

http://dx.doi.org/10.5772/intechopen.72762

Abstract

Corrosion of metals is a serious environmental problem. Most of the corrosion inhibitors are synthetic chemicals that are hazardous to environments and expensive. Today, there has been an increasing search for green corrosion inhibitors, due to the toxicity of some inhibitors. Green corrosion inhibitors are biodegradable and free of heavy metals or other toxic compounds. There is an intensive effort underway to develop new plant origin corrosion inhibitors for metal subjected to various environmental conditions. Salts of rare earth elements can be also used as inorganic alternative nontoxic corrosion inhibitors. Sol-gel coatings can be recommended as green protective films for metal corrosion. The origin of sol-gel process, its advantages and limitations, and the type of coating method are presented in this chapter, as well as a brief summary of recent literature concerning the use of hybrid inorganic-organic sol-gel and nanostructured sol-gel protective coatings for metal corrosion in different media.

Keywords: corrosion, green inhibitors, plant extracts, protective films, sol-gel process

1. Introduction

Metallic materials are the pre-eminent important materials used in structural and decorative applications and are still the most widely used group of materials particularly in both mechanical engineering and transportation industry [1]. The usefulness of these metals is constrained by one common problem known as corrosion. The corrosion, deterioration or destruction of metals is an unavoidable but controllable process by means of optimum corrosion management practices.

One of the established methods of protection against corrosion damage is the use of corrosion inhibitors. A corrosion inhibitor is a solid or liquid substance when added in a suitable

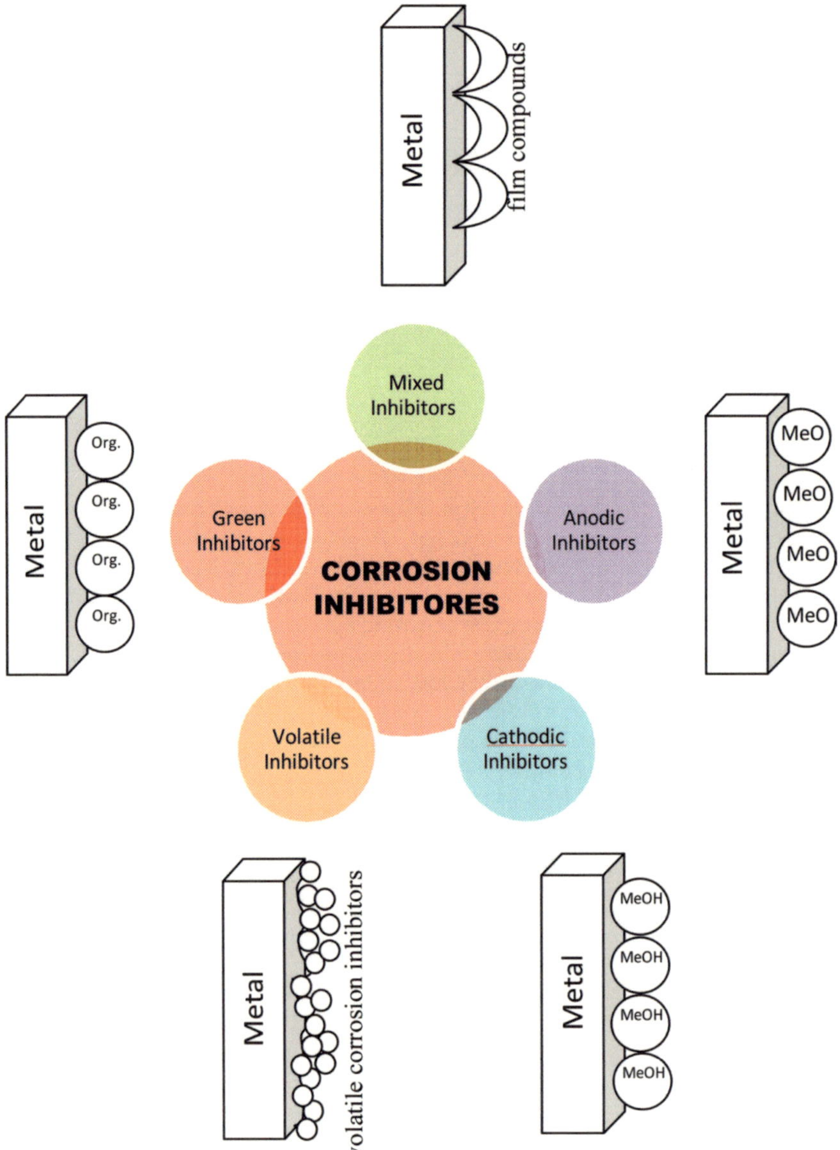

Figure 1. Types of corrosion inhibitors and their possible mechanisms.

concentration to an environment reduces the corrosion rate by reducing the corrosivity of the environment [2]. Corrosion inhibitors can be divided into two broad categories: passive film formation and adsorbing on the metal surface. In both cases, a barrier was created on the surface that prevents access of corrosive agents to the metal surface [3]. Corrosion inhibitors have always been considered to be the first line of defiance against corrosion. According to their action, there are four types of corrosion inhibitors: anodic inhibitors, cathodic inhibitors, mixed inhibitors and volatile corrosion inhibitors (VCI). Anodic inhibitors usually cause a large anodic shift of the corrosion potential to reach passivation region by forming a protective oxide film on the metal surface. Cathodic inhibitors act by selectively precipitating on cathodic areas. This blocking of cathodic areas limits the diffusion of reducing species to the surface. Also, they can slow the cathodic reaction itself. Mixed inhibitors work by reducing both the cathodic and anodic reactions by forming film compounds that block both anodic and cathodic sites. Volatile corrosion inhibitors (VCI) compounds have sufficient vapor pressures to allow vaporization and subsequent condensation and adsorption on metal surfaces as a monomolecular passivating layer and control the corrosion attack. In closed vapor spaces, such as shipping containers, volatile solids such as salts of dicyclohexylamine, cyclohexylamine, and hexamethyleneamine are used [4–6]. When choosing an inhibitor, there are several factors including cost and amount, easy availability and, most importantly, safety to environment need to be considered.

Another newly established method for corrosion control, covered in this chapter, is employing green sol-gel coatings. Sol-gel science, early discovered in the 1800s and studied in detail since the 1930s, has attracted researchers' interest to a great extent. As inorganic gels can be formed at low temperatures and converted to glasses without the need of melting process.

Sol-gel materials encompass a wide range of inorganic and organic-inorganic hybrid materials which share a common preparation strategy: the sol-gel process. It involves the generation of colloidal suspensions, "sols" [7]. During the process, inorganic networks grow by sol (colloidal suspension) formation then the sol undergoes gelation (gel). Metal alkoxides, mainly alkoxysilane, are most commonly used precursors for the sol synthesis as they react easily with water. Deposition of a protective layer can be a proper way to enhance the corrosion resistance of a substrate. This can be done by a variety of ways such as chemical [8] or physical [9] vapor deposition, laser cladding [10], thermal spraying [11] and sol-gel [12]. Upon comparison, the sol-gel method has advantages of being simple, deposition can be achieved at low temperatures. The low-temperature reaction makes it possible to incorporate organic modifiers into the inorganic network, thus providing new materials which have combined reactants' properties [13–15]. In conclusion, types of corrosion inhibitors with their possible mechanisms are summarized in **Figure 1**.

2. Green inhibitors

Green corrosion inhibitors or "eco-friendly inhibitors" are biodegradable compounds and do not contain heavy metals or other toxic compounds. They are biocompatible with nature. Similar to the general classification of "inhibitors," "green inhibitors" can also be grouped into two categories, namely organic green inhibitors and inorganic green inhibitors.

2.1. Organic green inhibitors

The organic green inhibitors are the quinine, alkaloids, phenol, phenylpropanoid flavonoids, organic acid and other natural products obtained from natural sources like plants. It also includes synthetic compounds with negligible toxicity. Herein, plant extracts, gums, drugs and oils have been considered as green corrosion inhibitors. The use of wastes from plants as corrosion inhibitors can be another way of extending the beneficial use of these plants and so enhance municipal waste management. Some of the notable developments on organic green inhibitors, especially plant extracts, are discussed here.

Aquatic plants contain some organic compounds that provide definite physiological actions on the human body and these bioactive substances include alkaloids, terpenoids, steroids, phenolic compounds, tannins and flavonoids and most of them are known to have corrosion inhibition efficiency.

Plant material extracts contain many active principles. They contain polar or hetero atoms such as S, N, O, P and aromatic rings, favoring the adsorption by donor-acceptor interaction between π electrons of donor atoms S, N, O, P and aromatic rings of inhibitors, and the vacant d-orbitals of metal surface atoms [16]. Because of adsorption of inhibitor molecules on the metal surface, protective film is formed that isolates the metal from the aggressive environment. Thus, corrosion is controlled.

Large numbers of organic extracts of plant materials have been used to prevent a variety of metals such as mild steel [17–28], various types of steel [29, 30], cast iron [31] aluminum and its alloys [32–34], zinc and its alloys [35–38], copper [39–41], nickel [42, 43] and tin [44, 45].

According to corrosive medium, extracts of plant materials have been used to prevent corrosion of metals in various medium. Most studies were conducted in acidic medium [31] than in basic [34] and neutral medium [35]. Different parts of plants that include roots, leaves, bark, flowers and fruits (shell, juice and seed) were used to reduce the corrosion. Many methods have been employed such as hydrogen evolution, weight loss method, thermometric methods, electrochemical studies and so on to evaluate corrosion inhibition efficiencies of inhibitors extracted from plants. The protective film has been analyzed by various surface analyses techniques such as fourier-transform infrared spectroscopy (FTIR) spectroscopy, UV-visible spectroscopy, scanning electron microscope (SEM), energy dispersive X-rays analysis (EDXA) and atomic force microscope (AFM).

Overtime, plants extracts have been used as corrosion inhibitors, alone or in combination with other inhibitors. The mixture of plants extracts shows better inhibition efficiency than individual members, which is called synergistic effect. Many additives have been used to improve the inhibition efficiency of plant extracts. For example Zn^{2+}, methylene dichloride, n-butanolic extract [46–51], KI [22, 52, 53], KCl, KBr [54–58], formaldehyde, diethylamine [59], ethyl acetate [60], and dimethyl sulfoxide [61].

2.1.1. Advantages and limitations of plant extract use

Plant extracts are environmental friendly, nontoxic, low cost and easily biodegradable. Despite these features as environmentally acceptable, readily available and renewable source for a wide range of green inhibitors, they have some main drawbacks. The drawback of most reports on plant extracts as corrosion inhibitors is that the active ingredient has not been identified. Also, they are easily biodegradable and contaminated by microorganisms which limit the storage of liquid extract or even the solid extract for a long term. According to a number of research works published in the last two decades on these groups of products, plant extracts occupy the first place by 68% for aluminum and its alloys corrosion inhibition (**Figure 1**) [16].

It is important to consider the type of plant extract (water, alcohol, acid). The quality of the ingredients present in plant extracts consequently affects the effectiveness of the extracts in inhibiting corrosion.

2.1.2. Mechanism of action of green inhibitors

Many theory mechanisms have been put by several workers. These mechanisms are being summarized in the following points:

- Organic substances may form onium ions in acidic solutions and are adsorbed on the cathodic sites of the metal surface.
- Organic substances in some plant extracts contain an alkaloid berberine which has a long chain of aromatic rings, an N-atom in the ring, and O atoms as H atoms substitution. The free electrons on the O and N atoms form bonds with the electrons on the metal surface.
- Allyl propyl disulfide, which is an S-containing unsaturated compound, exists in many plant extracts and affects the potential cathodic process of metals such as steel.
- Pyrrolidine in neutral and alkaline aqueous media ionizes, and the N-atom acquires a negative charge, and the higher charge on free electrons on N results in stronger bond between the pyrrolidine and metals.
- The N-atom in alkaloid ricinine is in the ring attachment of the $–OCH_3$ favoring the interaction with the metal surface.

2.2. Inorganic green inhibitors

Green inhibitors display substantially improved environmental properties as compared to the existing inhibitors whose toxic level is above 50%. The action of inorganic green inhibitors is to improve the passivity of a metal and alloys. According to that, most inorganic green inhibitors are compounds or metal ions that have more positive redox potential than the metal surface and also improve the passivity of a metal by favoring the electropositive metal to be reduced and deposited on the surface. There are several inorganic salts that are less toxic than chromate inhibitors. Among the metals used for corrosion inhibitor are: mercury (Hg) [62], lanthanide salts (La) [63, 64], samarium (Sm) [63], cerium salts [64–68] and platinum (Pt) [69] (**Table 1**).

Moreover, there are inorganic anions providing passivation protection to metal surfaces through their incorporation into the oxide layer; the most widely used of these are: chromate (CrO_4^{2-}), nitrate (NO_2^-), molybdate (MoO_3^-), phosphate ($H_2PO_3^-$) and silicates [70].

Inhibitor medium	Metal and medium	Reference
$La(NO_3)_3$, $Sm(NO_3)_3$, $LaCl_3$, $SmCl_3$	AISI 434 SS in NaCl	[65]
$CeCl_3$	Aluminum alloy in NaCl	[66]
$CeCl_3$	AA5083 alloy and galvanized steel in NaCl	[67]
$CeCl_3$	Sn-Fe alloy in NaCl	[68]
$CeCl_3$	AA6964 Al-Zn alloy in NaCl	[69]
$(Ce(NO_3)_3{\cdot}6H_2O)$	Tinplate in citric-citrate	[70]
Tryptamine (TA)	ARMCO iron in 0.5 M deaerated H_2SO_4	[71]

Table 1. Model inorganic green inhibitors.

2.2.1. *Mechanism of inhibitors*

In neutral and alkaline media, the oxygen reduction reaction is more favored as cathodic reaction Eq. (1).

$$O_2 + 2\,H_2O + 2e \rightarrow 4\,OH^- \tag{1}$$

The cathodic reaction in Eq. (1) of oxygen reduction increases local pH by generating a high concentration of OH^- ions. These OH^- ions formed over the cathodic sites react with metal ions present in the solution according to Eqs. (2) and (3).

$$Ce^{3+} + 3\,OH^- \rightarrow Ce\,(OH)_3 \tag{2}$$

$$La^{3+} + 3\,OH^- \rightarrow La\,(OH)_3 \tag{3}$$

These reactions give rise to the precipitation of lanthanide oxides or hydroxides over cathodic sites and block them. Blocking the cathodic active sites by these compounds decreases the available cathodic current and therefore reduces the corrosion process.

2.3. Sol-gel protective coatings

The sol-gel process involves several stages as shown in **Figure 2** [71]:

- Hydrolysis of the precursors with the formation of hydroxide species;
- Condensation with the formation of oxide species;
- Gelation with the formation of a "spanning cluster" yielding a network which entraps the remaining solution;
- Ageing with the formation of further crosslinks which change the structure in pore sizes and pore wall strengths;

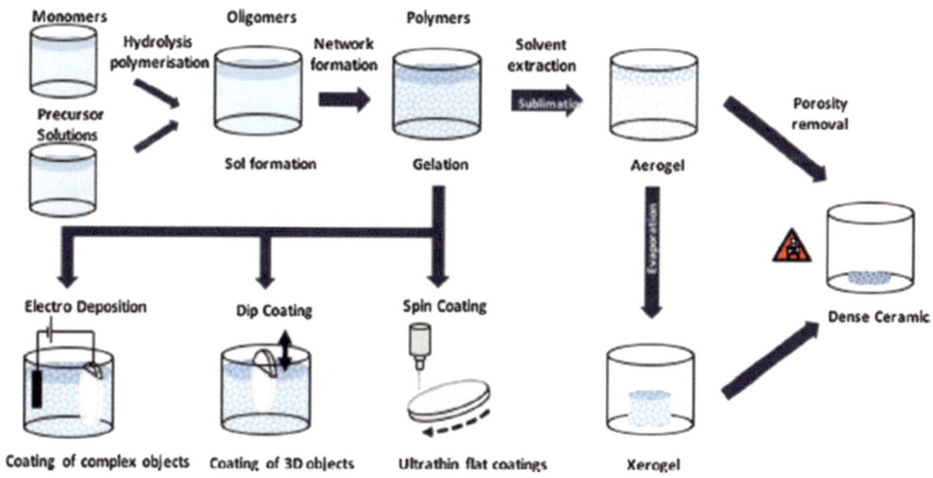

Figure 2. Sol-gel process and possible coatings routes [71].

- Drying with the loss of solvent and the associated development of capillary stress; and
- Densification with the collapse of the open structure and formation of a dense material.

2.3.1. *Advantages of sol-gel coatings*

There are many advantages using sol-gel coatings, and some of the most important features are listed as follows [72]:

1. Excellent chemical stability.
2. Enhanced corrosion resistance for metal substrates.
3. An environmentally friendly technique of surface protection "green coating" (Replacement of toxic pretreatments and coatings which have traditionally been used for increasing corrosion resistance of metals).
4. Sol-gel process can be achieved at room temperature, thus degradation or thermal volatilization of incorporated organic species can be avoided.
5. Cast coating is possible without the need for complicated processes. Since used precursors are in the liquid form.
6. These are economical and fast and can be used for all types of surfaces to control the size and porosity of the resultant thin film.
7. Precursor such as metal alkoxides and mixed alkyl/alkoxides are frequently volatile and easily purified to very high levels using techniques developed for the microelectronics industry.

8. Highly porous materials and nanocrystalline materials may be prepared in this way.
9. Since liquid precursors are used, it is possible to cast ceramic materials in a range of complex shapes and to produce thin films or fibers as well as monoliths, without the need for machining or melting.
10. Hybrid silica sol-gel coatings are of growing interest as corrosion protection materials for metals as these coatings have good adhesion to metallic substrates and offer good corrosion resistance by combining properties of organic polymers and properties of ceramics [73, 74].

2.3.2. Limitations of sol-gel coatings

Common sol-gel limitations can be summarized in the following points [7]:

- Precursors are sensitive to moisture;
- Precursors are expensive, limiting large-scale manufacture;
- The process is time-consuming; and
- The dimensional change can occur on densification, shrinkage, cracking on drying, require attention.

2.3.3. Applications of sol-gel coatings

2.3.3.1. Spin-coating

Spin coating has been used for several decades for the application of thin films. In this process, a small drop of the coating material is loaded onto the center of a substrate, which is then spun at a controlled high speed. In the spin-coating process, the substrates spin around an axis which should be perpendicular to the coating area. As a result, the coating material spreads forming a thin film on the surface. Film properties, such as thickness, will be affected by several factors such as: viscosity, surface tension, drying ease and the rotation speed [75].

2.3.3.2. Dip coating

This method can be achieved by immersing a metal into a tank containing the coating material then withdrawing it to obtain a homogeneous film. Compared to the spin coating, the film obtained by dip coating is thicker [76].

2.3.3.3. Spray coating technique

Spray coating technique is a process in which molten or softened particles are applied by impact onto a substrate to produce a coating. This is done by using a vessel under pressure that pumps loading through a spray gun. The operation occurs in a totally enclosed system to minimize radiation of coating and product wastes [75].

http://dx.doi.org/10.5772/intechopen.72762

2.3.3.4. Flow-coating process

In flow-coating process, the liquid-coating system is more or less poured over the substrate to be coated. The coating thickness depends on the angle of inclination of the substrate, the coating-liquid viscosity and the solvent evaporation rate. Thus, nonplanar large substrates can be readily coated [76].

There are many studies, investigate the ability of the sol-gel method to protect metals from corrosion, will be discussed in the following section.

2.3.4. Corrosion control by using sol-gel coatings

Sol-gel-coated layers can exhibit good adhesion to the metal substrate due to the formation of chemically adsorbed M—O—Si bonds, as shown in **Figure 3**, resulting in a good anticorrosion performance. The following section summarizes a brief collection of recent and novel studies concerning the use of sol-gel coatings for corrosion control in metals and metal alloys.

Al-Cu-Fe alloy powder, a highly effective lubrication-oil additive, was coated by conventional sol-gel method employing tetraethyloxysilane (TEOS) as the precursor for purpose of improving its corrosion resistance [78]. The sol-gel method was modified by employing ultrasound to study the effects of both ultrasonic power and irradiation time on the coating process. The coated powder had a high corrosion resistance in the acid media of pH 1. Vignesh et al. used 3-mercaptopropyltrimethoxysilane (MPTMS) as the precursor for sol-gel coating on aluminum metal [79]. The results brought to light the potential of MPTMS as a precursor in the development of sol-gel coating over Al metal. The FTIR studies revealed the mechanism of the gelation of MPTMS and the formation of Si—O—Al linkage on the metallic substrate. The formed film was found to be stable up to 350°C as evident from thermogravimetric analysis. The X-rays diffraction (XRD) data also confirmed the formation of sol-gel coating on the metal surface. The electrochemical impedance spectroscopy (EIS) analysis of MPTMS coating on aluminum showed the inhibition of the corrosion by acting as a barrier to the electrolyte by impeding its contact with the metal

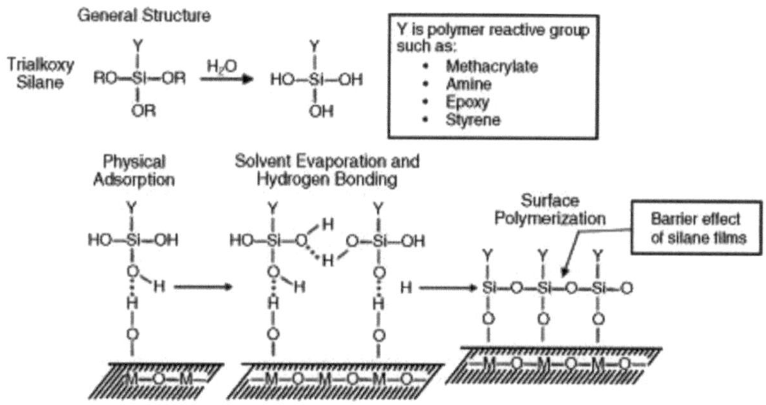

Figure 3. Bonding of silane precursors to the metallic surface [77].

surface. The polarization studies revealed that MPTMS sol-gel coating on aluminum blocked the anodic sites of the metal and thereby reduced the metal dissolution into the electrolyte solution, while SEM confirmed the protective nature of the MPTMS coating over aluminum substrate.

Seifzadeh et al. have studied the novel and crack-free nanocomposites based on sol-gel process for corrosion protection of copper [80]. The best results have been obtained after addition of sodium montmorillonite nanoclay at 0.01% concentration. Also, the phenyltrimethoxysilane sol-gel coating becomes more hydrophobic after addition of nanoclay at low concentrations (0.01 and 0.1%). Results indicated that phenyltrimethoxysilane coating provided a high protection from corrosion when compared with copper substrate. The addition of nanoclays resulted in a crack-free sol-gel coating and enhanced protective properties of the coating. A conversion coating was formulated and applied on mild steel prior to sol-gel Al_2O_3 coating to improve the corrosion resistance of mild steel reported by Tiwari et al. [81]. The comparison was made between conversion coated and conversion followed by sol-gel Al_2O_3-coated mild steel. The modified surfaces, produced by conversion coating and conversion followed by sol-gel Al_2O_3 coating, were characterized by XRD and X-ray photoelectron spectroscopy (XPS) and SEM. The corrosion resistance of the sol-gel coating was evaluated in 3.5 wt.% NaCl solution by one-component plasma (OCP), polarization and EIS measurement. By examining the surface morphology of solgel after Al_2O_3 coating (0.36 ± 0.03 μm thickness), the surface seems smooth with featureless topography. The sol-gel Al_2O_3 coating is free from any defect and is homogeneous. No change in the coating morphology is seen after long immersion (240 h) in 3.5 wt.% NaCl solution and repeated anodic polarization. The conversion coating by using silica sol and aluminum oxyhydroxide followed by heat treatment at 500°C produced mild steel surface suitable for sol-gel Al_2O_3 coating. The conversion coating formed a composite oxide containing Al_2O_3, SiO_2, α-Fe_2O_3 and Fe_3O_4 on the surface. The Al_2O_3 coating is amorphous and stable for 240 h in 3.5 wt.% NaCl solution. The protective nature of a coating composed of a hybrid sol-gel and a phosphate-based conversion layer for magnesium-alloy corrosion was investigated by Li et al. [82]. The sol-gel coating itself has a good anticorrosion performance; however, the presence of an additional conversion layer highly enhanced the protection ability. The corrosion process was examined by EIS which showed that the presence of the conversion layer increases the system resistance upon immersion in a 0.05 M NaCl solution, as compared to the single sol-gel coating. Hassannejad et al. studied the effects of cerium doping on corrosion resistance of amorphous silica-titania coating [83]. Potentiodynamic polarization and EIS spectroscopy have been used to study the corrosion behavior of composite coating in 3.5% NaCl solution. Results showed that ceria doping presented higher corrosion resistance than undoped silica-titanium composite layer and also the passive region of the doped films in potentiodynamic tests expanded with respect to undoped silica-titanium composite layer. The film corrosion resistance is noticed due to the inhibition action of cerium. There is a critical cerium concentration, above it cracking occurs, this can be related to the high expansion coefficient of cerium oxide. Mei et al. examined the effects of cerium salts on corrosion behaviors of Si-Zr hybrid sol-gel coatings [84]. The Si-Zr hybrid sol-gel coatings on aluminum substrate were prepared through hydrolysis and condensation of GPTMS and zirconium(IV) n-propoxide, used as inhibitors for corrosion, three types of cerium salts ($Ce(NO_3)_3$, $CeCl_3$ and $Ce(CH_3COO)_3$) were doped into the sol-gel coatings. The corrosion process is retarded

http://dx.doi.org/10.5772/intechopen.72762

by the addition of cerium salts due to its self-healing abilities. The sol-gel coating doped with $Ce(CH_3COO)_3$ has the highest corrosion resistance due to the utilized hydrolysis and condensation provided by CH_3COO^-. Ali et al. investigated the corrosion behavior of zinc substrates in 0.5 M HCl solution pretreated with various mono- and bilayered sol-gel films [85]. It was found that the bilayer morphology is similar to that of the outer layer. The protection abilities of mono- and bi-films are discussed in terms of film structure and functionality. The hybrid bifunctional organic coating (i.e. contains two different organic functional groups) is more efficient than monofunctional one. Hybrid monolayer has higher protection ability than its corresponding bilayers. Rahimi et al. reported the corrosion- and wear-resistance characterization of environmentally friendly sol-gel hybrid nanocomposite coating on aluminum. The hybrid nanocomposite coatings have been synthesized from TEOS and GPTMS precursors. The multilayer coatings were prepared by dip-coating technique. Effects of different layers on corrosion and wear resistance were studied. The coatings were found to increase the electrochemical corrosion resistance of aluminum. The maximum corrosion resistance properties, as determined by using electrochemical cyclic polarization testing, were observed for triple sol-gel-coated layer. The results suggested that crack-free plain organic-inorganic hybrid nanocomposite coatings with Si-O-Si structure backbone and CH_2 group incorporated into a silica network could be obtained on aluminum substrate by sol-gel synthesis. Atomic-force microscopy height images showed smooth nanostructural surface with root mean square less than 4 nm and the existence of organic and inorganic components in coatings could be a result of excellent wear and mechanical properties. Shi et al. studied that the sol-gel coatings were prepared using GPTMS and TEOS as precursors, diethylentriamine as curing agent. Inhibition effect of 2-methyl piperidine on magnesium alloy in 0.005, 0.05 and 0.5 wt.% NaCl solution is investigated.

The protection abilities of unmodified and modified GPTMS-VTMS sol-gel films with different Ti- and Al-alkoxides concentrations against zinc corrosion in 0.5 M HCl solution are reported by Ali et al. [85]. It was shown that the presence of the modifiers, either Ti or Al, facilitates the formation of the network structure and increases slightly the thermal stability of sol-gel materials. Weight-loss and electrochemical measurements proved that the highest protection ability of sol-gel coatings is assigned with low modifiers concentrations. The protection efficiency is increased from 69.2 to 96.2% and 95.8% upon modification with 0.005 M of Ti and Al modifiers, respectively. On the other hand, higher modifier concentrations may cause a galvanic corrosion which results in lower protection efficiency values. Corrosion inhibitors and nanoparticles can be added in the silane sol-gel network to improve corrosion resistance [86]. Balan et al. examined the effect of adding activated cerium nitrate nanoparticles into a coated sol-gel matrix on low-carbon steel substrates. Hybrid sol-gel films have been prepared from mixture of TEOS and GPTMS. Silica, alumina and their equimolar mixtures are used as nanoparticles with a concentration of 250 ppm. The SEM confirmed crack-free surface for prepared substrates with different pretreatments. Inorganic corrosion inhibitor was introduced into the coating via encapsulation in the nanoparticles reported by Tavandashti et al. [87]. Organosiloxane sol was prepared by mixing GPTMS, TEOS and isopropanol to obtain a sol with 75 mol.% organic content. The high corrosion-resistance performance of such coatings is due to the presence of encapsulated cerium nitrate corrosion inhibitor that can be released at the defects within the coating, hindering the corrosion reactions at defective sites. Asadi et al. investigated the cloisite concentration effect on the protective performance of sol-gel coating, prepared from GPTMS, TEOS and methyltriethoxysilane on mild steel substrate [88]. The optimum concentration of nanoparticles was determined. The data indicated superiority of the hybrid film embedded with 1000 ppm of cloisite nanoparticles. Confirming the electrochemical data, FTIR spectra and water contact angles showed that the nanoparticles well dispersed in the silane film might improve barrier properties moreover, a good trend correlation was observed between the noise resistance and low-frequency *IZI* data. Naderi et al. studied the synergistic effect of incorporation of nanoclay along with cerium nitrate as a corrosion inhibitor into an eco-friendly silane layer consisting of GPTMS, TEOS and methyltriethoxysilane applied on pure Al [89]. EIS and polarization results showed that the high protective nature of film containing cerium is due to film deposition on cathodic regions, thus preventing entrance of aggressive species to the surface. FTIR spectra proved the presence of the inhibitor in the film structure. N-nanoparticles silane coating in the presence of cerium salt provided higher corrosion protection compared to the film with only incorporated nanoclay.

Author details

Khadija M. Emran[1]*, Shimaa M. Ali[1,2] and Hamedh A. Al Lehaibi[1]

*Address all correspondence to: kabdalsamad@taibahu.edu.sa

1 Department of Chemistry, Collage of Science, Taibah University, Madinah, Kingdom of Saudi Arabia

2 Department of Chemistry, Collage of Science, Cairo University, Giza, Egypt

References

[1] Roberge PR. Handbook of Corrosion Engineering. Columbus, OH: McGraw-Hill Companies; 2000

[2] Malik MA, Hashim MA, Nabi F, Al-Thabaiti SA, Khan Z. Anti-corrosion ability of surfactants: A review. International Journal of Electrochemical Science. 2011;**6**:1927-1948

[3] Abd El-Lateef HM, Abbasov VM, Aliyeva LI, Ismayilov TA. Corrosion protection of steel pipelines against CO_2 corrosion—A review. Chemical Journal. 2012;**2**(2):52-63

[4] Zhang D, An Z, Pan Q, Gao L, Zhou G. Comparative study of bis-piperidiniummethyl-urea and mono-piperidiniummethyl-urea as volatile corrosion inhibitors for mild steel. Corrosion Science. 2006;**48**(6):1437-1448. DOI: 10.1016/j.corsci.2005.06.007

[5] Chen Z, Huang L, Zhang G, Qiu Y, Guo X. Benzotriazole as a volatile corrosion inhibitor during the early stage of copper corrosion under adsorbed thin electrolyte layers. Corrosion Science. 2012;**65**:214-222. DOI: 10.1016/j.corsci.2012.08.019

[6] Belarbi Z, Farelas F, Singer M, Nešić S. Role of amines in the mitigation of CO_2 top of the line corrosion. Corrosion. 2016;**72**(10):1300-1310

[7] Brinker CJ, Hurd AJ, Schunk PR, Frye GC, Ashley CS. Review of sol-gel thin film formation. Journal of Non-Crystalline Solids. 1992;**147-148**:424-436. DOI: 10.1016/S0022-3093(05)80653-2

[8] Zheludkevich ML, Salvado IM, Ferreira MGS. Sol–gel coatings for corrosion protection of metals. Journal of Materials Chemistry. 2005;**15**(48):5099-5111

[9] Wang H, Akid R, Gobara M. Scratch-resistant anticorrosion sol–gel coating for the protection of AZ31 magnesium alloy via a low temperature sol–gel route. Corrosion Science. 2010;**52**(8):2565-2570. DOI: 10.1016/j.corsci.2010.04.002

[10] Makhlouf AH, Tiginyanu I. Nanocoatings and Ultra-Thin Films: Technologies and Applications. New Delhi: Woodhead Publishing Limited; 2011

[11] Attia SM, Wang J, Wu G, Shen J, Ma J. Review on sol-gel derived coatings: Process, techniques and optical applications. Journal of Materials Science and Technology. 2002;**18**(3):211-218

[12] Brinker CJ, Clark DE. Better Ceramics through Chemistry. 3rd ed. Philadelphia, PA: Materials Research Society; 1988

[13] Liang B, Wang G, Zhang B, Zhang X. Corrosion resistance of Al-Cu-Fe alloy powder coated with silica using an ultrasound-assisted sol-gel method. Corrosion Science. 2013;**73**:292-299. DOI: 10.1016/j.corsci.2013.04.018

[14] Alvarez D, Collazo A, Perez C. The use of sol–gel film as pre-treatment for tinplate used in the canning industry. Progress in Organic Coatings. 2013;**76**(12):1819-1826. DOI: 10.1016/j.porgcoat.2013.05.023

[15] Li YS, Ba A, Mahmood MS. An environmentally friendly coating for corrosion protection of aluminum and copper in sodium chloride solutions. Electrochimica Acta. 2008;**53**(27):7859-7862. DOI: 10.1016/j.electacta.2008.05.062

[16] Xhanari K, Finšgarv M, Hrnčič MK, Maver U, Knez Ž, Seiti B. Green corrosion inhibitors for aluminium and its alloys: A review. RSC Advances. 2017;**7**(44):27299-27330

[17] Chauhan LR, Gunasekaran G. Corrosion inhibition of mild steel by plant extract in dilute HCl medium. 2007;**49**(3):1143-1161

[18] Okafor PC, Ikpi ME, Uwah IE, Ebenso EE, Ekpe UJ, Umoren SA. Inhibitory action of *Phyllanthus amarus* extracts on the corrosion of mild steel in acidic media. Corrosion Science. 2008;**50**(8):2310-2317

[19] Okafor PC, Ebenso EE, Ekpe UJ. *Azadirachta indica* extracts as corrosion inhibitor for mild steel in acid medium. International Journal of Electrochemical Science. 2010;**5**(7):978-993

[20] Singh MR. A green approach: A corrosion inhibition of mild steel by *Adhatoda vasica* plant extract in 0.5 MH_2SO_4. Journal of Materials and Environmental Science. 2013;**4**(1): 119-126

[21] Ji G, Shukla SK, Dwivedi P, Sundaram S, Prakash R. Inhibitive effect of *Argemone mexicana* plant extract on acid corrosion of mild steel. Industrial and Engineering Chemistry Research. 2011;**50**:11954-11959. DOI: 10.1021/ie201450d

[22] Eduok UM, Umoren SA, Udoh AP. Synergistic inhibition effects between leaves and stem extracts of Sidaacuta and iodide ion for mild steel corrosion in 1M H_2SO_4 solutions. Arabian Journal of Chemistry. 2012;**5**:325-337

[23] Benali O, Benmehdi H, Hasnaoui O, Selles C, Salghi R. Green corrosion inhibitor: Inhibitive action of tannin extract of *Chamaerops humilis* plant for the corrosion of mild steel in 0.5 M H_2SO_4. Journal of Material and Environmental Science. 2013;**4**:127-138

[24] Mourya P, Sitashree B, Singh MM. Corrosion inhibition of mild steel in acidic solution by *Tagetes erecta* (Marigold flower) extract as a green inhibitor. Corrosion Science. 2014;**85**:352-363

[25] Ji G, Anjum S, Sundaram S, Prakash R. *Musa paradisica* peel extract as green corrosion inhibitor for mild steel in HCl solution. Corrosion Science. 2015;**90**:107-117

[26] Mathina A, Rajalakshmi R. Corrosion inhibition of mild steel in acid medium using canna Indica as green corrosion inhibitor. Rasayan Journal of Chemistry. 2016;**9**:56-66

[27] Ibisi NE, Ihejirika AV. Inhibitory potentials of *Bamboosa vulgaris* leaves extract on corrosion inhibition of mild steel in 1M HCl solution. Saudi Journal of Engineering and Technology. 2017;**2**:211-218

[28] Ogwo KD, Osuwa JC, Udoinyang IE, Nnanna LA. Corrosion inhibition of mild steel and aluminium in 1 M hydrochloric acid by leaves extracts of *Ficus sycomorus*. Physical Science International Journal. 2017;**14**:1-10

[29] Faustin M, Lebrini M, Robert F, Roos C. Corrosion studies of C38 steel by alkaloids extract of a tropical plant type. International Journal of Electrochemical Science. 2011;**6**:4095-4113

[30] Faustin M, Maciuk A, Salvin P, Roos C, Lebrini M. Corrosion inhibition of C38 steel by alkaloids extract of *Geissospermum laeve* in 1M hydrochloric acid: Electrochemical and phytochemical studies. Corrosion Science. 2015;**92**:287-300

[31] Emran KM, Al-Ahmadi AO, Ahmed NM, Torjoman BA, Sheekh SN. Corrosion and corrosion inhibition of cast iron in hydrochloric acid (HCl) solution by cantaloupe (*Cucumis melo*) as green inhibitor. African Journal of Pure and Applied Chemistry. 2015;**9.3**:39-49

[32] Ating EI, Umoren SA, Udousoro II, Ebenso EE, Udoh AP. Leaves extract of *Ananas sativum* as green corrosion inhibitor for aluminium in hydrochloric acid solutions, green. Chemical Letters Review. 2010;**3**:61-68

[33] Fetouh HA, Abdel-Fattah TM, El-Tantawy MS. Novel plant extracts as green corrosion inhibitors for 7075-T6 aluminium alloy in an aqueous medium. International Journal of Electrochemical Science. 2014;**9**:1565-1582

[34] Singh A, Ahamad I, Quraishi MA. *Piper longum* extract as green corrosion inhibitor for aluminium in NaOH solution. Arabian Journal of Chemistry. 2016;**9**:S1584-S1589

[35] Abd-El-Naby BA, Abdullatef OA, Abd-El-Gabr AM, Shaker MA, Esmail G. Effect of some natural extracts on the corrosion of zinc in 0.5 M NaCl. International Journal of Electrochemical Science. 2012;**7**:5864-5879

[36] Bright A, Michlin Ruphina Maragatham S, Malar vizhi I, Selvaraj S. Corrosion behavior of zinc in 1.0 n hydrochloric acid with *Cnidoscolus chayamansa*: A green approach. International Journal of Recent Scientific Research. 2015;**6**:3594-3601

[37] El-Housseiny S. Inhibition of the acid corrosion of zinc by cannabis plant extract. American Journal of Chemistry. 2017;**7**:6-15

[38] Patel BB, Vashi RT. *Azadirachta indica* (neem) extract as green inhibitor for corrosion of brass in nitric acid media. Journal of Applied Chemistry. 2017;**6**:340-349

[39] Sangeetha TV, Fredimoses M. Inhibition of mild copper metal corrosion in HNO_3 medium by acid extract of *Azadirachta indica* seed. Journal of Chemistry. 2011;**8**(S1):S1-S6

[40] Deyab MA. Egyptian licorice extract as a green corrosion inhibitor for copper in hydrochloric acid solution. Journal of Industrial and Engineering Chemistry. 2015;**22**:384-389

[41] Fouda AS, Elmorsi MA, Abou-Elmagd BS. Adsorption and inhibitive properties of methanol extract of *Eeuphorbia heterophylla* for the corrosion of copper in 0.5 M nitric acid solutions. Polish Journal of Chemical Technology. 2017;**19**:95-103. DOI: 10.1515/pjct-2017-0014

[42] El-Etre AY, Abdallah M, El-Tantawy ZE. Corrosion inhibition of some metals using Lawsonia extract. Corrosion Science. 2005;**47**(2):385-395

[43] El-Etre AY, El-Tantawy Z. Inhibition of metallic corrosion using Ficus extract, Portugaliae. Electrochimica Acta. 2006;**24**:347-356

[44] Ibrahim ME, El-Khrisy AM, Al-Abdal-lah EM, Baraka A. Evaluation of the inhibitor action of certain natural substances used as corrosion inhibitors. 1. In the dissolution of tin in nitric acid. Mo Metalloberflanche Beschichten won Metall und Kunstsoff. 1981;**35**(4):134-136

[45] Sharma SK, Peter A, Obot IB. Potential of *Azadirachta indica* as a green corrosion inhibitor against mild steel, aluminum, and tin: A review. Journal of Analytical Science and Technology. 2015;**6**:26-1-16

[46] Djeddi N, Benahmed M, Akkal S, Laouer H, Makhloufi E, Gherraf N. Study on methylene dichloride and butanolic extracts of *Reutera lutea* (Desf.) Maire (Apiaceae) as effective corrosion inhibitions for carbon steel in HCl solution. Research on Chemical Intermediates. 2015;**41**:4595-4616

[47] Rajenderan S, Shanmugapriya S, Rajalakshmi T, Amal Raj AJ. Corrosion inhibition by an aqueous extract of rhizome powder. Corrosion. 2005;**61**:685-692

[48] Priya SL, Chitra A, Rajenderan S, Anuradha K. Corrosion behaviour of aluminium in rain water containing garlic extract. Surface Engineering. 2005;**21**(3):229-231

[49] Njoku VO, Oguzie EE. Inhibitory properties of *Baphia nitida* extract in acid mild steel corrosion. EUROCORR. 2009. pp. 849-870

[50] John Amalraj A, Wilson Sahayaraj J, Peter Pascal Regis A, Sahayaraj A, Pandian P, Johnraj P, Kasthuri A, Kumar C. Synergistic effect of Allium cepa-Zn^{2+} system on the corrosion of carbon steel in ground water. Engineering. 2013;**5**:178-183

[51] Kavitha N, Manjula P, Anandhakumar N. Synergistic effect of *C. papaya* leaves extract-Zn^{2+} in corrosion inhibition of mild steel in aqueous medium. Research Journal of Chemical Science. 2014;**4**:88-93

[52] Arab ST, Al-Turkustani AM, Al-Dhahiri RH. Synergistic effect of *Azadirachta indica* extract and iodide ions on the corrosion inhibition of aluminium in acid media. Journal of the Korean Chemical Society. 2008;**52**:281-294

[53] Obot IB, Obi-Egbedi NO, Umoren SA, Ebenso EE. Synergistic and antagonistic effects of anions and *Ipomoea invulcrata* as green corrosion inhibitor for aluminium dissolution in acidic medium. International Journal of Electrochemical Science. 2010;**5**:994-1007

[54] Sharma S, Parihar PS, Nair RN, Verma PS, Sharma A. Influence of ion-additives on inhibitory action of extract of *Trigonella foenum graceums* seeds for AA6063 in acid medium. Rasayan Journal of Chemistry. 2012;**5**:16-23

[55] Ridhwan AM, Rahim AA, Shah AM. Synergistic effect of halide ions on the corrosion inhibition of mild steel in hydrochloric acid using mangrove tannin. International Journal of Electrochemical Science. 2012;**7**:8091-8104

[56] Krishnaveni K, Ravichandran J. Effect of aqueous extract of leaves of *Morinda tinctoria* on corrosion inhibition of aluminium surface in HCl medium. Transactions of Nonferrous Metals Society of China. 2014;**24**(8):2704-2712

[57] Chidi OBI, Stella Wegwu N. Corrosion inhibitive dynamics of *Pennisetum purpureum* (elephant grass) on aluminum metal in acidic media. International Journal of Innovation Science and Engineering. 2016;**4**(3):1-15

[58] Njoku VO, Oguzie EE, Obi C, Ayuk AA. *Baphia nitida* leaves extract as a green corrosion inhibitor for the corrosion of mild steel in acidic media. Advanced Chemistry. 2014; **2014**: Article ID 808456, 10 pages

[59] Zhang J, Zhang Q, Chen G, Zhao JR, Tang DY. Modification of *Punica granatum* Linne husk extract as eco-friendly corrosion inhibitors in oil fields. Applied Mechanics and Materials. 2014;**488**:273-276

[60] Dekmouche M, Saidi M, Hadjadj M, Ghiaba Z, Yousfi M. Green approach to corrosion inhibition by ethyl acetate extract from *Pistacia atlantica* gals in hydrochloric acid solution. International Journal of Electrical Science. 2014;**9**(7):3969-3978

[61] FelipeM BMC, Silva DR, Martinez-Huitle CA, Medeiros SRB, Maciel MAM. Effectiveness of *Croton cajucara* Benth on corrosion inhibition of carbon steel in saline medium. Material Corrosion. 2013;**64**(6):530-534

[62] Cui Y, Zhao D, Qu H, Qu G, Liu J, Ren P. The study of composite substituting mercury corrosion inhibitors for zinc in zinc-manganese battery. Advanced Material Research. 2014;**881-883**:102-107

[63] Bernal S, Botana FJ, Calvino JJ, Marcos M, Perez-Omil JA, Vidal H. Lanthanide salts as alternative corrosion inhibitors. Journal of Alloys and Compounds. 1995;**255**:638-641

[64] Hinton BRW, Arnott DR, Ryan NE. The inhibition of aluminum alloy corrosion by cerous cations. Metals Forum. 1984;**7**:211-217

[65] Arenas A, Bethencourt M, Botana FJ, de Damborenea J, Marcos M. Inhibition of 5083 aluminium alloy and galvanized steel by lanthanide salts. Corrosion Science. 2001;**43**: 157-170

[66] Arenas MA, Conde A, de Damborenea JJ. Cerium: A suitable green corrosion inhibitor for tinplate. Corrosion Science. 2002;**44**:511-520

[67] Shi H, Han E, Liu F. Corrosion protection of aluminium alloy 2024-T3 in 0.05 M NaCl by cerium cinnamate. Corrosion Science. 2011;**53**:2374-2384

[68] Mora N, Cano E, Polo JL, Puente JM, Bastidas JM. Corrosion protection properties of cerium layers formed on tinplate. Corrosion Science. 2004;**46**(3):563-578

[69] Moretti G, Guidi F, Grion G. Tryptamine as a green iron corrosion inhibitor in 0.5 M deaerated sulphuric acid. Corrosion Science;**46**(2):387-403

[70] Bethencourt M. Lanthanide compounds as environmental friendly corrosion inhibitors of aluminium alloys: A review. Corrosion Science. 1998;**40**(11):1803-1819

[71] Owens GJ, Singh RK, Foroutan F, Alqaysi M, Han C, Mahapatra C, Kim H, Knowles JC. Sol–gel based materials for biomedical applications. Progress in Materials Science. 2016;**77**:1-79

[72] Seifzadeh D, Golmoghani-Ebrahimi E. Formation of novel and crack free nanocomposites based on sol gel process for corrosion protection of copper. Surface and Coatings Technology. 2012;**210**:103-112. DOI: 10.1016/j.surfcoat.2012.08.073

[73] Lakshmi RV, Bharathidasan T, Basu BJ. Superhydrophobic sol-gel nanocomposite coatings with enhanced hardness. Applied Surface Science. 2011;**257**(24):10421-10426. DOI: 10.1016/j.apsusc.2011.06.122

[74] Lakshmi RV, Yoganandan G, Mohan AVN, Basu BJ. Effect of surface pre-treatment by silanization on corrosion protection of AA2024-T3 alloy by sol–gel nanocomposite coatings. Surface and Coatings Technology. 2014;**240**:353-360. DOI: 10.1016/j.surfcoat.2013.12.051

[75] Tiwari SK, Sahu RK, Pramanick AK, Singh R. Development of conversion coating on mild steel prior to sol gel nanostructured Al_2O_3 coating for enhancement of corrosion resistance. Surface and Coatings Technology. 2011;**205**(21-22):4960-4967. DOI: 10.1016/j.surfcoat.2011.04.087

[76] Murillo-Gutierrez NV, Ansart F, Bonino JP, Menu MJ, Gressier M. Protection against corrosion of magnesium alloys with both conversion layer and sol-gel coating. Surface and Coatings Technology. 2013;**232**:606-615. DOI: 10.1016/j.surfcoat.2013.06.036

[77] Zaferani SH. The science of anti-corrosion thin films. Corrosionpedia. 2016

[78] Poznyak SK, Zheludkevich ML, Raps D, Gammel F, Yasakau KA, Ferreira MGS. Preparation and corrosion protective properties of nanostructured titania-containing hybrid sol-gel coatings on AA2024. Progress in Organic Coatings. 2008;**62**(2):226-235. DOI: 10.1016/j.porgcoat.2007.12.004

[79] Ali SM, Allehaibi H. Protective sol-gel coatings for zinc corrosion: Precursor type effect. Surface and Coatings Technology. 2017;**311**:172-181. DOI: 10.1016/j.surfcoat.2017.01.010

[80] Li YS, Tran T, Xu Y, Vecchio NE. Spectroscopic studies of trimetoxypropylsilane and bis(trimethoxysilyl)ethane sol-gel coatings on aluminum and copper. Spectrochimica Acta: Part A Molecular and Biomolecular Spectroscopy. 2006;**65**(3-4):779-786. DOI: 10.1016/j.saa.2005.12.040

[81] Nemeth S, Liu YC. Mechanical properties of hybrid sol-gel derived films as a function of composition and thermal treatment. Thin Solid Films. 2009;**517**(17):4888-4891. DOI: 10.1016/j.tsf.2009.03.099

[82] Li Y, Lu W, Wang Y, Tran T. Studies of (3-mercaptopropyl)trimethoxylsilane and bis(trimethoxysilyl)ethane sol-gel coating on copper and aluminum. Spectrochimica Acta Part A. 2009;**73**(5):922-928. DOI: 10.1016/j.saa.2009.04.016

[83] Shi H, Liu F, Han E. Corrosion protection of AZ91D magnesium alloy with sol-gel coating containing 2-methyl piperidine. Progress in Organic Coatings. 2009;**66**(3):183-191. DOI: 10.1016/j.porgcoat.2009.07.004

[84] Shi H, Liu F, Han E. Corrosion behaviour of sol–gel coatings doped with cerium salts on 2024-T3 aluminum alloy. Materials Chemistry and Physics. 2010;**124**(1):291-297. DOI: 10.1016/j.matchemphys.2010.06.035

[85] Ali SM, Allehaibi H. Nano-structured sol–gel coatings as protective films against zinc corrosion in 0.5 M HCl solution. Journal of Saudi Chemical Society. 2017;**21**(4):473-480. DOI: 10.1016/j.jscs.2016.12.001

[86] Balan P, Shelton MJ, Ching DOL, Han GC, Pal niandy LK. Modified silane films for corrosion protection of mild steel. Procedia Materials Science. 2014;**6**:244-248. DOI: 10.1016/j.mspro.2014.07.030

[87] Tavandashti NP, Sanjabi S. Corrosion study of hybrid sol–gel coatings containing boehmite nanoparticles loaded with cerium nitrate corrosion inhibitor. Progress in Organic Coatings. 2010;**69**(4):384-391. DOI: 10.1016/j.porgcoat.2010.07.012

[88] Asadi N, Naderi R, Saremi M. Determination of optimum concentration of cloisite in an eco-friendly silane sol-gel film to improve corrosion resistance of mild steel. Applied Clay Science. 2014;**95**:243-251. DOI: 10.1016/j.clay.2014.04.018

[89] Naderi R, Fedel M, Deflorian F, Poelman M, Olivier M. Synergistic effect of clay nanoparticles and cerium component on the corrosion behavior of eco-friendly silane sol-gel layer applied on pure aluminum. Surface and Coatings Technology. 2013;**224**:93-100. DOI: 10.1016/j.surfcoat.2013.03.005

Chapter 4

The Impact of Drugs as Corrosion Inhibitors on Aluminum Alloy in Coastal-Acidified Medium

Ojo Sunday Isaac Fayomi,
Paul Apeye Lucky Anawe and Ayodele Daniyan

Additional information is available at the end of the chapter

http://dx.doi.org/10.5772/intechopen.72942

Abstract

The use of corrosion inhibitors has proven to be one of the effective methods of corrosion protection of metals and alloys. Aluminum alloys are good candidate materials for structural components in major industrial application owing to their excellent corrosion resistance. However in high intensity coastal and acidified medium the challenge of micro porosity and hydrogen embrittlement distributed along the interface cannot be neglected as suitable corrosion inhibitive compounds is needful without significantly reacts with the environmental components. The challenges of most effective organic and inorganic Inhibitors when dissolved in aqueous environments are increasingly due to their toxicity. Drug as inhibitive compound have been seen as suitable replacement for this high toxic organic inhibitors since both react by adsorption on a metallic surface. They are usually compounds that form film and cause the formation of precipitates on the metal surface, thereby blocking both cathodic and anodic and sites. Therefore, this study covers the general overview of impact of some drugs as corrosion inhibitive compound on aluminum for industrial applications and their environmental impact.

Keywords: corrosion inhibitors, protection, coastal, acidified, drug and impact

1. Introduction

Aluminum is a silvery white material and a member of boron group. It is one of the most abundant metal in the Earth's crust, and the third most abundant element within, after oxygen and silicon. It is yielding, long-lasting, lightweight, malleable metal with physical appearance varying from silvery coloration to approximate gray, which depend on the roughness of the surface. Aluminum is not a Ferro-magnetic material. It is also not soluble in alkanols, despite the fact that, in certain form, it can be water-soluble. It has an appreciable yield strength of

S/N	Research work	Result	Reference
1	Meclizine (Antihistaminic Drug) as Corrosion Inhibitors of Aluminum in HCl Environment	The result showed that the drug excellently inhibited the aluminum in the acid medium at 303 K	[2]
2	Clotrimazole and Fluconazole (Antifungal drugs) as Corrosion Inhibitors of Aluminum in HCl Environment	The results pointed out that both drug server as excellent inhibitors in the Medium	[3]
3	Amine Compound as Corrosion Inhibitors of Aluminum in HCl Environment	The result indicated 92% inhibitive efficiency at 30°C and 83% at 60°C using 100 ppm inhibitor concentration	[4]
4	Voltaren (expired) Drug to inhibit Aluminum in HCl Environment	Corrosion potential (Ecorr) was shifted to the direction that is more noble by the Inhibitor molecules	[5]
5	Domiana Extract as inhibitor of Aluminum in Acidic medium	From the result, the inhibition efficiency displayed direct variation with inhibitor's concentration but demonstrated inverse relation with temperature	[6]
6	Corrosion Inhibitor of Aluminum using Chloroquine in HCl Environment Media	It was deduced that chloroquine excellently inhibited the aluminum in the acidic environment and the efficiency of the inhibitor increased with concentration of inhibitor but not favored by temperature	[7]
7	Corrosion Inhibitor of Aluminum using *Aloe Vera* in NaCl medium	The result showed that Aloe Vera is potential green corrosion inhibitor in saline environment	[8]
8	Corrosion inhibition of Aluminum Alloy in NaOH Medium using Anti-inflammatory drug	The result showed the potential of the Omeprazole drug in inhibiting corrosion in the basic environment	[9]
9	Corrosion inhibition of Aluminum using Fluconazole & Clotrimazole in HCl Medium	The result displayed the antifungal drug as potential inhibitor of aluminum in the acidic environment	[10]
10	Corrosion inhibition of Aluminum using green corrosion inhibitor in acidic medium	The result showed the extract as eco-friendly inhibitor in the acidic environment	[11]

Table 1. Analysis of inhibitive drugs and their performance.

about 10 MPa for the pure metal, while that of its alloys can range from about 150 to 650 MPa. Aluminum is about 33% dense and stiff in comparing to steel.

The machinability and formability of aluminum is excellent with outstanding corrosion resistance due to its ability to form an effective and tenacious thin surface layer of its oxide in atmosphere. It is also noteworthy that the strongest alloys aluminum are less corrosion resistant because of the possibility internal galvanic cell that may be developed in service [1]. This corrosion resistance is also often greatly reduced in the presence of various salts or dissimilar metals and often corrosion of aluminum can be mitigated using corrosion inhibitors, even the use of inhibitive drugs (**Table 1**).

2. Passivity of pure aluminum and its alloys

2.1. Corrosion of aluminum

Al has a reactive metal, can react with acidic, basic or salty foods to release itself into the food product. Also, an aluminum container that is worn out or pitted may contaminate the food by releasing excess aluminum into it [12]. Therefore, temperature does have an effect on the corrosion of aluminum. The aluminum subjected to the higher temperature does corrode because of the increase in particles' collision [13]. Corrosion of aluminum depends on environment, design, alloy and preventive measures put into considerations. A dirt-free aluminum surface is highly reactive and will spontaneously react with air or water to form aluminum oxide. This oxide is usually stable and possess excellent adhesion to the surface of metal therefore protects the material (aluminum) from further corrosion or oxidation. Thus aluminum has excellent corrosion resistance in stable oxide layer environments [14]. This oxide layer will be weaken in environments where there is low or high pH particularly, in the presence of aggressive ions. That is, at a pH below 4 and beyond a pH of about 8.5, there will generally be an increase in the aluminum corrosion rate, though this depend on the ions that are present in that particular environment. The oxide layer will be broken down by the aggressive ions locally and localized corrosion attacks start. Of all the ions that display high aggression, chloride ion (Cl^-) is of the most practical importance [15]. From thermodynamically point of consideration aluminum is a metal with high activity. Nevertheless, in aerated environment, the metal is rapidly covered with an intense oxide layer [16]. This oxide layer is basically inert, and prevents further oxidation. The thickness of this layer depends largely on the temperature, environment and alloy elements. The oxide films that are formed in air at ordinary room temperature are about 2.5 nm of thickness on pure aluminum. Heating to about 400°C may results in films up to about 25 nm (this is about 10 times that at room temperature [17]. If the oxide film is destroyed, e.g. by a scratch, new oxide will spontaneously and immediately form on the uncovered metal [18]. Thus pure and even many of its alloys display excellent performance in corrosion protection.

3. Aluminum alloy

Major reason for alloying aluminum is to increase hardness, strength and resistance to fatigue, stress relaxation, wear or creep. The Effects of different alloying elements and combinations of them on these properties are specific and related to the phase diagrams of constituent alloy elements.

Figure 1 shows the relationships between some of the more commonly used alloys in the *6xxx* series. The elements that are most frequently used in commercial aluminum alloys to provide better strength especially, when related to strain hardening via cold working or heat treatment, or both are; zinc, copper, silicon, magnesium and manganese (**Figure 2**). All these elements have high solid solubility in aluminum, and usually, the solubility increases as temperature increases.

Aluminum and its alloys possess significant industrial and economic value due to its light weight, good thermal and electrical conductivity as well as low cost [20]. The most important property of aluminum is its ability to effectively resist atmospheric and several aqueous solution because of the quick formation of protective oxide of thin film which adhere tenaciously to the surface, that serves as barrier between the corrosive environment and the metal.

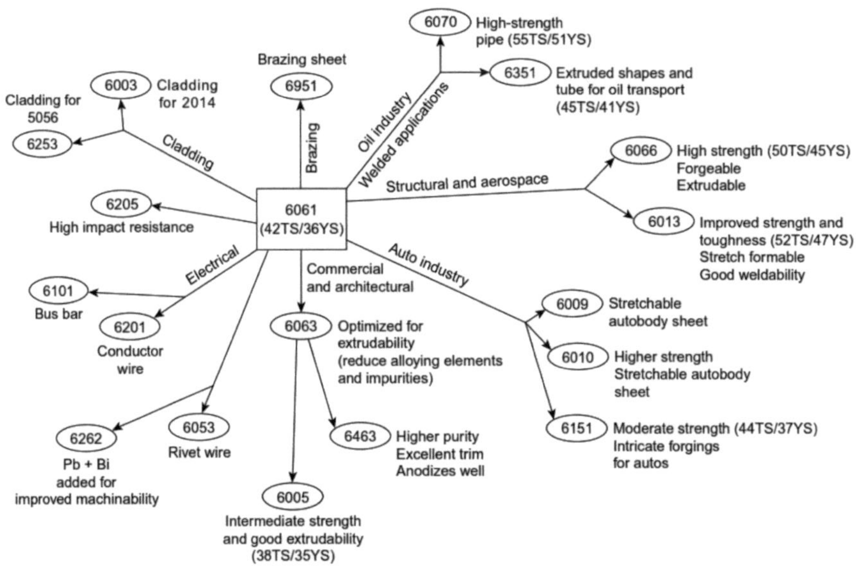

Figure 1. Relationships that exist among commonly used alloys in the *6xxx* series (Al-Si-Mg). Their yield strength (YS) and tensile strength (TS) in ksi units [19].

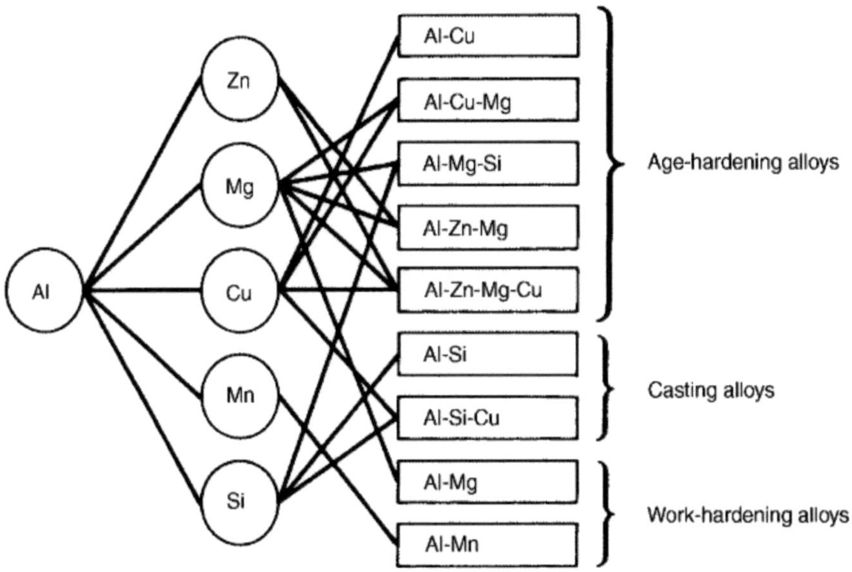

Figure 2. The principal aluminum alloys [19].

Increasing strength as oxidizing agent ↑

Element	Half Reaction		E°(V)
Gold	$Au^{3+}_{(aq)} + 3e^{-}$ ⟶	$Au_{(s)}$	+1.50
Platinium	$Pt^{2+}_{(aq)} + 2e^{-}$ ⟶	$Pt_{(s)}$	+1.18
Silver	$Aq^{+}_{(aq)} + e^{-}$ ⟶	$Ag_{(s)}$	+0.80
Iron	$Fe^{3+}_{(aq)} + e^{-}$ ⟶	Fe^{2+}	+0.77
Copper	$Cu^{2+}_{(aq)} + 2e^{-}$ ⟶	$Cu_{(s)}$	+0.34
Tin	$Sn^{4+}_{(aq)} + 2e^{-}$ ⟶	$Sn^{2+}_{(aq)}$	+0.13
Nickel	$Ni^{2+}_{(aq)} + 2e^{-}$ ⟶	$Ni_{(s)}$	-0.25
Iron	$Fe^{2+}_{(aq)} + 2e^{-}$ ⟶	$Fe_{(s)}$	-0.44
Chromium	$Cr^{3+}_{(aq)} + 3e^{-}$ ⟶	$Cr_{(s)}$	-0.74
Zinc	$Zn^{2+}_{(aq)} + 2e^{-}$ ⟶	$Zn_{(s)}$	-0.76
Manganese	$Mn^{2+}_{(aq)} + 2e^{-}$ ⟶	$Mn_{(s)}$	-1.18
Titanium	$Ti^{2+}_{(aq)} + 2e^{-}$ ⟶	$Ti_{(s)}$	-1.37
Aluminium	$Al^{3+}_{(aq)} + 3e^{-}$ ⟶	$Al_{(s)}$	-1.66
Magnesium	$Mg^{2+}_{(aq)} + 2e^{-}$ ⟶	$Mg_{(s)}$	-2.27
Cerium	$Ce^{3+}_{(aq)} + 3e^{-}$ ⟶	$Ce_{(s)}$	-2.34
Praseodymium	$Pr^{3+}_{(aq)} + 3e^{-}$ ⟶	$Pr_{(s)}$	-2.35
Sodium	$Na^{+}_{(aq)} + e^{-}$ ⟶	$Na_{(s)}$	-2.71
Calcium	$Ca^{2+}_{(aq)} + 2e^{-}$ ⟶	$Ca_{(s)}$	-2.87

Increasing strength as reducing agent ↓

Figure 3. Aluminum in the redox strength of metals.

The reaction is in line with the position of aluminum in the redox strength of metals (**Figure 3**). It has been observed that this film (oxide) is stable at a range of about 5–9 range of pH [21].

4. Pitting corrosion of aluminum

Aluminum usually suffers pitting corrosion in the presence of chloride ions. Pitting corrosion is a dangerous form of localized corrosion that selectively attacks areas of a metal surface where there is; (i) a surface scratch or a mechanically induced break in an otherwise protective film, (ii) an emerging dislocation or slip step caused by applied or residual stresses and (iii) a compositional heterogeneity such as an inclusion, segregate or precipitate [22]. Pitting is one of the most insidious forms of corrosion. It is very destructive to many engineering materials since it causes perforation of the metal. As a result of its localized nature, it often causes sudden unexpected failures. Pitting is often difficult to detect because corrosion products may cover small pits. General illustration of pitting corrosion on aluminum alloys is shown in **Figure 4**.

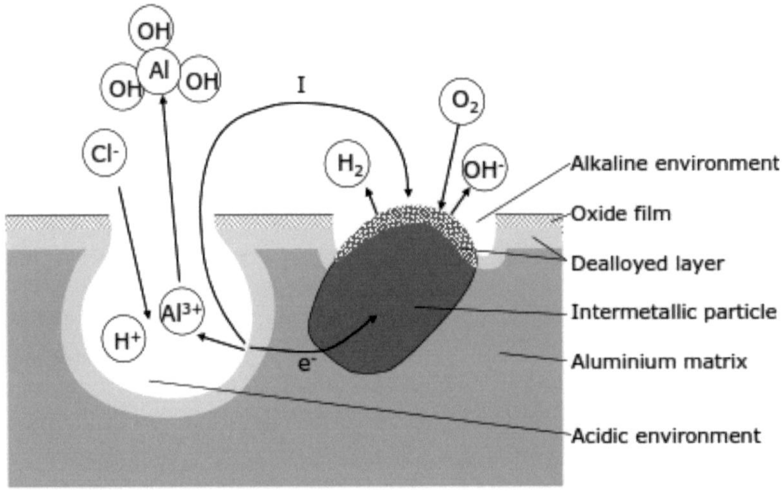

Figure 4. General illustration of pitting corrosion on aluminum alloys [18].

5. Corrosion inhibitors

Corrosion inhibitors have been established to be effective and easier means of corrosion control. Using chemical inhibitors to reduce the rate of corrosion processes is somewhat amenable. Corrosion inhibitors found applications in various industries, examples are; oil and gas, chemical industries, production, refineries, heavy manufacturing, water treatment and etc. [23]. In extraction the oil, chemical and processing industries, the use of corrosion inhibitors are usually considered to be the primary line of protection. Therefore, a large numbers of scientific research have been concentrated to the issue of corrosion inhibitors [24–29]. **Table 1** shows application of corrosion inhibitors in some specific system environment. An inhibitor is a chemical substance or mixture of substances which when added in very little concentrations to a corrosive environment, it effectively retards or prevents corrosion without significantly reacts with the environmental components.

Inhibitors can be classified as either organic or inorganic compounds and they are usually dissolved in aqueous environments. Some of the most effective inorganic inhibitors are carbonates, chromates, nitrites, silicates and phosphates. The organic inhibitors comprise amines, nitrogen compounds, sulfur compounds such as thioethers, thioalcohols, thioamides, thiourea and hydrazine. Chromates and zinc salts are used increasingly less due to their toxicity and are currently largely replaced by organic inhibitors.

Also, with reference to their influence on the electrochemical reaction relating to metal, and their environment corrosion inhibitors (the inorganic inhibitors) can also be classified as anodic, cathodic or mixed corrosion inhibitors (**Figure 5**). **Figure 6** displays the adsorption mechanism of an organic inhibitors on aluminum. These have been in application for quite some years and the most common examples of their uses are in films and coatings on metals where chromate, nitrate, borates, phosphate, benzoates, and oxides are incorporated, Nitrites

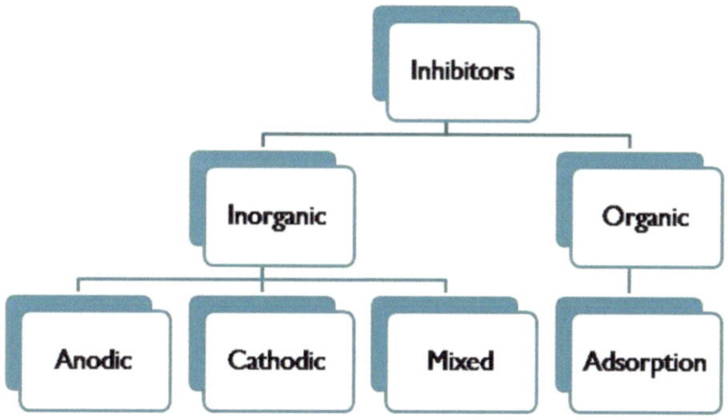

Figure 5. Classification of corrosion inhibition.

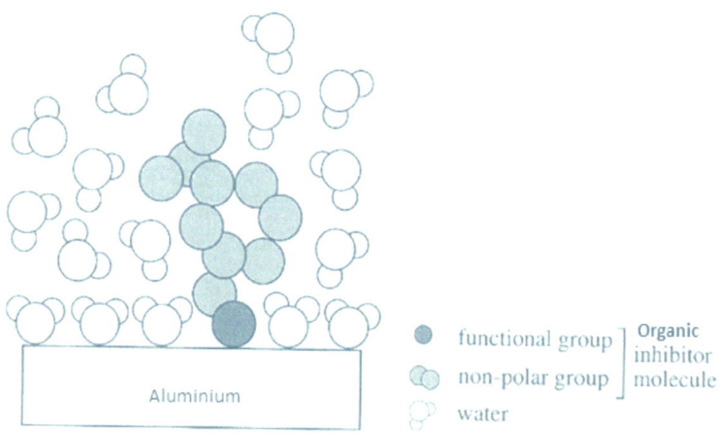

Figure 6. Adsorption mechanism of an organic inhibitor on an aluminum surface in an aqueous environment.

are being used as inhibitor in concrete reinforcement. While the organic inhibitors react by adsorption on a metallic surface.

Anodic inhibitors generally act by forming a protective oxide film on the surface of the metal to be protected resulting in a large anodic shift of the E_{corr} (corrosion potential). This shift usually forces the surface of metal into the region of passivation. That is, they slow down the anodic dissolution reaction rate of and produce products of reaction which form a thin film over anode. Examples are; Chromates, nitrates, tungstate and molybdates.

Cathodic inhibitors usually work by either slowing down the cathodic reaction or precipitating (selectively) on cathodic regions to inhibit the diffusion of reducing parts to the surface. That is, they interrupt the flow of electrons from the anode to the cathode and produce products of reaction which precipitate selectively at cathodic sites. Examples are zinc sulfate, nickel sulfate, aminoethylene and polyphosphate.

Mixed inhibitors act by reducing both the anodic and cathodic reactions. They are usually compounds that form film and cause the formation of precipitates on the metal surface, thereby blocking both cathodic and anodic and sites. Common examples are the phosphates and silicates.

Therefore, this study deals with general overview of aluminum, corrosion inhibitors, their classifications, Inhibitive drugs, considerations in employing them, their efficiency industrial applications and their environmental impact.

6. Properties of a good inhibitor

In order to prevent or mitigate the corrosion of materials (metallic), inhibitors in question must be able to satisfy these following measures [26]:

i. It must be effective in corrosion protection at a very little concentration of the inhibitor.

ii. It must offer protection to all the part of the component that are exposed from the corrosion attack of corrosion.

iii. It must maintain its effectiveness at extreme operating conditions (high velocity and temperature).

iv. During application, over or under dosage of inhibitor, the rate of corrosion must not drastically increased.

v. The inhibitor or inhibitor's products must not form any deposit on the surface metal especially at regions of heat transfer.

vi. It must mitigate both localized and uniform corrosion.

vii. The range of effectiveness should be long.

viii. It must not pose health hazard or pollution potential.

7. Drug inhibitors

Typical examples of organic eco-friendly inhibitors are fast green and carmine. The molecules of fast green have aromatic ring with electroactive nitrogen and oxygen atoms while molecules of carmine have aromatic rings (electron rich) with electroactive oxygen atoms. These molecular structures are shown in **Figure 7**. These compounds are adsorbed at the metal surface and they block the active site and consequently, decreasing the corrosion rate and consequently increase the inhibition efficiency [30]. Nevertheless, Most of organic inhibitors are poisonous and harmful to the environment. The increasing health and ecological awareness have drawn the attentions toward research and development of very effective and efficient eco-friendly inhibitors.

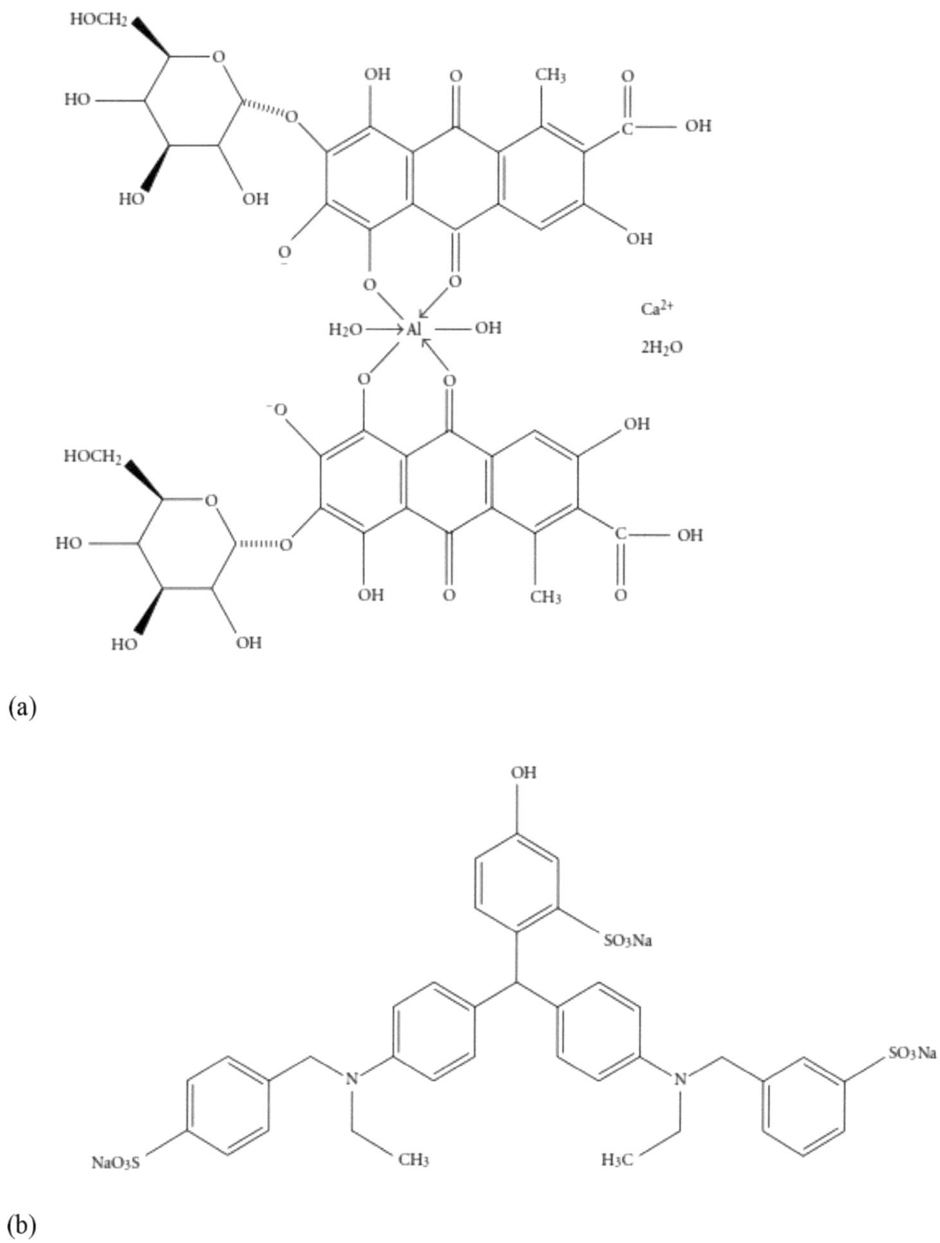

Figure 7. Structure of (a) carmine and (b) fast green [31].

Therefore it is vital and essential to develop cheap and environmentally secure corrosion inhibitors [32, 33]. Recently drugs have now been used as corrosion inhibitors. According to Gece [34], the use of drugs as corrosion inhibitors of metals corrosion possesses some merits over using some inorganic/organic inhibitors because of their little or insignificant impact on the environment. Drugs are nontoxic, cheap, negligible negative effects on environment, so it suggested replacing the traditional toxic corrosion inhibitors [35]. Many Researchers in this field generally agree that drugs are corrosion inhibitors that can compete favorably with discovered green corrosion inhibitors and that most of these drugs can be synthesized from natural products. The choice of some drugs used as corrosion inhibitors is based on the following: (a) drug molecules contain oxygen, nitrogen and sulfur as the main active centers, (b) drugs are reported to be environmental friendly and important in biological reactions and (c) some drugs are easy to produced. Therefore, it has been established from research findings that, since majority of these drugs are natural product or plants extracts have been discovered to accomplish the aim of employing a cheap, environmentally suitable, abundantly available and effective high inhibition efficiency and little or insignificant environmental impact. **Tables 2** and **3** reveal application of corrosion in specific system environment and examples of drugs (quinolones) commonly used as corrosion inhibitors in our common environment respectively [34]. Nowadays, studies are now focusing on using pharmaceutical drugs nontoxic corrosion inhibitors to slow down the corrosion of metals and their alloys. For instance, Abdallah [36] used rhodanine azosulpha drugs to inhibit the corrosion of 304 stainless steel in HCl solution, [37] also anti-bacterial drug to control the deterioration of aluminum in HCl solution. Similarly, Obot and Obi–Egbedi [38] used antihypertensive drugs to slow down the corrosion of aluminum and aluminum alloys in aqueous solutions.

The presence of benzene ring in the molecules and heterocycles like pyridine, thiophenes, isoxazoles, others that present in drugs has sensitized researchers all over the world to intensifying their investigation of drug compounds as green corrosion inhibitors as save alternative to the conventional toxic corrosion inhibitors [39].

A number of studies have reported the use of drugs particularly antibacterial as corrosion inhibitors [40]. For instance, Eddy and Ebenso [40] investigated on Corrosion Inhibition and Adsorption Characteristics of Tarivid on Mild Steel in H_2SO_4 using thermometric and gasometric methods. Their finding showed that tarivid drug inhibits the deterioration of mild steel in the acid environment (H_2SO_4). The inhibition efficiency of tarivid was established to increase in values as its concentration increased however; it decreased as the temperature increases. The mechanism of physical adsorption was proposed from the acquired kinetic and thermodynamic factors. Also, Langmuir isotherm adsorption model was followed. Abdallah [36] studied the effect of Rhodanine azosulpha drugs as corrosion inhibitors on the corrosion of 304 stainless steel in HCl environment using gravimetry and potentiodynamic methods. Parallel adsorption technique was used for the inhibition process on the steel surface because of the presence of multiactive center within the molecules of the inhibitor.

Solmaz et al. [41] have shown that Rhodanine is a good corrosion inhibitor for low carbon steel in hydrochloric acid. They used electrochemical techniques for the corrosion findings.

The effect of a number of antibiotic drugs, like; spectinomycin, paromomycin and streptomycin on the corrosion performance of zinc metal in hydrochloric acid environment was studied by [42], using some electrochemical and gravimetric techniques. The percentage of inhibition efficiency was discovered to be increasing as the concentration of drug increases but decreases with temperature. It has been discovered that majority of the drugs used display vital roles in biological reactions due to their antibacterial, anticonvulsant, antidiabetic, and other properties

http://dx.doi.org/10.5772/intechopen.72942

[10, 11]. The criteria used in selection of these drugs as corrosion inhibitors depend on the following (i) whether the molecules contain sulfur, oxygen and nitrogen as active centers, (ii) the ease of production as well as purification. (iii) They are eco-friendly.

System	Inhibitor	Metals	Concentration
Acids			
HCl	Ethylaniline	Fe	0.5%
..	Mercaptobenzotriazole	..	1%
..	Pyridine + phenylhydrazine	..	0.5% + 0.5%
..	Rosin amine + ethylene oxide	..	0.2%
Sulfuric	Phenylacridine	..	0.5%
Phosphoric	Sodium iodide	..	200 ppm
Others	Thiourea	..	1%
..	Sulfonated castor oil	..	0.5-1.0%
..	Arsenic Oxide	..	0.5%
..	Sodium arsenate	..	0.5%
Water			
Potable	Calcium bicarbonate	Steel, cast iron	10 ppm
..	Polyphosphate	Fe, Zn, Cu, Al	5-10 ppm
..	Calcium hydroxide	Fe, Zn, Cu	10 ppm
..	Sodium silicate	..	10-20 ppm
Cooling	Calcium bicarbonate	Steel, cast iron	10 ppm
..	Sodium chromate	Fe, Zn, Cu	0.1%
..	Sodium nitrite	Fe	0.05%
..	Sodium phosphate monobasic	..	1%
..	Morpholine	..	0.2%
Boilers	Sodium phosphate monobasic	Fe, Zn, Cu	10 ppm
..	Polyphosphate	..	10 ppm
..	Morpholine	Fe	variable
..	Hydrazine	..	O2 scavenger
..	Ammonia	..	neutralizer
..	Octadecylamine	..	variable
Engine coolants	Sodium chromate	Fe, Pb, Cu, Zn	0.1-1%
..	Sodium nitrite	Fe	0.1-1%
..	Borax	..	1%
Glycol/water	Borax + MBT	All	1% + 0.1%
Oil field brines	Sodium silicate	Fe	0.01%
..	Quaternaries	..	10-25 ppm
..	Imidazoline	..	10-25ppm
Seawater	Sodium silicate	Zn	10 ppm
..	Sodium nitrite	Fe	0.5%
..	Calcium bicarbonate	All	pH dependent
..	Sodium phosphate monobasic + Sodium nitrite	Fe	10 ppm + 0.5%

Table 2. Application of corrosion inhibitors in specific system environment [24].

S/N	Drug (used as corrosion inhibitor)	The molecular structure	The clinical usage	Class of inhibitor	Type of environment (corrosive medium)
1.	Ciprofloxacin	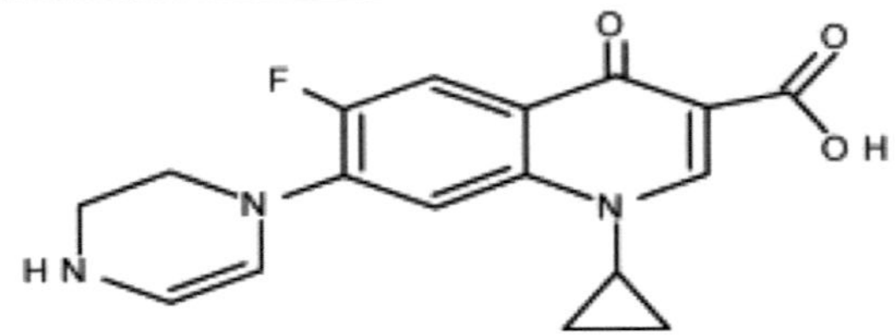	Typhoid fever, Prostatitis, Diarrhea	Mixed	Saline (NaCl)
2.	Ofloxacin	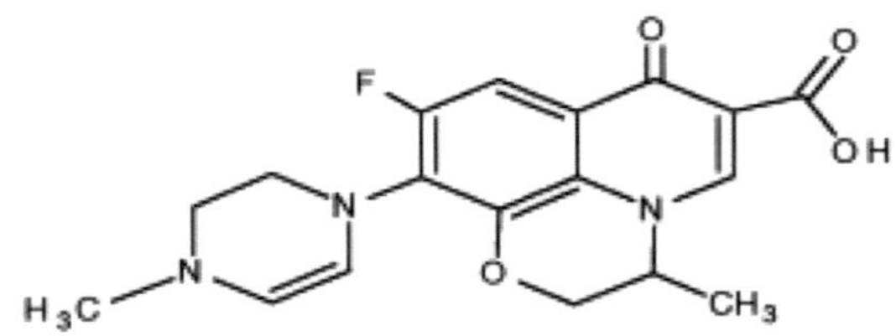	Urethritis, Gonorrhea, Cervicitis	Mixed	Acidic (HCl)
3.	Quinoline	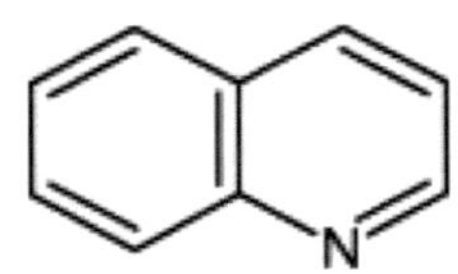	Antimalarial	Mixed	Acidic (HCl)
4.	Quinaldic acid	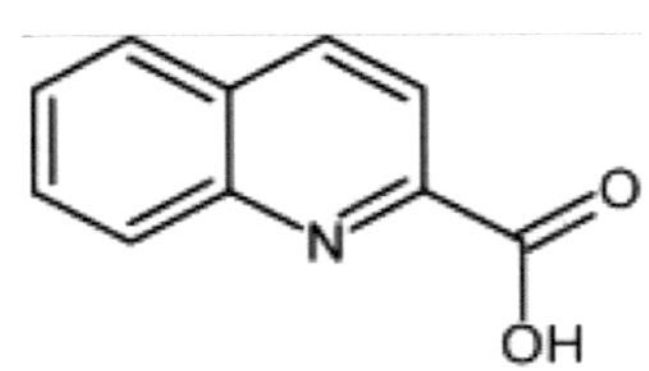	Antimalarial	Mixed	Acidic (HCl)
5.	Quinaldine	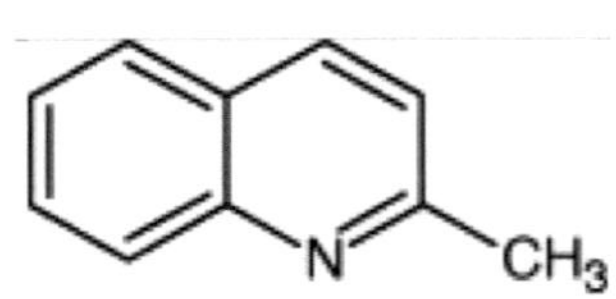	Antimalarial	Mixed	Acidic (HCl)

S/N	Drug (used as corrosion inhibitor)	The molecular structure	The clinical usage	Class of inhibitor	Type of environment (corrosive medium)
6		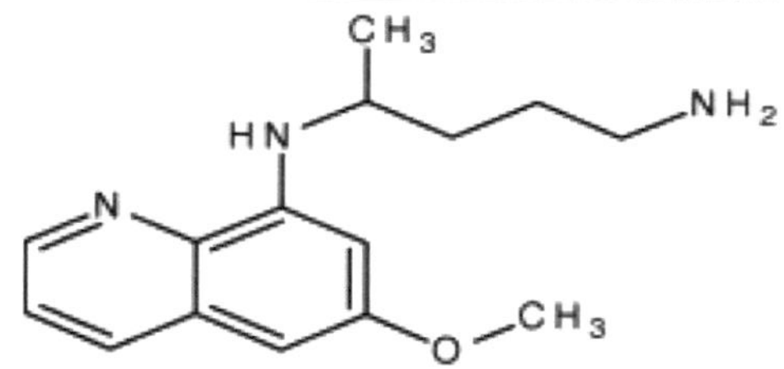	Antimalarial pneumonia	Mixed	Acidic (HCl)
7.	Amifloxacin	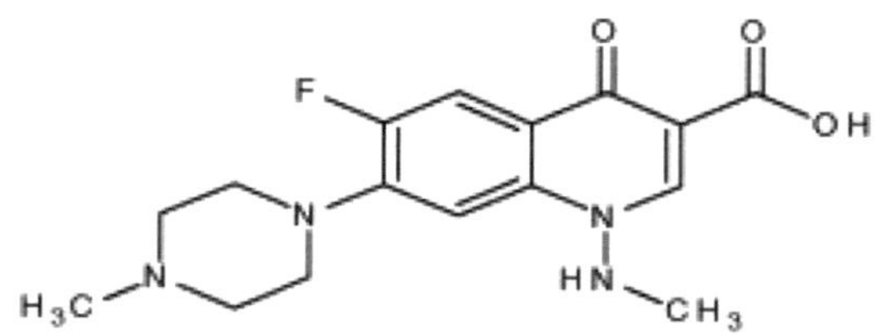	Typhoid fever, Diarrhea (infectious)	Mixed	Saline (NaCl)
8.	Sparfloxacin	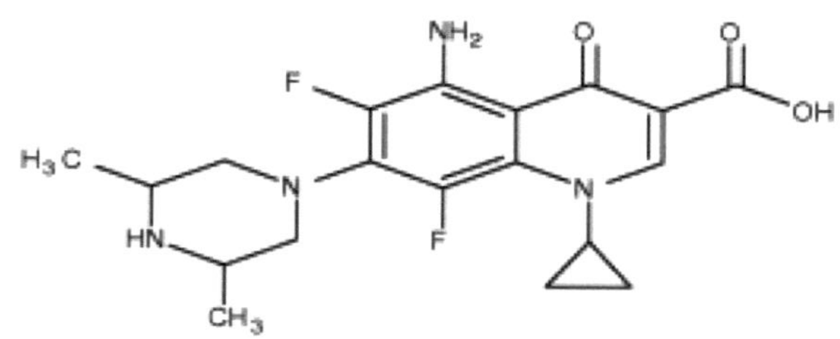	Urinary tracts, Typhoid Fever, Bones	Adsorption	Acidic (HCl and H_2SO_4)
9.	Enrofloxacin	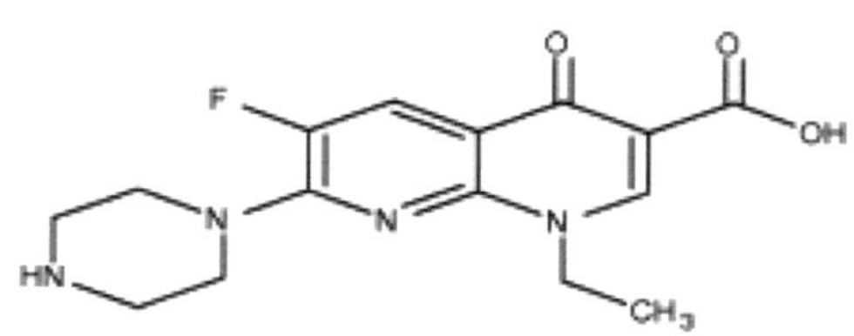	Infections in urinary tracts	Mixed	Saline (NaCl)

Table 3. Examples of drugs (quinolones) commonly used as corrosion inhibitors [35].

Author details

Ojo Sunday Isaac Fayomi[1,2]*, Paul Apeye Lucky Anawe[3] and Ayodele Daniyan[4]

*Address all correspondence to: ojosundayfayomi3@gmail.com

1 Department of Chemical, Metallurgical and Materials Engineering, Tshwane University of Technology, Pretoria, South Africa

2 Department of Mechanical Engineering, Covenant University, Ota, Ogun State, Nigeria

3 Department of Petroleum Engineering, Covenant University, Ota, Ogun State, Nigeria

4 Department of Materials Science and Engineering, Obafemi Awolowo University, Ile-Ife, Nigeria

References

[1] Das S. Development of aluminium alloy composites for engineering applications. Transactions of Indian Institute of Metallurgy. 2004;**57**(4):325-334

[2] Ishwara Bhat J, Alva VDP. A study of aluminium corrosion inhibition in acid medium by an antiemitic drug. Transactions of the Indian Institute of Metals. 2011;**64**:377. DOI: 10.1007/s12666-011-0102-9

[3] Obot IB, Obi-Egbedi NO, Umoren SA. Antifungal drugs as corrosion inhibitors for aluminium in 0.1 M HCl. Corrosion Science. 2009;**51**(8):1868-1875

[4] Oki M, Oki K, Otaigbe J, Otikor S. Corrosion inhibition of aluminium in HCl by amine modified epoxy resin. Journal of Materials. 2013;**2013**:5. Article ID 479728. DOI: 10.1155/2013/479728

[5] Abdel Hameed RS, Ismail EA, Abu-Nawwas AH, Al-Shafey HI. Expired Voltaren drugs as corrosion inhibitor for aluminium in hydrochloric acid. International Journal of Electrochemical Science. 2015;**10**:2098-2109

[6] El-Etre AY, Shahera M, Shohayeb MA, Elkomy S, Abdelhamed S. Inhibition of acid aluminium corrosion in presence of aqueous extract of Domiana. Journal of Basic and Environmental Sciences. 2016;**3**:25-36

[7] Adejoro IA, Akintayo DC, Ibeji CU. The efficiency of chloroquine as corrosion inhibitor for aluminium in 1M HCl solution: Experimental and DFT study. Jordan Journal of Chemistry. 2016;**11**(1):38-49

[8] Bikić F, Kasapović D, Delijić K, Radonjić D. Application of Aloe Vera as green corrosion inhibitor for aluminum alloy types AA8011 and AA8006 in 3.5% NaCl. Bulletin of the Chemists and Technologists of Bosnia and Herzegovina. 2017;**48**:35-40

[9] Gupta NK, Quraishi MA, Singh P, Srivastava V, Srivastava K, Verma C, Mukherjee AK. Curcumine longa: Green and sustainable corrosion inhibitor for aluminum in HCl medium. Analytical & Bioanalytical Electrochemistry. 2017;**9**(2):245-265

[10] Abood HA. The study of the inhibitory properties of Omeprazole on the corrosion of Aluminum 6063 in alkaline media. Basrah Journal of Science. 2011;**28**:74-93

[11] Uma K, Rekha S. Investigation and inhibition of aluminium corrosion in methane sulphonic acid solution by organic compound. Journal of Chemical and Pharmaceutical Research. 2015;**7**(4):165-169

[12] Hackerman N, Hurd RM. Corrosion Inhibition and Molecular Structure, First International Congress on Metallic Corrosion. London: Butterworth; 1962. pp. 166-172

[13] Norayan R. An Introduction to Metallic Corrosion and Its Prevention. New Delhi: Oxford & IBH Publishing Co; 1983

[14] West JM. Basic Corrosion and Oxidation. Chichester UK: Ellis Horwood; 1980

[15] Jones DA. Principles and Prevention of Corrosion. 2nd ed. New Jersey, United States: Upper Saddle River; 1996

[16] Pourbaix M. Atlas of Electrochemical Equilibria in Aqueous Solutions. Huston: NACE Cebelcor; 1974

[17] Shimizu K, Furneaux RC, Thompson GE, Wood GC, Gotoh A, Kobayashi K. On the nature of "easy paths" for the diffusion of oxygen in thermal oxide films on aluminium. Oxidation of Aluminium. 1991;**35**(5/6):427-439

[18] Davis JR et al., editor. Metals Handbook. Vol. 13. 9th ed. Ohio: ASM International; 1987. pp. 104-122 and 583-609

[19] Davis JR. Alloying: Understanding the Basics. Ohio, USA: ASM International; 2001. pp. 351-416

[20] Fouda AS, Abdallah M, Eissa M. Corrosion inhibition of Aluminum in 1 M phosphoric acid solutions using some Chalcones derivatives and synergistic action with halide ions. African Journal of Pure and Applied Chemistry. 2013;**7**(12):394-404. DOI: 10.5897/AJPAC2013. 0534

[21] Mazhar AA, Badaway WA, Abou-Romia MM. Impedance studies of corrosion resistance of aluminium in chloride media. Surface and Coatings Technology. 1986;**29**:335-345

[22] Baker MA, Castle JE. The initiation of pitting corrosion at MnS inclusion. Corrosion Science. 1993;**34**(4):667-682

[23] Van Delinder LS. Corrosion Inhibitors Basics: An Introduction. Houston, Texas, USA: NACE International; 1984

[24] Corrosion inhibitors www.chemia.odlew.agh.edu.pl/dydaktyka/Dokumenty/Corrosion/Inhibitors.pdf [Accessed: 28-06-17]

[25] Bregman JJ. Corrosion Inhibitors. New York/London: MacMillan; 1963

[26] Yurt A, Balaban A, Kandemir SU, Bereket G, Erk B. Investigation on some Schiff bases as HCl corrosion inhibitors for carbon steel. Materials Chemistry and Physics. 2004;**85**:420-426

[27] Musa AY, Jalgham RTT, Mohamad AB. Molecular dynamic and quantum chemical calculations for phthalazine derivatives as corrosion inhibitors of mild steel in 1M HCl. Corrosion Science. 2012;**56**:176-183

[28] Rosenfeld IL. Corrosion Inhibitors. New York: McGraw-Hill; 1977

[29] Thomas JGN. The Mechanism of Corrosion. Oxford, UK: Butterworths Heinemann; 1994

[30] Hegazy MA, Abdallah M, Awad MK, Rezk M. Part I: Experimental results. Corrosion Science. 2014;**81**:54-64

[31] Prabhu RA, Shanbhag AV, Venkatesha TV. Influence of tramadol [2-[(dimethylamino) methyl]-1-(3-methoxyphenyl) cyclohexanol hydrate] on corrosion inhibition of mild steel in acidic media. Journal of Applied Electrochemistry. 2007;**37**:491-497

[32] Loto RT, Loto CA, Popoola API. Corrosion inhibition of thiourea and thiadiazole derivatives: A review. Journal of Materials and Environmental Science. 2012;**3**:885-894

[33] Ladha DG, Naik UJ, Shah NK. Investigation of cumin (*Cuminum cyminum*) extract as an eco-friendly green corrosion inhibitor for pure Aluminium in acid medium. Journal of Materials and Environmental Science. 2013;**4**:701-708

[34] Gece G. Drugs: A review of promising novel corrosion inhibitors. Corrosion Science. 2011;**53**(12):3873-3898

[35] Mahdi AS. Amoxicillin as green corrosion inhibitor for concrete reinforced steel in simulated concrete pore solution containing chloride. International Journal of Advanced Research in Engineering and Technology. 2014;**5**:99-107

[36] Abdallah M. Rhodanine azosulpha drugs as corrosion inhibitors for corrosion of 304 stainless steel in hydrochloric acid solution. Corrosion Science. 2002;**44**(4):717-728

[37] Abdallah M. Antibacterial drugs as corrosion inhibitors for corrosion of aluminium in hydrochloric solution. Corrosion Science. 2004;**46**(8):1981-1996

[38] Obot IB, Obi-Egbedi NO. Fluconazole as an inhibitor for aluminium corrosion in 0.1 M HCl. Colloids and Surfaces A: Physicochemical and Engineering Aspects. 2008;**330**: 207-212

[39] Khan G, Newaz KS, Basirun WJ, Ali HB, Faraj FL, Khan GM. Application of natural product extracts as green corrosion inhibitors for metals and alloys in acid pickling processes - A review. International Journal of Electrochemical Science. 2015;**10**:6120-6134

[40] Eddy NO, Ebenso EE. Corrosion inhibition and adsorption characteristics of tarivid on mild steel in H_2SO_4. E-Journal of Chemistry. 2010;**7**(1):442-448

[41] Solmaz R, Kardas G, Yazici B, Erbil M. Inhibition effect of rhodanine for corrosion of mild steel in hydrochloric acid solution. Protection of Metals. 2005;**41**(6):581-585

[42] Abdallah M, Zaafarany BA, Jahdaly AL. Corrosion inhibition of zinc in hydrochloric acid using some antibiotic drugs. Journal of Materials and Environmental Science. 2016;**7**(4): 1107-1118

Chapter 5

Corrosion Inhibitors for Reinforced Concrete: A Review

Han-Seung Lee, Velu Saraswathy,
Seung-Jun Kwon and Subbiah Karthick

Additional information is available at the end of the chapter

http://dx.doi.org/10.5772/intechopen.72572

Abstract

The objective of the topic is to review the recent trends in corrosion inhibitors for reinforced concrete and their application in the laboratory and field conditions. Inhibitors are chemical substances which when added to the concrete in small concentrations will inhibit or prolong the time to initiation of corrosion in concrete structures. This chapter focuses on the type of inhibitors used in concrete based upon their mode of action and the way of application. The section deals with anodic, cathodic, mixed inhibitors; performance of admixed inhibitor vs. migrating/surface applied inhibitor and their evaluation studies; and electrochemical injection of corrosion inhibitor (EICI) in concrete and electrochemical chloride extraction techniques has been reviewed.

Keywords: anodic inhibitors, cathodic inhibitors, mixed inhibitors, organic inhibitors, migrating inhibitors, electrochemical injection, reinforced concrete

1. Introduction

Corrosion of steel in concrete structures plays a significant role in affecting the service life of the concrete structures. Various methods have been developed with the intent of preventing the corrosion and to enhance the service life. The methods include the coating to the concrete surface, the coating to the reinforcement, cathodic protection, electrochemical methods, alternative reinforcement, and corrosion inhibitors. Among all the available techniques, the use of corrosion inhibitors is one of the most appropriate and efficient methods for corrosion protection of reinforced concrete structures due to the easy operation, low cost, and excellent corrosion resistance effect [1–8]. Inhibitors which are added to the concrete in small concentrations are intended to delay and slow the onset of corrosion in reinforced concrete. Most of the inhibitors act by stabilizing the steel surface by forming the protective film, and some inhibitors react with concrete forming the complex thus reducing the permeability of the concrete. Corrosion inhibitors are generally used as admixtures in concrete for new construction, but they can also be utilized for repairs by admixed with concrete for patches, sprayed onto the surface of the concrete or applied by saturation treatment [9–13].

As per NACE international, "a corrosion inhibitor is a substance when added to an environment, either continuously or intermittently to prevent corrosion by forming a passive film on the metal." In other processing industries, inhibitors are the first line of defense against corrosion [14].

Inhibitors slow down the corrosion process by:

- increasing the anodic or cathodic polarization behavior;
- reducing the movement of ions of the metallic surface;
- increasing the electrical resistance of the metallic surface.

1.1. Classification of corrosion inhibitors

Corrosion inhibitors for concrete are classified based on [15–18] (**Figure 1**):

- Electrochemical mechanism of action (anodic or cathodic, or both).
- Type of chemicals used (organic and inorganic).
- The way of application; either mixed with concrete or applied on hardened concrete (migrating inhibitors).

1.2. Anodic (passivating) inhibitors

This type of inhibitors forms an insoluble protective film on anodic surfaces to passivate the steel. An anodic inhibitor shifts the potential to the passivation zone causing the formation of a thin passive film on the anodic sites, which increases the potential of the anode and decrease the corrosion in rate. There are two types of passivating inhibitors.

- Oxidizing anions, such as chromates, nitrates, and nitrites, passivate the steel in the absence of oxygen.
- Nonoxidizing ions, such as phosphate, tungstate, and molybdates, require the presence of oxygen to passivate the steel.

Anodic inhibitors are effective only when present in sufficiently high concentrations [15]. Typically, the concentration required is determined by the level of chloride present in the environment in which the steel is exposed. The most commonly used anodic inhibitor for concrete is calcium nitrite [18]. Calcium nitrates show similar inhibiting properties in concrete. Some anodic inhibitors, such as nitrites, cause accelerated corrosion and pitting if the

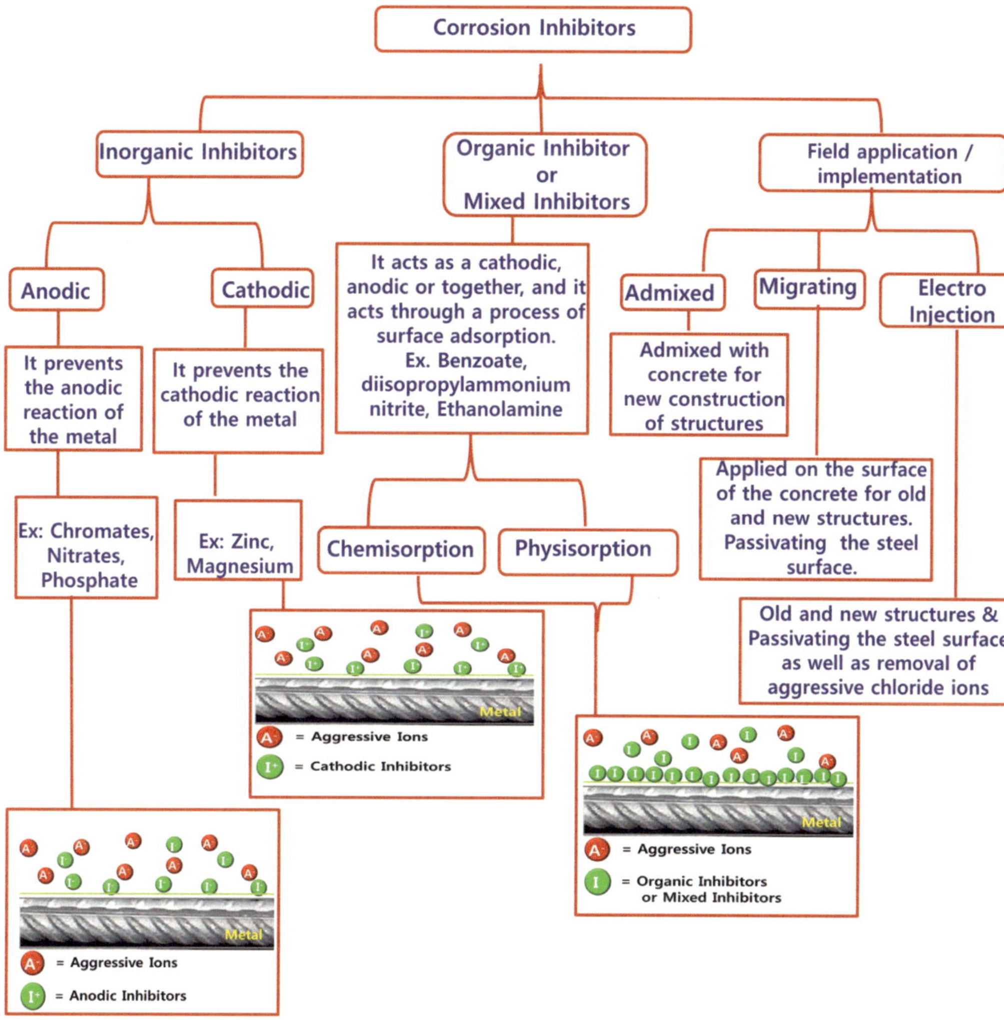

Figure 1. Classification of corrosion inhibitors.

concentration drops below the threshold/critical level [19]. Nitrates, benzoates, chromates, molybdates, and orthophosphates are used as anodic inhibitors.

1.3. Cathodic inhibitors

Cathodic inhibitors slow down corrosion by reducing the rate of cathodic reactions in the corrosion environment. A cathodic inhibitor causes the formation of insoluble compounds precipitating on the cathodic sites in the form of a barrier film. These inhibitors are called precipitation inhibitors [20]. Zinc and magnesium salts are cathodic inhibitors that form precipitates of zinc hydroxide and magnesium hydroxide at the cathode [21, 22]. Phosphates precipitate as ferrous and ferric phosphates on the steel substrate [23]. Cathodic inhibitors are not as effective as anodic inhibitors, but they are not likely to cause pitting [24].

1.4. Mixed inhibitors

Mixed inhibitors quash both the anodic and cathodic reactions, and they reduce the corrosion rate by forming a thin protective hydrophobic film on entire surface of the metal through adsorption mechanism [25]. The material with the hydrophobic group that has polar groups, such as N, S, and OH, is effective. These inhibitors are called adsorption or film-forming inhibitors [26–28]. The effectiveness of the film depends on the chemical composition, molecular structure, and their affinities for the metal surface. Cationic inhibitors, such as amines, and anionic inhibitors, such as sulfonates, get adsorbed on the metal surface, whether it is charged positively or negatively. The strength of the bond is the dominant factor for organic inhibitors. Organic inhibitors can also act as pore blockers, reducing the permeability of the concrete. Amines and amino alcohols are the commonly used mixed inhibitors, which displace the chloride ion and form a durable passivating film [29].

Organic inhibitors are further classified into two categories, based upon their application: inhibitor admixed concrete and migrating corrosion inhibitors that are applied on the hardened surface of the concrete [10, 19]. Corrosion inhibitors may be a good alternative for repair concrete due to their lower cost and ease of application. The main application methods for corrosion inhibitors are: added to fresh concrete as an admixture, applied on the hardened concrete and damaged structures migrating corrosion inhibitors (MCI), added to repair mortars, used as a surface treatment to rebars before concreting [10].

1.5. Effect of anodic, cathodic, and mixed inhibitors in concrete

Corrosion of steel in concrete structures is the primary concern when the structures are exposed to the coastal marine environment. The use of corrosion inhibitors can delay the onset of chloride-induced corrosion, prolong the time to initiation of corrosion, and thereby reduce the corrosion rate [30, 31]. Calcium nitrite is the most commonly used anodic corrosion inhibitor for reinforced concrete. This inhibitor is a passivating inhibitor, which forms a passive film on the surface of steel and significantly reduces the corrosion rate of steel in chloride-contaminated concrete [1, 32]. The inhibitive action of steel with calcium nitrite is as follows [1, 10]:

$$2Fe^{2+} + 2OH^{-} + 2NO_2^{-} \rightarrow 2NO\uparrow + Fe_2O_3 + H_2O \tag{1}$$

$$Fe^{2+} + OH^{-} + NO_2^{-} \rightarrow NO\uparrow + \gamma FeOOH \tag{2}$$

Different authors have reported different threshold value threshold $[Cl^-]/[NO_2^-]$ ratio depends on the environment. It was reported that the re-passivation of the corroded steel was observed at the $[Cl^-]/[NO_2^-]$ ratio from 0.5 to 1 [32]. For chloride-contaminated concrete or mortar, the $[Cl^-]/[NO_2^-]$ ratio ranges from 1 to 3, and for simulated alkaline solutions and chloride-contaminated concrete, the $[Cl^-]/[NO_2^-]$ ratio ranges from 1 to 4 [3, 31]. Several laboratory studies have established the

System	E_{corr} (mV vs. SCE)	Tafel slopes (mV dec^{-1})		Corrosion rate (mmpy)	Efficiency (%)
		ba	bc		
Plain	−544	53	84	0.0010	—
Plain +1% Cl^-	−566	58	88	0.0042	—
Plain +2% Cl^-	−576	50	87	0.0062	—
Plain +3% Cl^-	−580	58	85	0.0091	—
Plain +0.5% NO_2	−538	50	87	0.0001	92
Plain +0.5% NO_2 + 1% Cl^-	−547	53	87	0.0004	91
Plain +0.5% NO_2+ 2% Cl^-	−556	51	85	0.0007	89
Plain +0.5% NO_2+ 3% Cl^-	−566	51	81	0.0017	81

Table 1. Potentiodynamic polarization parameters for steel in ternary cement extract [45].

performance of calcium nitrite as an efficient corrosion inhibitor [1, 10, 15, 27, 33]. But, at the same time, possible increase of corrosion rate in case of low dosage or the presence of cracks [16]. Also, increase in dosage resulted in a decrease of the setting time and lowers the compressive strength of concrete and corrosion process may get accelerated [10, 32]. However, their application in corrosion protection is limited due to its toxicity and carcinogenicity [6, 34]. Chromates are carcinogenic, and hence its use is also prohibited [35, 36]. Haleem et al. [37] investigated the pitting corrosion behavior of phosphates, tungstate, and molybdates in chloride-contaminated $Ca(OH)_2$ solution and found that pitting corrosion current reaches steady state values, which depend on the way of introducing the inhibiting anions in the solution. Studies revealed that sodium phosphate forms a protective layer on the surface of the steel in chloride-contaminated synthetic concrete pore solution. Besides pretreatment of

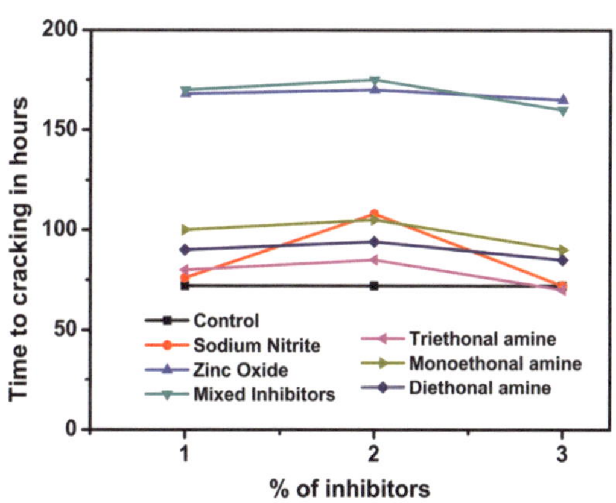

Figure 2. Impressed voltage test [1].

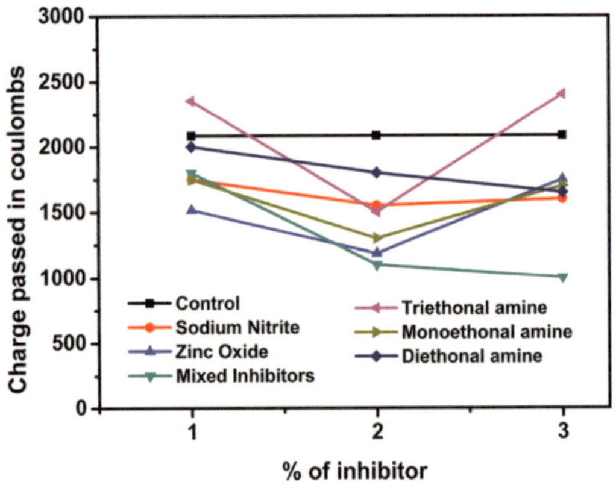

Figure 3. Rapid chloride penetration test [1].

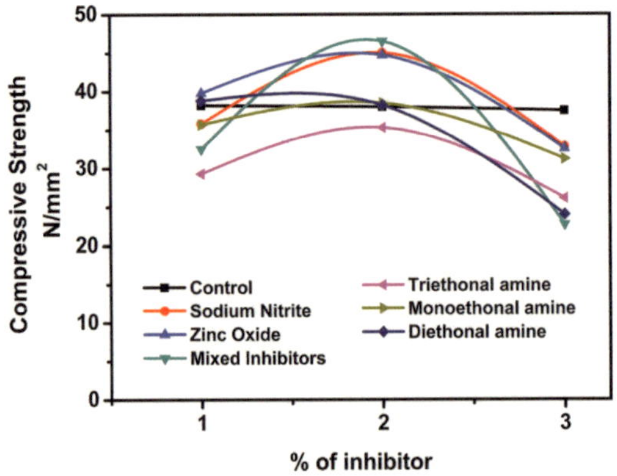

Figure 4. Compressive strength vs. percentage of inhibitors.

Systems	Corrosion rate (mmpy)		
	1%	2%	3%
Control	0.0093	—	—
$NaNO_2$	0.0023	0.0093	0.0250
ZnO	0.0022	0.0023	0.0024
Mixed	0.0009	0.0012	0.0037
Triethanolamine	0.0117	0.0038	0.0071
Monoethanolamine	0.0171	0.0042	0.0096
Diethanolamine	0.0203	0.0034	0.0083

Table 2. Corrosion rate of rebar exposed to various types of inhibitors admixed with chloride [1].

Systems	E_{corr} (mV)	Tafel slopes ($mV\ dec^{-1}$)		I_{corr} (μAcm^{-2})	Corrosion rate (mmpy)
		bc	ba		
OPC - Control	–580	80	350	20.5	0.240
$NaNO_2$ 1%	–440	110	260	2.51	0.029
2%	–450	120	230	2.72	0.031
3%	–480	140	240	2.94	0.034
ZnO 1%	–440	115	300	1.35	0.015
2%	–490	140	280	1.29	0.014
3%	–480	150	260	1.79	0.020
Mixed 1%	–420	180	225	0.90	0.010
2%	–485	120	200	1.20	0.013
3%	–490	135	220	1.30	0.015
Triethanolamine 1%	–470	110	292	4.52	0.052
2%	–530	90	284	7.81	0.090
3%	–545	75	248	9.73	0.112
Monoethanolamine 1%	–510	83	306	10.5	0.121
2%	–560	87	300	10.2	0.118
3%	–550	82	301	12.7	0.147
Diethanolamine 1%	–500	90	400	13.4	0.155
2%	–570	85	356	14.8	0.172
3%	–565	80	345	12.6	0.146

Table 3. Potentiodynamic polarization parameters for mild steel in concrete containing different inhibitors [15].

steel with 0.5 M inhibitor solution for 72 h forms a protective layer which further enhanced the protection efficiency [38–41]. Nahali et al. [42] studied the effect of Na_3PO_4 addition in chloride-contaminated mortar and concluded that the addition of phosphate inhibitor in concrete decreases the chloride ions effect on localized corrosion.

Yohai et al. [43, 44] reported that the phosphate ions behaved as a mixed type inhibitor. Soluble phosphate compounds, such as disodium phosphate or mono fluoro phosphate, can be mixed with mortar or applied on concrete [42] by immersion or by surface application [45]. Song et al. [46] evaluated the role of alkaline nitrites on the inhibition of corrosion of steel in binary (OPC + PPC) and ternary cement (OPC + PPC + PSC) and reported that ternary system provides 81% protection efficiency (**Table 1**) for steel in concrete in the presence of 3% chloride concentration. Muralidharan et al. [47] evaluated the effect of various inhibitive ions (hydroxide, citrate, and stannate) in chloride-contaminated (OPC + FA) extract and mortar and found that the addition of inhibitive ions decreased the corrosion rate of the steel rebar in simulated concrete environments.

Furthermore, the effectiveness of corrosion inhibitors was enhanced by using hybrid corrosion inhibitors; these inhibitors had good corrosion resistance and low toxicity, but the mechanism of protection is not clear [48]. The combined effect of pitting and uniform corrosion on the surface of the rebar was studied by using sodium molybdate (Na_2MoO_4), cerium nitrate hexahydrate, 2,5 Dimercapto-1,3,4-thiadiazole (DT), 2-Mercaptobenzothiazole (MBT), and 1H-benzotriazole (BTA). All the inhibitors except DT and Na_2MoO_4 are toxic. From the study, they concluded that Na_2MoO_4 is the non-toxic inhibitor, which re-passivates the pits formed on the surface of the steel rebar by forming an insoluble $FeMoO_4$ compound [36]. Saraswathy et al. [1] and Song et al. [15] studied the effect of anodic, cathodic, and mixed inhibitors in concrete by conducting various short- and long-term accelerated techniques. They concluded that the mixed inhibitor had shown higher time to cracking (**Figure 2**), lower coulomb values (**Figure 3**), higher compressive strength (**Figure 4**), and lesser corrosion rate (**Tables 2** and **3**).

2. Effect of organic inhibitors in concrete

2.1. Admixed vs. migrating corrosion inhibitors (MCI)

Recently, organic inhibitors gained more attraction in the construction industry due to their promising application as admixtures in reinforced concrete with improved protection efficiency and low cost [49]. Organic inhibitors are used as an admixed corrosion inhibitor [13, 18] or migrating corrosion inhibitors (MCI) [18, 50–52]. Migrating inhibitors are also called as surface applied inhibitors. Both admixed and migrating corrosion inhibitors prevent steel corrosion by forming a thin layer of protective barrier film on the surface of the rebar through adsorption mechanism. Amines and alkanol amines [1, 52–54] and their salts are used as organic inhibitors in concrete. It has been reported that the alkanol amine-based inhibitors could be able to decrease the corrosion rate of the carbonation induced corrosion effect only for the chloride-induced corrosion of reinforcement [55]. Furthermore, the results indicated that penetrating corrosion inhibitors when applied to the surface of existing or new structures can able to reduce the corrosion rate below 0.1 μA/cm^2. Organic inhibitors can adsorb by chemisorption or physisorption. Chemisorption leads to the chemical reaction between active centers of corrosion onto metal and organic inhibitor, where charges from the polar group of inhibitor and charges from metal and metal oxides formed in concrete are shared. Then, organic inhibitor forms the thin coating film blocking the metal surface from the aggressive species by strong chemical bonds. Physisorption leads to the formation of a mechanical barrier, made by organic inhibitors [56]. This inhibitor can halt both the anodic and cathodic reactions, thus reducing the corrosion rate of the reinforcing steel [57]. It has been reported that carboxylate and amino alcohol-based corrosion inhibitors show dual actions in concrete as the amines and alkanol amines adsorb on the metal surface and form a protective film against chloride [58, 59], whereas the carboxylate ester compound reacts with calcium hydroxide, precipitates and blocks the pores of the concrete. This pore blocking property is said to be a secondary protection mechanism against reinforcement corrosion [16, 18, 44, 58]. Furthermore, another study reported that some organic inhibitors adsorb better on the iron surface in the active state than in the passive state [60]. Ormellese et al. [61] reported that amines and alkanol amines (**Table 4**) had poor corrosion inhibition effect on steel in chloride

Amines [61]		Carboxylic acids [66]
Methylamine	CH_3–NH_2	Sodium gluconate
Dimethylamine	$(CH_3)_2$–NH	D-Saccharic acid, monopotassium salt
Ethylamine	CH_3CH_2–NH_2	Calcium α-D Eptagluconate
Propylamine	$CH_3CH_2CH_2$–NH_2	Phthalic acid, monopotassium salt
Cyclohexylamine	C_6H_{11}–NH_2	Lactic acid
Triethylentetramine	NH_2–CH_2CH_2–$(NHCH_2CH_2)_2$–NH_2	Maleic acid
Hexamethylentetramine	$C_6H_{12}N_4$	Suberic acid
Alkanolamines		Adipic acid
Monoethanolamine	$(OHCH_2CH_2)$–NH_2	Sodium benzoate
Dimethylethanolamine	$(CH_3)_2(OHCH_2CH_2)$–N	
Triethanolamine	$(OHCH_2CH_2)_3$–N	
Methyldiethanolamine	$(CH_3)(OHCH_2CH_2)_2$–N	

Table 4. Various amines and alkanolamines investigated by Ormellese et al. [61] and Monticelli et al. [66].

containing alkaline solutions. They further observed that polycarboxylates are more efficient against pitting corrosion. Comparison of carboxylates with amines/alkanol amines and amino acids were investigated in the same study, and the results are summarized in Predeferri's diagram showing pitting potential vs. chloride concentration (**Figure 5**). From the figure, it is evident that amines/alkanol amines had only moderate effectiveness, while carboxylates showed the best behavior at low and medium chloride levels. It was reported that amines were reported to reduce the compressive strength of cement paste and the reducing effect increased with increasing dosage [62]. Heren et al. [63] showed that the decline in concrete strength was enhanced by an increase in corrosion inhibitor concentration.

Surface-applied migrating corrosion inhibitors (MCI) are suitable for repairing the chloride-contaminated concrete. In concrete, chloride and inhibitor ions work on three mechanisms: natural

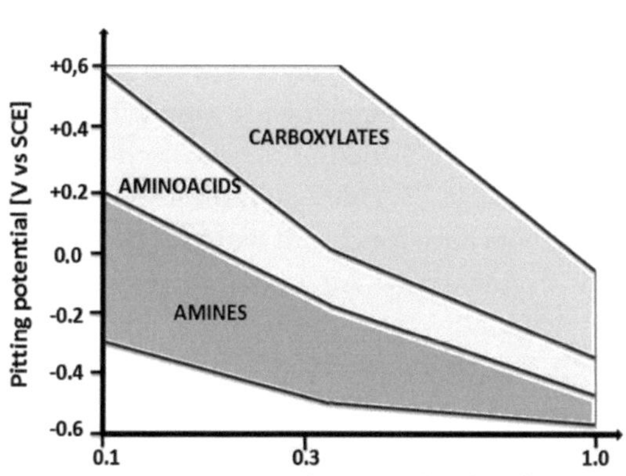

Figure 5. Pedeferri type diagram showing pitting potential vs. chloride ions and the range of corrosion organic inhibitors [61].

diffusion, electrical migration, and transport by movement of pore solution due to the capillary suction or pressure gradient. MCIs are typically based on commercial organic compounds. Aminoalcohols rapidly penetrates through the pores of the concrete, and it protects the steel rebar by forming a hydrophobic layer by physical or chemical adsorption. Another function of this inhibitor is, it reacts with the cement particles and forms an insoluble compound blocking the pores of the concrete [58]. This type of inhibitors is an appropriate strategy to rehabilitate reinforced concrete structures because of its ease of operation, low cost, and safety. MCIs are topically applied to the chloride-contaminated concrete surfaces before placement of the patches and overlays [64]. Zheng et al. [7] studied the durability of the surface applied inhibitor in concrete by conducting various accelerated tests and concluded that the durability of the concrete is improved. Criado et al. [65] examined the surface applied corrosion inhibitors in mortars with fly-ash additives. Another study has shown that amino alcohol-based inhibitors are most effective on the reference mortar specimens than on the specimens with ladle furnace slag. Further, it was noted that the corrosion rate was lowered in the samples with lesser chloride percentage, but was not effective with higher chloride ion containing mortar specimens [66].

Organic MCIs are mostly mixed inhibitors. It creates a hydrophobic layer that helps to repel moisture away from the steel rebar [67]. Monticelli et al. [68] investigated the different type of inhibitors based on amine/alkanol amine and carboxylate (dicarboxylic and hydrocarboxylic acids) on steel corrosion behavior in saturated calcium hydroxide solution containing 0.1 M chloride and chloride admixed mortar. They concluded that among the inhibitors studied, dicyclohexylammonium nitrite seems to be a better inhibitor than dicyclohexylamine. Further, they found that maleic acid was efficient in increasing the pitting potential of the steel in chloride-contaminated alkaline solutions. Of all the acids tested, maleic acid failed to hinder the corrosion process during longer exposure in mortars as like sodium nitrite. However, there is some controversy regarding their use. Elsener et al. [69] reported that amines/alkanol amines are not effective against chloride-induced corrosion. Most of the problems related to low effectiveness have been associated with leaching and evaporation of AMAs volatile components [70, 71] leading to decrease the inhibitor effective concentration in concrete.

Thangavel et al. [72] studied the effect of migrating vs. admixed corrosion inhibitors for steel in Portland, pozzolana, and slag cement concretes under macrocell condition. Macrocell corrosion parameters such as anode potential, macrocell current, and the total integrated current were monitored over a period of 1 year. The corrosion rate of the steel rebar embedded in different types of cement concretes was assessed by gravimetric weight loss method. The admixed inhibitor consists of 0.5% sodium citrate +0.5% sodium stannate and migrating inhibitor consisted of 0.5% amino alcohol, 0.5% amines and 0.5% nitrites. **Table 4** shows the various types of inhibitor combinations studied. The authors have concluded that MCI performed better than the admixed system in all the three types of cement used. Among the three types of cement, PSC showed lowest corrosion rate (**Table 5**). MCI showed 60% reduction in macrocell current (**Figures 6–8**) and admixed system showed 40% reduction in macrocell current in PSC concrete.

2.2. Electrochemical injection of corrosion inhibitors (EICI)

Several methods of repair treatments are adopted in the rehabilitation of concrete structures. They are a partial replacement of concrete cover, electrochemical re-alkalization, electrochemical chloride extraction, cathodic protection, and surface application of corrosion inhibitors [73–75].

Systems	Alkalinity (pH)	Free chloride contents (ppm)	Corrosion rate (mmpy)
OPC			
Control	11.45	4667	0.2044
Migrating	12.16	3750	0.1878
Admixed	12.08	3813	0.1892
PPC			
Control	11.46	3257	0.1810
Migrating	12.40	2862	0.1621
Admixed	12.33	3229	0.1751
PSC			
Control	11.46	2019	0.1396
Migrating	12.40	1813	0.1027
Admixed	12.31	2017	0.1188

Table 5. Corrosion rate of steel in OPC, PPC and PSC concretes [73].

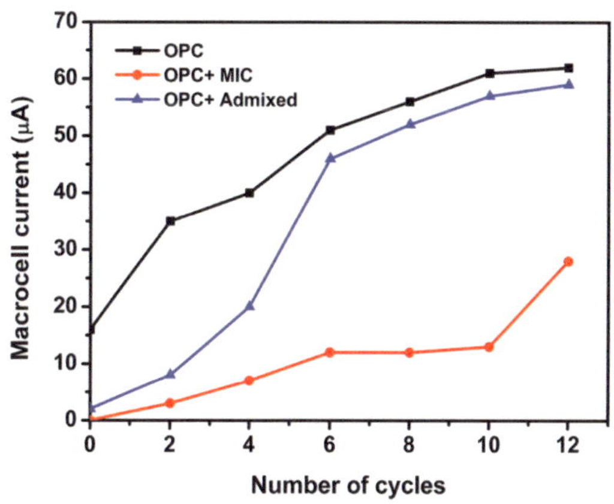

Figure 6. Macrocell current vs. number of cycles of exposure for steel in OPC concrete under macrocell condition [72].

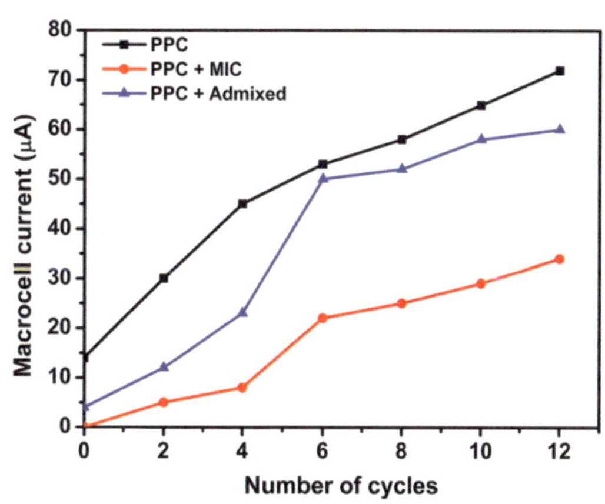

Figure 7. Macrocell current vs. number of cycles of exposure for steel in PPC concrete under macrocell condition [72].

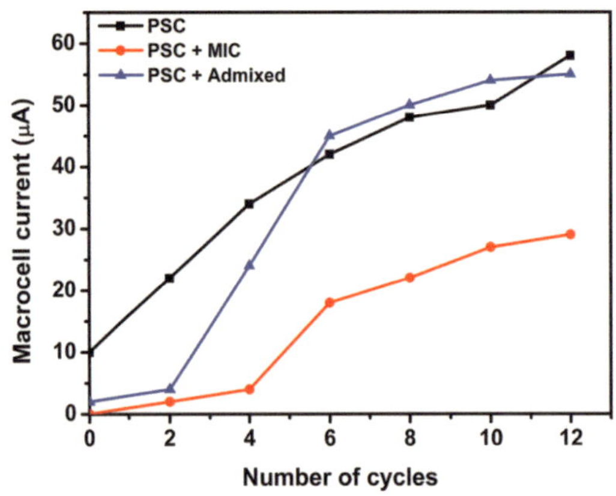

Figure 8. Macrocell current vs. number of cycles of exposure for steel in PSC concrete under macrocell condition [72].

Electrochemical injection of corrosion inhibitors (EICI) was found to be an efficient corrosion mitigation technique for carbonated and chloride-contaminated reinforced concrete structures to improve the durability. In EICI, a current density of 1–5 A m^{-2} is usually applied between the embedded steel cathode and an externally placed anode on the concrete surface in an aqueous solution containing inhibitors for a few weeks [76]. During the application of current, the cationic species of the corrosion inhibitor migrate into the concrete cover to the cathode, whereas the chloride ions in the concrete migrate out of the concrete towards the external anode [73, 76]. EICI was found be an effective method of preventing corrosion in existing structures; and this method serves as a rehabilitative measure to retard or reduce corrosion. Kubo et al. [77] applied the EICI technique in a 40-year-old carbonated reinforced concrete railway viaduct in Tokyo. They tried the electrochemical injection of the organic base corrosion inhibitor, ethanolamine into an existing aged carbonated concrete was found to allow penetration of an adequate concentration of the inhibitor near the steel. It was found that pH near the steel was changed due to the buffering action of the inhibitor. It was also found that due to the variation in the concrete cover (as is usually the case in real concrete structures) the current density distribution was affected resulting in an uneven injection treatment. Further, it was reported that detailed information on the distribution of concrete cover within the structure is therefore needed to establish a reasonable treatment plan for successful inhibitor injection.

Mangayarkarasi et al. [78] studied the efficiency of the electrochemical injection of inhibitors in chloride admixed concrete with OPC, PPC, and PSC concrete by using electrochemical techniques. The inhibitor consists of 0.1 M guanidine, 0.1 M thiosemicarbazide, 2 M triethanolamine, and 2 M ethyl acetate. They concluded the multi-component inhibitor injection showed more than 95% efficiency (**Table 6**) in terms of reduction in corrosion rate irrespective of chloride levels in different concretes at a current density of 0.5 A/m^2. Karthick et al. [2] investigated the EICI by using the same type of inhibitor in a chloride-contaminated old concrete slab (**Figure 9**) at a current density of 0.5 A/m^2. The efficiency of the inhibitor formulation was evaluated through various electrochemical tests, and the mechanism of inhibition was established by FTIR, SEM, EDAX, and MIP studies. They found that FTIR studies (**Figure 10**) indicated that the functional group of the inhibitors namely –NH_2, C–H, and C≡C move towards the cathode of embedded rebar and adsorbed on the metal surface and thereby provided

System	Cement	E_{corr} (mV vs. SCE)	I_{corr} (mA.cm^{-2}) × 10^{-5}	Corrosion rate (mmpy) × 10^{-3}	Inhibitor efficiency (%)
Without electro injection	OPC	−591	91.910	10.650	—
	PPC	−567	82.080	9.512	—
	PSC	−478	30.690	3.556	—
With electro injection	OPC	−282	6.034	0.699	93.43
	PPC	−345	2.276	0.263	97.22
	PSC	−295	2.024	0.234	93.40

Table 6. Polarization parameters for the corrosion of rebar embedded in OPC, PPC, and PSC concretes with and without electro injection process [79].

inhibition. MIP studies (**Figure 11**) proved that nearly 50% reduction in porosity was observed for concrete samples after EICI. SEM analysis (**Figure 12**) indicated that the hybrid inhibitor formulation act as a pore blocking agent. They also found that the EICI treatment can shift the rebar potential (+300 mV) towards the passive region in the potentiodynamic polarization curve (**Figure 13**) with the result the corrosion rate of the rebar also reduced considerably.

Pan et al. [79] evaluated the inhibition efficiency of eight organic inhibitors by monitoring the corrosion rate steel in aggressive, simulated concrete pore solution through electrochemical injection method. The data showed that tetrabutylammonium bromide (TBA-B) and tetramethylammonium chloride (TMA-C) outperformed other chemicals as corrosion inhibitors in the simulated concrete pore solution, by reducing the corrosion rate of steel by 85 and 75%, respectively. Nguyen et al. [80] performed the EICI rehabilitation treatments in repair mortar with tetra butyl ammonium bromide salt at a current density of 5 A/m^2, by using two electrolytes of 0.1 M NaOH and Na_3BO_3 for 1 and 4 weeks, respectively. The results reveal that the EICI treatment with 0.1 M Na_3BO_3 was more efficient in improving the chloride penetration resistance and the compressive strength of the mortar, relative to 0.1 M NaOH as the electrolyte.

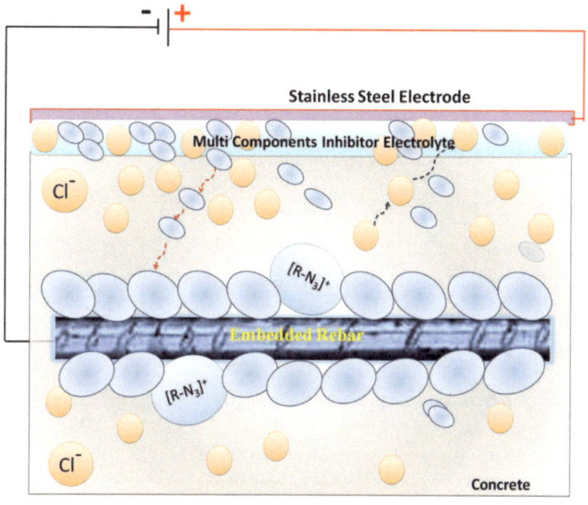

Figure 9. Schematic representation of electrochemical EICI process in concrete [2].

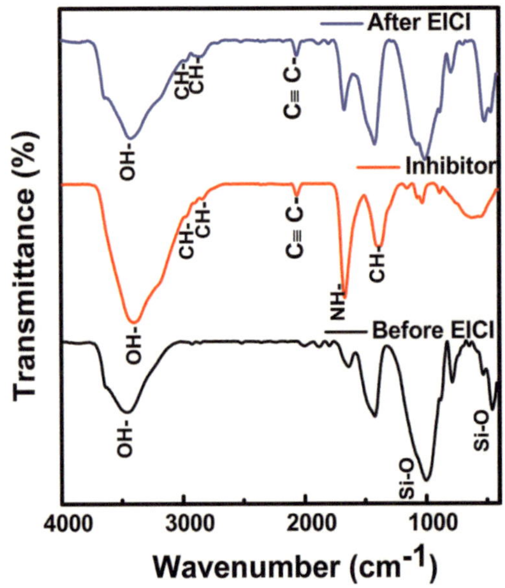

Figure 10. FTIR spectroscopy for concrete—before and after EI [2].

Liu and Shi [81] presented a state of the art report on electrochemical chloride extraction (ECE) and EICI, which covers the laboratory studies and computational models to predict the kinetics of ECE and EICI. Sawada et al. [76] evaluated ethanolamine and guanidine, the two organic inhibitors for steel, in aqueous media by immersing the carbonated and non-carbonated concrete and impressing a current density of 1–5 A/m^2 for 3–14 days. It was found that the efficiency was far higher in carbonated concrete than in non-carbonated concrete and that, in the carbonated specimens, the inhibitors became concentrated near the embedded steel. In non-carbonated concrete, guanidine penetration was accelerated, but the ethanolamine penetration was not significantly enhanced by the application of current density.

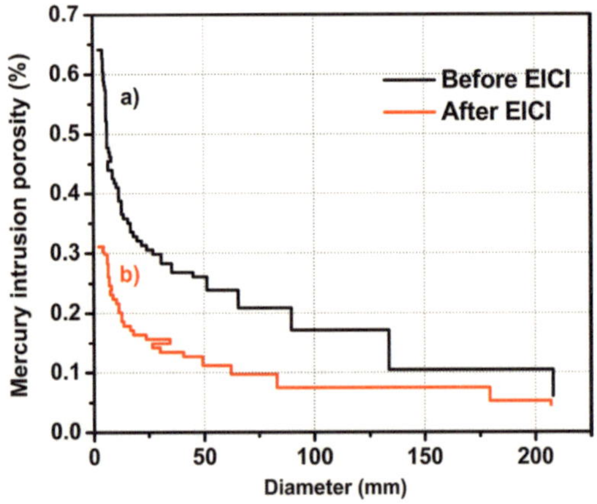

Figure 11. Mercury intrusion porosity for before and after EI process for concrete [2].

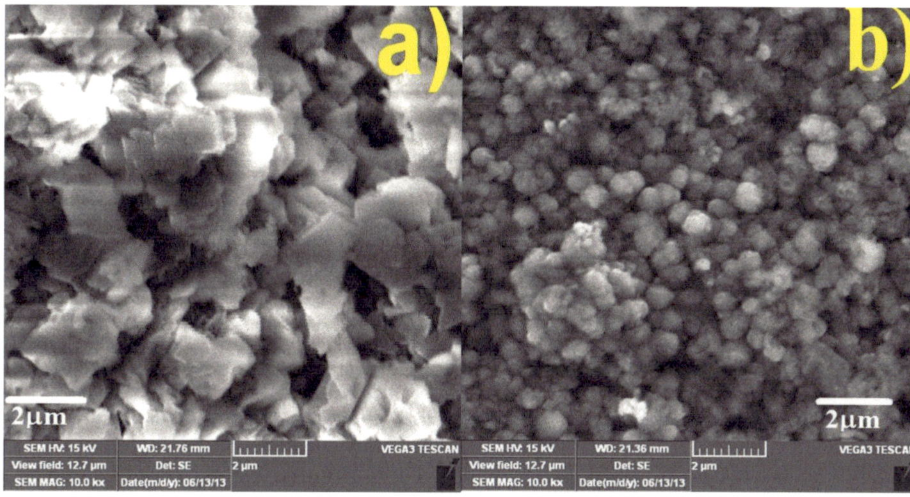

Figure 12. SEM images for concrete before and after EI [2].

Xu et al. [54] adopted a novel method called bi-directional electromigration rehabilitation [BIEM] for injecting the corrosion inhibitor (triethylenetetramine) into chloride-contaminated concrete specimens. In this process, an electric field was applied between the embedded steel cathode and external anode to inject the inhibitor from an external electrolyte to the concrete specimens and extract the chloride ions from the cover zone concrete. After the treatment, the specimens were drilled to determine the concentration profiles of the corrosion inhibitor, chloride, and hydroxyl ions within the concrete. ECE was compared with as a control experiment using saturated $Ca(OH)_2$ solution as an external electrolyte. They found that the chloride content decreased, and alkalinity increased after treatment. The concentration of the inhibitor injected around the embedded steel bars was adequate to provide corrosion protection. Among

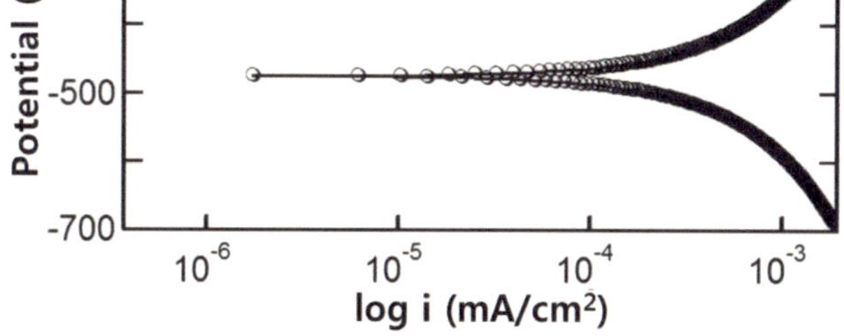

Figure 13. Potentiodynamic polarization curve for rebar in concrete before and after EI [2].

the electrochemical rehabilitation techniques available, ECE is the most widely used method of repairing chloride-contaminated structures. However, this approach cannot completely remove the chlorides in the structure and is only a temporary solution because chloride ions tend to come back after treatment is stopped [75]. Simultaneous effect of chloride removal and re-passivation of concrete by using nitrite (migrating) inhibitor was tried by the application of ECE treatment. The electrochemical response of the embedded rebar after the ECE treatment was found that the migration of nitrite ions re-passivated the steel surface and the chloride removal efficiency was also increased [82].

3. Summary of inhibitors used

Tables 7 and **8** show the review of the most commonly used corrosion inhibitor dosage in concrete. From the table, it is evident that recently organic inhibitors are mainly focused and most of the studies are carried out in simulated concrete pore solution containing chloride. Only a few studies are conducted in mortar and concrete. Very fewer field studies are undertaken.

Reference	Inhibitor used	Dosage	Environment
Jiang et al. [6]	Deoxyribonucleic acid as a biological corrosion inhibitor	0.0025%	Simulated concrete pore solution
Zheng et al. [7]	Organic inhibitors—surface applied		Concrete
Diamanti et al. [22]	Organic inhibitors	0.0001, 0.001, 0.01, and 0.1 M	Concrete pore solution
Xu et al. [33]	Anodic, cathodic, and mixed inhibitors	1, 2% by vol.	Simulated environment
Nahali et al. [41]	Na_3PO_4		Mortar $+Cl^-$
Yohai et al. [42]	Na_3PO_4		Mortar $+Cl^-$
Fei et al. [47]	Cathodic inhibitor		Pore solution
Verbuggen et al. [48]	Organic inhibitors	10^{-4} M	Concrete pore solution
Ormellese et al. [50]	Organic inhibitors/Migrating	0.3 mol/L	Concrete pore solution
Shi et al. [52]	Migrating corrosion inhibitor		Mortar
Xu et al. [54]	Organic inhibitors—electro-injection		
Criado et al. [66]	Organic inhibitors	0.05 M	Activated fly ash mortar
Prieto et al. [67]	Organic inhibitors—surface applied	0.500 kg/m^2	Mortar
Cabrini et al. [68]	Organic inhibitors	0.1 M	Pore solution
Kubo et al. [78]	Electrochemical injection (Amine based)	0.5–2.0 mol/l	Real concrete bridge (40-year-old railway viaduct)
Sanchez et al. [82]	Electrochemical removal and migration of inhibitor	Nitrite	Reinforced concrete

Table 7. Review of inhibitors in concrete (2010–2017).

Type of inhibitor	Exposure condition	Inhibitors	pH	Dosage	Efficiency (%)	References
Anodic inhibitors	Concrete admixed 1.96% NaCl	$Ca(NO_3)_2$	6.0–6.5	0.49	26.51	[32]
				0.98	60.54	[32]
				1.96	88.65	[32]
	Sat. $Ca(OH)_2$ + 3% NaCl	Na_3PO_4	10.5–11.8	1 mol/l	60.02	[38]
				2.5 mol/l	64.85	[38]
				5 mol/l	78.2	[38]
				7.5 mol/l	80.0	[38]
				10 mol/l	78.1	[38]
	Concrete immersed in 3% NaCl	$NaNO_2$	7.5–8.5	1%	87.9	[15]
				2%	87.08	[15]
				3%	85.83	[15]
Cathodic Inhibitors	Concrete immersed in 3% NaCl	ZnO	6.95–7.8	1%	93.75	[15]
				2%	94.16	[15]
				3%	91.66	[15]
Mixed Inhibitors or organic inhibitors	Concrete immersed in 3% NaCl	$NaNO_2$ + ZnO	7.2–8.5	1%	95.8	[15]
				2%	94.58	[15]
				3%	93.75	[15]
	Concrete immersed in 3% NaCl	Triethanolamine	6.0–7.0	1%	77.04	[15]
				2%	62.5	[15]
				3%	53.33	[15]
	Concrete immersed in 3% NaCl	Monoethanolamine	11–11.8	1%	49.58	[15]
				2%	50.8	[15]
				3%	38.75	[15]
	Concrete immersed in 3% NaCl	Diethanolamine	9.8–10.2	1%	35.42	[15]
				2%	28.33	[15]
				3%	39.16	[15]
	OPC Concrete immersed in 1% NaCl	Disodium β-glycerol phosphate Pentahydrate + sodium 3-aminobenzoate	Adjusted to 7.0	0.05 M + 0.05 M	63	[65]
		Disodium β-glycerol phosphate pentahydrate + sodium N-phenylanthranilate	adjusted to 7.0	0.05 M + saturated	81	[65]

Table 8. Different type of inhibitor dosage in concrete.

4. Conclusions

The following conclusions were drawn:

- The review focused only on the use of various types of inhibitors in concrete under laboratory and field condition.
- The corrosion inhibitors are effective in preventing reinforcement from corrosion within concrete structures.
- The corrosion inhibitor forms a protective film around the embedded steel bars.
- The inhibitors for concrete are classified as an anodic, cathodic, and mixed type of inhibitors based on the mode of action.
- Recently, mixed and organic types of inhibitors are the most commonly used inhibitors in concrete due to their synergistic effect.
- Organic corrosion inhibitors were ineffective in preventing steel corrosion when their concentrations were too low [19]. The corrosion inhibitors can provide adequate protection for reinforcement only when the concentration of the inhibitor is higher than that of the chloride ions in the pore solution [54].
- Some inhibitors get leached out over an extended period of exposure.
- Over dosage of inhibitors will cause pitting, leaching, and retarding effect.
- While under dosage of inhibitor may lead to inefficient protection of the steel rebars.
- The efficiency of the inhibitor strongly depends on the initial chloride ions concentration in concrete.
- The performance of the inhibitor strongly depends on the quality of concrete.
- Amines-, alkanol amines-, and carboxylate-based mixed inhibitors are mostly used in reinforced concrete.
- Amines and alkanol amines are capable of diffusing through concrete when applied on the surfaces of the structures by capillary action.
- Migrating inhibitors play a significant role in the repair and rehabilitation of damaged and chloride-contaminated concrete structures.
- Migrating or surface applied inhibitors are found to be the cost-effective treatment for field implementation due to the ease of application. For this type of application, penetrating type amine/carboxylic-based inhibitors are preferable.
- The surface-applied inhibitors can penetrate up to the depth of the embedded steel reinforcement; thus, adequate concentrations are necessary to provide corrosion protection when the concrete cover is too thick, or the concrete compaction is too high [83].
- Electrochemical injection of corrosion inhibitors (EICI) is found to be an effective corrosion mitigation technique for carbonated and chloride-contaminated reinforced concrete structures to improve the durability.

- In this type of repair method, continuous monitoring of the condition of steel is essential not only in the electrochemically treated area but also in its adjacent regions.
- Still, the mechanism of mixed/organic inhibitors is not very well understood in long-term applications. Hence, it needs more extensive investigation in real concrete structures by considering the various aspects, such as chloride content, types of cement, types of inhibitors, etc.

Acknowledgements

This research was supported by the Basic Science Research Program through the National Research Foundation of Korea (NRF) funded by the Ministry of Science, ICT, & Future Planning (No. 2015R1A5A1037548). One of the authors (V.S.) thanks the Director, CECRI, and CSIR for the permission to pursue my fellowship at Hanyang University, Korea.

Author details

Han-Seung Lee[1], Velu Saraswathy[1,2]*, Seung-Jun Kwon[3] and Subbiah Karthick[1]

*Address all correspondence to: corrsaras@gmail.com

1 Department of Architectural Engineering, Hanyang University, Sangrokgu, Ansan, Korea

2 Corrosion and Materials Protection Division, CSIR-Central Electrochemical Research Institute, Karaikudi, Tamil Nadu, India

3 Department of Civil Engineering, Hannam University, Daejeon, Korea

References

[1] Saraswathy V, Song H-w. Improving the durability of concrete by using inhibitors, Building and Environment. Building and Environment. 2007;**42**(1):464-472. DOI: 10.1016/j.buildenv.2005.08.003

[2] Karthick SP, Madhavamayandi A, Murlidharan S, Saraswathy V. Electrochemical process to improve the durability of concrete structures. Journal of Building Engineering. 2016;**7**:273-280. DOI: 10.1016/j.jobe.2016.07.005

[3] Krolikowski A, Kuziak J. Impedance study on calcium nitrite as a penetrating corrosion inhibitor for the steel in concrete. Electrochimica Acta. 2011;**56**(23):7845-7853. DOI: https://doi.org/10.1016/j.electacta.2011.01.069

[4] Vamisi Priya P, Siva Prasad Rao N. Comparative study on various methods used for corrosion protection of rebar in concrete. International journal of Applied Research. 2017;**3**(1):426-431. DOI: http://www.allresearchjournal.com/archives/2017/vol3issue1/PartF/3-1-110-857.pdf

[5] Christodoulou C, Glass G, Webb J, Austin S, Goodier C. Assessing the long-term benefits of impressed current cathodic protection. Corrosion Science. 2010;**52**(8):2671-2679. DOI: https://doi.org/10.1016/j.corsci.2010.04.018

[6] Jiang SB, Jiang LH, Wang ZY, Shuya Bai MJ, Song S, Yan X. Deoxyribonucleic acid as an inhibitor for chloride-induced corrosion of reinforcing steel in simulated concrete pore solutions. Construction and Building Materials. 2017;**150**:238-247. DOI: 10.1016/j.conbuildmat.2017.05.157

[7] Zheng H, Li W, Ma F, Kong Q. The effect of a surface applied corrosion inhibitor on the durability of concrete. Construction and Building Materials. 2012;**37**:36-40. DOI: 10.1016/j.conbuildmat.2012.07.007

[8] El-Hacha R, Mirmiran A, Cook A, Rizkalla S. Effectiveness of surface-applied corrosion inhibitors for concrete bridges. Journal of Materials in Civil Engnieering. 2011;**23**(3):271-280. DOI: 10.1061/(ASCE)MT.1943-5533.0000163

[9] De Schutte Gr, Luo L. Effect of corrosion inhibiting admixtures on concrete properties. Construction and Building Materials. 2004;**18**(7):483-489. DOI: 10.1016/j.conbuildmat 2004.04.001

[10] Soylev TA, Richardson MG. Corrosion inhibitors for steel in concrete; state-of-the art report. Construction and Building Materials. 2008;**22**(4):609-622. DOI: 10.1016/j.conbuildmat.2006.10.013

[11] Martínez I, Andrade C, Castellote M, De Viedma GP. Advancements in non-destructive control of efficiency of electrochemical repair techniques. Corrosion Engineering, Science and Technology. 2008;**44**(2):108-118. DOI: 10.1179/174327808X286266

[12] Pritzl MD, Tabatabai H, Ghorbanpoor A. Long-term chloride profiles in bridge decks treated with the penetrating sealer or corrosion inhibitors. Construction and Building Materials. 2015;**101**(1):1037-1046. DOI: 10.1016/j.conbuildmat.2015.10.158

[13] Angst UM, Büchler MM, Schlumpf J, Marazzani B. An organic corrosion-inhibiting admixture for reinforced concrete: 18 years of field experience. Materials and Structures. 2016;**49**(7):2807-2818. DOI: 10.1617/s11527-015-0687-2

[14] http://corrosion-doctors.org/Corrosion-History/Course.htm#Inhibitors

[15] Song H-W, Saraswathy V. Analysis of corrosion resistance behavior of inhibitors in concrete using electrochemical techniques. Metals and Materials International. 2006;**12**(4):323-329. DOI: 10.1007/BF03027549

[16] Vyrides I, Rakanta E, Zafeiropoulou T, George BG. Efficiency of amino alcohols as corrosion inhibitors in reinforced concrete. Open Journal of Civil Engineering. 2013;**3**(2A):1-8. DOI: 10.4236/ojce.2013.32A001

[17] Ormellese M, Berra M, Bolzoni F, Pastore T. Corrosion inhibitors for chlorides induced corrosion in reinforced concrete structures. Cement and Concrete Research. 2006;**36**(3):536-547. DOI: 10.1016/j.cemconres.2005.11.007

[18] Gaidis MJ. Chemistry of corrosion inhibitors. Cement and Concrete Composites. 2004;**26**(3):181-189. DOI: 10.1016/S0958-9465(03)00037-4

[19] Justnes H. Preventing chloride induced rebar corrosion by anodic inhibitors—Comparing calcium nitrate with calcium nitrite. In: 29th Conference on our World in Concrete & Structures; 25-26 August, 2004; Singapore. CI-Premier PTE LTD; 2004. pp. 44-58. DOI: www.cipremier.com/100029006

[20] Koch GH, MPH B, Thompson NG, Virmani YP, Payer JH. Corrosion costs and preventive strategies in the United States. Houston: NACE Publications; 2004. p. 12. DOI: https://www.nace.org/uploadedFiles/Publications/ccsupp.pdf

[21] Heiyantuduwa R, Alexander MG, Mackechnie JR. Performance of a penetrating corrosion inhibitor in concrete affected by carbonation induced corrosion. Journal of Materials in Civil Engineering. 2006;**18**(6):842-850. DOI: 10.1061/(ASCE)0899-1561(2006)18:6(842)

[22] Baiqing Z, Xiaowei W, Qin L, Yisheng P. Performance and mechanism of a water stabilizer for low hardness cooling water. Anti-Corrosion Methods and Materials. 2003;**505**:347-351

[23] Gallant D, Simard S. A study on the localized corrosion of cobalt in bicarbonate solutions containing halide ions. Corrosion Science. 2005;**47**(7):1810-1838. DOI: 10.1016/j.corsci.2004.08.008

[24] Davis JR. Corrosion: Understanding the Basics. Materials Park, OH, USA:ASM International; 2000. p. 563. ISBN: 978-0-87170-641-6

[25] Kondratova IL, Montes P, Bremner TW. Natural marine exposure results for reinforced concrete slabs with corrosion inhibitors. Cement and Concrete Composites. 2003;**25**(4-5): 483-490. DOI: 10.1016/S0958-9465(02)00088-4

[26] Elsener B. Mixed-in inhibitors. In: COST 521: Corrosion of Steel in Reinforced Concrete Structures. Prevention–Monitoring–Maintenance, Final Report; Weydert R. editor. 18-19 February; Luxembourg. 2002. pp. 43-55

[27] Soeda K, Ichimura T. Present state of corrosion inhibitors in Japan. Cement and Concrete Composites. 2003;**25**(1):117-122. DOI: 10.1016/S0958-9465(01)00058-0

[28] Hansson CM, Mammoliti L, Hope BB. Corrosion inhibitors in concrete—Part 1: The principles. Cement and Concrete Research. 1998; **28**:1775-1781. DOI: 10.5772/57255

[29] Jamil, HE, Shriri A, Boulif R, Montemor MF, Ferreira MGS. Corrosion behaviour of reinforcing steel exposed to an amino alcohol based corrosion inhibitor. Cement and Composites. 2005;**27**(6):671-678. DOI: 10.1016/j.cemconcomp.2004.09.019

[30] Elsener B, Angst U. 14 - Corrosion inhibitors for reinforced concrete. In: Aïtcin PC, Flatt RJ, editor. Science and Technology of Concrete Admixtures. 1st ed. Elsevier, Woodhead Publishing, 2016. pp. 321-339. ISBN 9780081006931, https://doi.org/10.1016/B978-0-08-100693-1.00014-X. (http://www.sciencedirect.com/science/article/pii/B9780081006931 00014X)

[31] Elsener B. Corrosion Inhibitors for Steel in Concrete-State of the Art Report. European Federation of Corrosion and Institute of Materials, The University of Virginia, European

Federation of Corrosion; 2001. p. 68. ISBN: 9781902653488, DOI: https://books.google.co.kr/books?id=yohGAAAAYAAJ

[32] Ann KY, Jung HS, Kim HS, Kim SS, Moon HY. Effect of calcium-nitrite based corrosion inhibitor in preventing corrosion of embedded steel in concrete. Cement and Concrete Research. 2006;**36**(3):530-535. DOI: 10.1016/j.cemconres.2005.09.003

[33] Xu JX, Jiang H, Wang WL, Tang L, Cui Li. Effectiveness of inhibitors in increasing chloride threshold value for steel corrosion. Water Science and Engineering. 2013;**6**(3):354-363. DOI: 10.3882/j.issn.1674-2370.2013.03.011

[34] Zhou X, Yang H, Wang F. [BMIM]BF4 ionic liquids as effective inhibitor for carbon steel in alkaline chloride solution. Electrochimica Acta. 2011;**56**(11):4268-4275. DOI: 10.1016/j.electacta.2011.01.081

[35] European Union Risk Assessment Report. ECHA Member State Committee Draft Support Document for Identification of Boric Acid as a Substance of Very High Concern Because of Its CMR Properties, SVHC Support [Internet]. Available from: https://echa.europa.eu/documents/10162/d51fd473-40ec-4831-bc2d-6f53bdf9cbbe [Accessed: 9 June, 2010]

[36] Verbruggen H, Terryn H, Graeve ID. Inhibitor evaluation in different concrete pore solution for the protection of steel rebars. Construction and Building Materials. 2016;**124**:887-896. DOI: 10.1016/j.conbuildmat.2016.07.115

[37] Abd El Haleem SM, Abd El Wanees S, Abd El Aal EE, Diab A. Environmental factors affecting the corrosion behaviour of reinforcing steel. Variation in pitting corrosion current in relation to the concentration of the aggressive and inhibitive ions. Corrosion Science. 2010;**52**(5):1675-1683. DOI: 10.1016/j.corsci.2010.01.021

[38] Sail L, Ghomari F, Khelidj A, Bezzar A, Benali O. The effect of phosphate corrosion inhibitor on steel in synthetic concrete pore solutions. Advances in Materials Research. 2013;**2**(3):155-172. DOI: http://dx.doi.org/10.12989/amr.2013.2.3.155

[39] Dhouibi L, Triki E, Salta M, Rodrigues P, Raharinaivo A. Studies on corrosion inhibition of steel reinforcement by phosphate and nitrite. Materials and Structures. 2003;**36**(8):530-540. DOI: 10.1007/BF02480830

[40] Etteyeb N, Sanchez M, Dhouibi L, Alonso C, Andrade C, Triki E. Corrosion protection of steel reinforcement by a pre-treatment phosphate solution. Assessment of passivity by electrochemical techniques. Corrosion Engineering Science and Technology. 2006;**41**(4):336-341. DOI: 10.1179/174327806X120775

[41] Etteyeb N, Dhouibi L, Takenouti H, Alonso MC, Triki E. Corrosion inhibition of carbon steel in alkaline chloride media by Na_3PO_4. Electrochimica Acta. 2007;**52**(27):7506-7512. DOI: 10.1016/j.electacta.2007.03.003

[42] Nahali N, Dhouibi L, Idrissi H. Effect of Na_3PO_4 addition in mortar on steel reinforcement corrosion behavior in 3% NaCl solution. Construction and Building Materials. 2015;**78**:92-101. DOI: 10.1016/j.conbuildmat.2014.12.099

[43] Yohai L, Vazquez M, Valcarce MB. Phosphate ions as corrosion inhibitors for steel in chloride-rich environments. Electrochimica Acta. 2013;**102**:88-96. DOI: 10.1016/j.electacta.2013.03.180

[44] Yohai L, Valcarce MB, Vazquez M. Testing phosphate ions as corrosion inhibitors for construction steel in mortars. Electrochimica Acta. 2016;**202**:316-324. DOI: 10.1016/j.electacta.2015.12.124

[45] Tritthart J. Transport of a surface-applied corrosion inhibitor in cement paste and concrete. Cement and Concrete Ressearch. 2003;**33**(6):829-834. DOI: 10.1016/S0008-8846(02)01067-0

[46] Song H-W, Saraswathy V, Muralidharan S, Lee CH, Thangavel K. Role of alkaline nitrites in the corrosion performance of steel in composite cements. Journal of Applied Electrochemistry. 2009;**39**(1):15-22. DOI: 10.1007/s10800-008-9632-1

[47] Muralidharan S, Saraswathy V, Thangavel K, Srinivasan S. Competitive role of inhibitive and aggressive ions in the corrosion of steel in concrete. Journal of Applied Electrochemistry. 2000;**30**(11):1255-1259. DOI: 10.1023/A:1026570120698

[48] Fei FL, Hu J, Wei JX, Yu QJ, Chen ZS. Corrosion performance of steel reinforcement in simulated concrete pore solutions in the presence of imidazole quaternary ammonium salt corrosion inhibitor. Construction and Building Materials. 2014;**70**:43-53. DOI: 10.1016/j.conbuildmat.2014.07.082

[49] Osial M, Wiliński D. Organic substances as corrosion inhibitors for steel in concrete—An overview. Journal of Building Chemistry. 2016;**1**:42-53. DOI: 10.17461/j.buildchem.2016.107

[50] Ormellese M, Bolzoni F, Goidanich LS, Pedeferri MP, Brenna A. Corrosion inhibitors in reinforced concrete structures. Part 3—Migration of inhibitors into concrete. Corrosion Engineering, Science and Technology. 2011;**46**(4):334-339. DOI: 10.1179/174327809X419230

[51] Dariva CG, Galio AF. Corrosion Inhibitors-Principles, Mechanisms and Applications. In: Aliofkhazraei M, editor. Developments in Corrosion Protection. UK: InTech; 2014. pp. 365-380. DOI: 10.5772/57010. https://www.intechopen.com/books/developments-in-corrosion-rotection/corrosion-inhibitors-principles-mechanisms-and-applications

[52] Shi W, Wang T-Z, Dong Z-H, Guo XP. Application of wire beam electrode technique to investigate the migrating behavior of corrosion inhibitors in mortar. Construction and Building Materials. 2017;**134**:167-175. DOI: 10.1016/j.conbuildmat.2016.12.036

[53] Liu JZ, Zhao D, Cai JS, Shi L, Liu JP. Aryl aminoalcohols as corrosion inhibitors for carbon steel in chloride-contaminated simulated concrete pore solution. International Journal of Electrochemical Science. 2016;**11**:1135-1151

[54] Xu C, Jin WL, Wang HL, HT W, Huang N, Li ZY, Mao JH. Organic corrosion inhibitor of triethylenetetramine into chloride contamination concrete by electro-injection method. Construction and Building Materials. 2016;**115**:602-617. DOI: 10.1016/j.conbuildmat.2016.04.076

[55] Paredes MA, Carvallo AA, Kessler R, Virmani YP, Sagues AA. Corrosion Inhibitor in Concrete—Second Interim Report, Federal Highway Administration, Department of

Transportation, State of Florida, 2010; p. 345. http://www.fdot.gov/materials/administration/resources/library/publications/researchreports/structures/10-531.pdf

[56] Ormellese M, Perez EA, Raffaini G, Ganazzoli F, Lazzari L. Inhibition mechanism in concrete by organic substances: An experimental and theoretical study. Journal of Materials Science and Engineering. 2010;**4**(10):66-77

[57] Wombacher F, Maeder U, Marazzani B. Aminoalcohol based mixed corrosion inhibitors. Cement and Concrete Research. 2004;**26**(3):209-216. DOI: 10.1016/S0958-9465(03)00040-4

[58] Soylev TA, McNally C, Richardson M. Effectiveness of amino alcohol-based surface-applied corrosion inhibitors in chloride-contaminated concrete. Cement and Concrete Research. 2007;**37**(6):972-977. DOI: 10.1016/j.cemconres.2007.03.010

[59] Jamil, HE, Shriri A, Boulif R, Bastos C, Montimor MF, Ferreira MGS. Electrochemical behaviour of amino alcohol-based inhibitors used to control corrosion of reinforcing steel. Electrochimica Acta. 2004;**49**(17-18):2753-2760. DOI: 10.1016/j.electacta.2004.01.041

[60] Olsson C-OA, Agarwal P, Frey M, Landolt D. An XPS study of the adsorption of organic inhibitors on mild steel surfaces. Corrosion Science. 2000;**42**(7):1197-1211. DOI: 10.1016/S0010-938X(99)00140-7

[61] Ormellese M, Lazzari L, Goidanich S, Fumagalli G, Brenna A. A study of organic substances as inhibitors for chloride-induced corrosion in concrete. Corrosion Science. 2009;**51**(12):2959-2968. DOI: 10.1016/j.corsci.2009.08.018

[62] Brown MC, Weyers RE, Sprinkel MM. Effects of corrosion-inhibiting admixtures on material properties of concrete. ACI Materials Journal. 2001;**98**(3):240-250. DOI: 10.14359/10279

[63] Heren Z, Olmez H. The influence of ethanolamines on the hydration and mechanical properties of Portland cement. Cement and Concrete Research. 1996;**26**(5):701-705. DOI: 10.1016/S0008-8846(96)85007-1

[64] Liu Z, Miao C, Zhou W, Liu J. Maintenance and improvement of durability of reinforced concrete using migrating corrosion inhibitors. Journal of the Chinese Ceramic Society. 2008;**36**(10):1497-1500

[65] Criado M, Monticelli C, Fajardo S, Gelli D, Grassi V, Bastidas JM. Organic corrosion inhibitor mixtures for reinforcing steel embedded in carbonated alkali-activated fly ash mortar. Construction and Building Materials. 2012;**35**:30-37. DOI: 10.1016/j.conbuildmat.2012.02.078

[66] Prieto MI, Cobo A, Rodríguez A, Gonzalez MN. The efficiency of surface-applied corrosion inhibitors as a method for the re passivation of corroded reinforcement bars embedded in ladle furnace slag mortars. Construction and Building Materials. 2014;**54**:70-77. DOI: 10.1016/j.conbuildmat.2013.12.014

[67] Cabrini M, Fontana F, Lorenzi S, Pastore T, Pellegrini S. Effect of organic inhibitors on chloride corrosion of steel rebars in alkaline pore solution. Journal of Chemistry. 2015;**2015**:521507. DOI: 10.1155/2015/521507

[68] Monticelli C, Frignani A, Trabanelli G. A study on corrosion inhibitors for concrete application. Cement and Concrete Research. 2000;**30**(4):635-642. DOI: 10.1016/S0008-8846(00)00221-0

[69] Elsener B, Buechler M, Stalger F, Böhni H. Migrating corrosion inhibitor blend for reinforced concrete: Part 2. Inhibitor as repair strategy. Corrosion. 2000;**56**(7):727-732. DOI: 10.5006/1.3280576

[70] Bavarian B, Reiner L. Migrating corrosion inhibitor protection of steel rebar in concrete. Materials Performance. 2003;**42**(2):3-5

[71] Jamil, HE, Montemor MF, Boulif R, Shriri A, Ferreira MGS. An electrochemical and analytical approach to the inhibition mechanism of an amino-alcohol-based corrosion inhibitor for reinforced concrete. Electrochimica Acta. 2003;**48**(23):3509-3518. DOI: 10.1016/S0013-4686(03)00472-9

[72] Thangavel K, Muralidharan S, Saraswathy V, Quraishi MA, Song H-W. Migrating vs admixed corrosion inhibitors for steel in Portland, Pozzolana and slag cement concretes under macro cell condition. The Arabian Journal for Science and Engineering. 2009;**34**(2C):81-93

[73] NACE SP0107-2007. Electrochemical Realkalization and Chloride Extraction for Reinforced Concrete [Internet]. Available from: https://www.nace.org/cstm/Store/Product.aspx?id=1c0efa23-0162-41c5-857e-1133af5398cd

[74] Fajardo G, Escadeillas AG. Electrochemical chloride extraction (ECE) from steel reinforced concrete specimens contaminated by artificia sea-water. Corrosion Science. 2006;**48**(1):110-125. DOI: 10.1016/j.corsci.2004.11.015

[75] Bennett JK, Schue TJ, Clear KC, Lankard DL, Hartt WH, Swiat WJ. Electrochemical Chloride Removal and Protection of Concrete Bridge Components: Laboratory Studies, Report # SHRP S-657 [Internet]. 1993. Available from: http://onlinepubs.trb.org/onlinepubs/shrp/SHRP-S-657.pdf

[76] Sawada S, Kubo J, Page CL, Page MM. Electrochemical injection of organic corrosion inhibitors into carbonated cementitious materials: Part I—Effects on pore solution chemistry. Corrosion Science. 2007;**49**(3):1186-1204. DOI: 10.1016/j.corsci.2006.06.020

[77] Kubo J, Tanaka Y, Page CL, Page MM. Application of electrochemical organic corrosion inhibitor injection to a carbonated reinforced concrete railway viaduct. Construction and Building Materials. 2013;**39**:2-8. DOI: 10.1016/j.conbuildmat.2012.05.010

[78] Mangayarkarasi G, Muralidharan S. Electrochemical protection of steel in concrete to enhance the service life of concrete structure. Procedia Engineering. 2014;**86**:615-622. DOI: 10.1016/j.proeng.2014.11.087

[79] Pan T, Nguyen TA, Shi X. Assessment of electrical injection of corrosion inhibitor for corrosion protection of reinforced concrete. Journal of the Transportation. Research Board. 2008;**2044**:51-60. http://dx.doi.org/10.3141/2044-06

[80] Nguyen TH, Nguyen TA, Nguyen TV, Le VK, Dinh TMT, Thai H, Shi X. Effect of electrical injection of corrosion inhibitor on the corrosion of steel rebar in chloride-contaminated repair mortar. International Journal of Corrosion. 2015;**2015**:862623. DOI: 10.1155/2015/862623

[81] Liu Y, Shi X. Electrochemical chloride extraction and electrochemical injection of corrosion inhibitor in concrete: State of the knowledge. Corrosion Reviews. 2009;**27**(1-2):53. DOI: 10.1515/CORRREV.2009.27.1-2.53

[82] Sanchez M, Alonso C. Electrochemical chloride removal in reinforced concrete structures: Improvement of effectiveness by simultaneous migration of calcium nitrite. Construction and Building Materials. 2011;**25**(2):873-878. DOI: 10.1016/j.conbuildmat.2010.06.099

[83] Sawada S, Page CL, Page MM. Electrochemical injection of organic inhibitors into concrete. Corrosion Science. 2005;**47**(8):2063-2078. DOI: 10.1016/j.corsci.2004.10.001

Chapter 6

Green Corrosion Inhibitors, Past, Present, and Future

Omnia S. Shehata, Lobna A. Korshed and Adel Attia

Additional information is available at the end of the chapter

http://dx.doi.org/10.5772/intechopen.72753

Abstract

Green corrosion inhibitors are of interest because there has been an increase in environmental awareness and a change in regulations that restrict regular corrosion inhibitors due to their toxicity. Natural products are a good source of green corrosion inhibitors, where most of their extracts containing the necessary elements such as O, C, N, and S, which are active in organic compounds, assist in adsorption of these compounds on metals or alloys to form a film that protects the surface and hinders corrosion. Numerous natural products and their application in different processes, especially in steel reinforcement embedded in concrete, are discussed. Development of green chemistry and green chemical technologies offers novel synthetic methods for ionic liquids, which are considered as new corrosion green inhibitors, and their mechanism of adsorption, how these green inhibitors act in different media, and their protective role for different metals and alloys are discussed. Finally, industrial applications of vapor-phase inhibitors and their mechanisms are presented.

Keywords: green inhibitors, corrosion protection, green chemistry, ionic liquid, vapor-phase inhibitors

1. Introduction

Corrosion is a natural phenomenon where metals and alloys try to revert to their more stable thermodynamics form due to reaction with the environment that surrounds them. Corrosion is expensive due to loss of materials or their properties, which leads to loss of time during maintenance, the shutting down of systems, and severe failure of some structures, which in some cases may be hazardous and cause injury.

To protect metals or alloys from corrosion, approaches such as isolating the structure from aggressive media (using coatings or film-forming chemicals) or compensating for the loss of electrons (corrosion is an oxidation process) from the corroded structure (e.g. cathodic protection by impressed current or by using active sacrificial anodes) are employed.

Corrosion inhibition may include organic or inorganic compounds that adsorb on the metallic structure to isolate it from its surrounding media to stop the oxidation-reduction process. Organic inhibitors create their inhibition by adsorbing their molecules on the metal or alloy surface to form a protective layer [1]. Alternatively, inorganic inhibitors act as anodic inhibitors and their metallic atoms are enclosed in the film to improve their corrosion resistance. Most investigated corrosion inhibitors are toxic and cause severe environmental hazards upon disposal. Therefore, their use has been limited by environmental regulations. Nevertheless, inhibitors still play a critical role in corrosion prevention. Inhibitors are classified according to:

- Three types of electrode process, namely, anodic, cathodic, and mixed;
- The chemical nature of the environment to acid inhibitors (organic or inorganic), neutral inhibitors, alkaline inhibitors, and vapor-phase inhibitors.

Hazards caused by toxicity of regular inhibitors have led to the use of new green corrosion inhibitors. Most of these inhibitors derived from natural products act as anticorrosion agents, which are eco-friendly and harmless upon disposal. Other green inhibitors such as rare earths and organic polymers are not focused on in this chapter and further information may be found elsewhere [2–4]. This chapter will focus on green inhibitors from natural products, ionic liquids, and safety vapor-phase inhibitors, with a brief description of their mechanisms of action.

Green inhibitors act when they are added in very low concentrations to treat the surface of metals or alloys in a corrosive environment. Plant extracts are deemed to be rich, naturally synthesized chemical compounds [5] that affect the corrosion rate by adsorption of effective species on metal surfaces when added to many industrial systems through:

- Changing the rate of anodic and/or cathodic reactions;
- Effecting the diffusion rate of aggressive ions interacting with metallic structures;
- Increasing electrical resistance of the metal surface by forming a film (coat) on it.

During corrosion, metal ions move into the solution at active areas (the anode) and pass electrons from the metal to an acceptor at less active areas (the cathode); the cathodic process requires the presence of an electron acceptor such as oxygen, oxidizing agents, or hydrogen ions. Corrosion can be minimized by retarding or completely stopping the anodic or cathodic reactions, or both. Inhibitors are adsorbed on the metal surface, forming a protective barrier, and interact with anodic and/or cathodic reaction sites to decrease the oxidation and/or reduction of corrosive reactions [6].

There are several different, common cathodic reactions in corrosive media:

$$2H^{+} + 2e^{-} \rightarrow H_2 \tag{1}$$

$$O_2 + 4H^{+} + 4e^{-} \rightarrow 2H_2O \tag{2}$$

Reduction reaction is hydrogen gas evolution as follows:

$$2H^{+} + 2e^{-} \rightarrow H_{ads} \rightarrow 2H_{2} \tag{3}$$

In Eq. (3) hydrogen ions adsorbed on the metal surface are catalyzed in combination with other hydrogen ions hold hidden giving evolved hydrogen gas on cathode surface, in presence inhibitor amount of hydrogen bubble refer to ability of inhibitor to prevent this reaction and protected metal from corrosion [7]. The action of inhibitors molecules occurs due to their adsorption on an exposed metal surface in the form of neutral molecules instead of hydrogen ions adsorbed from the metal surface.

$$Inhibitor + nH_{ads} \rightarrow Inhibitor_{ads} + H_{2} \tag{4}$$

or by displacement of water molecules on the surface, as described in Eq. (2). Green inhibitors have adsorptive properties, known as site-blocking elements [8, 9].

Figure 1 shows the interest shown in green inhibitors in the last few decades, as explained by the number of publications in that period. We summarize the research focused on using natural products that have been used as green inhibitors in various aggressive media for corrosion protection of different alloys and metals as depicted in **Table 1**. The table shows the numerous studies arranged in chronological order and discussed in the reference next to each type of inhibitor. Each inhibitor's percentage of inhibition efficiency is reported at the maximum

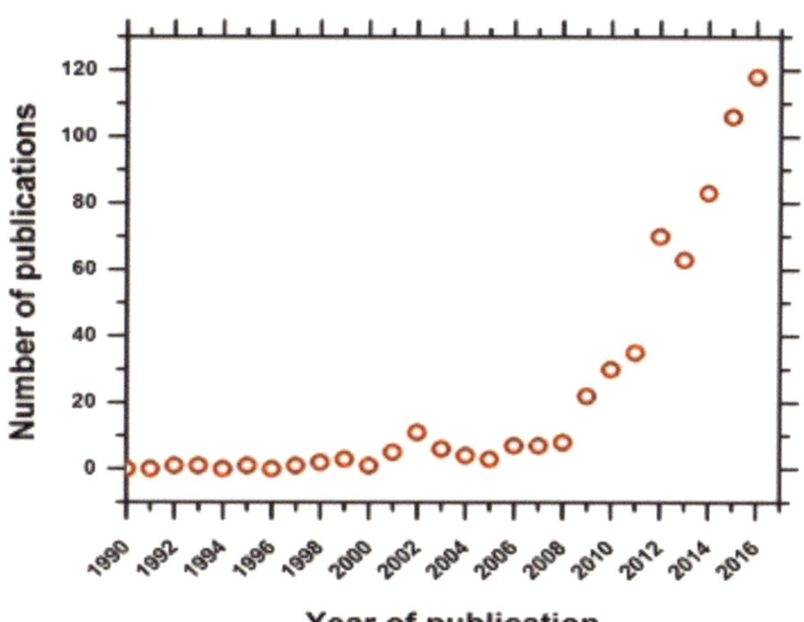

Figure 1. The interest in green inhibitors versus corrosion as indicated by number of publications. Keywords of corrosion and green inhibitors were used to retrieve the data from Web of Knowledge®.

Source of natural products	Metal/alloy studied	Agg. media studied	Efficiency (%)	References
Flour and yeast	Iron	Acid media	65–82	[10]
Furit—peels of pomegranate	Aluminum	Acid media	83	[11]
Hibiscus sabdariffa extract	Al and Zn	HCl	85	[12]
Tobacco leaves	Mild steel and Aluminum	Neutral then acidic media	87.5 (steel), 77.5 (Al)	[13]
Castor seeds			71.0 (steel), 45.9 (Al)	
Black pepper			87.5 (steel), 65.8 (Al)	
Soya bean			65.5 (steel), 27.1 (Al)	
Gum (acacia)			NA (steel), 21.8 (Al)	
Catechu			50.0 (steel), 6.3 (Al)	
Opium (0.1%)			NA (steel), 42.7 (Al)	
Opuntia extract	Steel, Aluminum, Zinc and Copper	Acid media	75 (steel), 70 (Al), 55 (Zn), 56 (Cu)	[14]
Aloe eru leaves			80 (steel), 64 (Al), 43 (Zn), 43 (Cu)	
Orange peels			80 (steel), 59 (Al), 59 (Zn), NA (Cu)	
Mango peels			82 (steel), 82 (Al), 80 (Zn), 30 (Cu)	
Pomegranate fruit shell			65 (steel), 81 (Al), 71 (Zn), 73 (Cu)	
Papaia	Mild steel	Acid media	94	[15]
Poinciana pulcherrima Cassia occidentalis			96	
			94	
Datura stramonium seeds *Calotropis procera*			93	
			98	
Azadirachta indica			84	
Auforpio turkiale sap			69	
Swertia angustifolia	Mild steel	Acid media	75–96	[16]
A. indica	Mild steel	Acid media	88–96	[17]
A. indica	Mild steel	NaCl	86.1	[18]
Punica granatum			79.2	
Momordica charantia			82.4	
Pongamia glabra, Annona squamosa	Mild steel	Acid media	89–95	[19]
Acacia arabica	Mild steel	Acid media	93–97	[20]
Natural honey	Carbon steel	NaCl	82–91	[21]
Rosmarinus officinalis	Al–Mg alloy	NaCl	75	[22]
Mimosa tannin	Carbon steel	Acid media	66–87	[23]
Vanillin	Carbon steel	Acid media	93–98	[24]
Dodecanohydrazide, cis-9-octadecanohydrazide and 10-undecanohydrazide derived from fatty acids	Mild steel	Acid media	85	[25]

Source of natural products	Metal/alloy studied	Agg. media studied	Efficiency (%)	References
Herbs (thyme, coriander, hibiscus, anis, black cumin, and garden cress)	Steel	Acid media	37–92	[26]
Reducing saccharides fructose and mannose	Aluminum and Zinc	Alkaline media	92	[27]
Opuntia extract	Aluminum	Acid media	76–96	[28]
Vernonia amygdalina	Al alloys	Acid media	49.5–72.5	[29]
Allium cepa, Allium sativum, M. charantia	Mild steel	Acid media	86–94	[30]
Guar gum	Carbon steel	Acid media	75–93.88	[31]
Zanthoxylum alatum	Mild steel	Acid media	76–95	[32]
Nypa fructicans	Mild steel	Acid media	75.11	[33]
Caffeine–Mn^{2+}	Carbon steel	Chloride ions	50	[34]
Caffeine and nicotine			80–90	[107]
Eugenol from cloves	Steel	Acid media	80	[35]
acetyleugeno			91	
Ricinus communis leaves	Mild steel	NaCl	43–84	[36]
Berberine extracted from *Coptis chinensis*	Mild steel	Acid media	79.7	[37]
Halfabar	Steel	Acid media	90.50	[38]
Chamomile			92.97	
Black cumin			88.43	
Kidney bean			88.43	
Sansevieria trifasciata	Aluminum	Acid and alkaline media + halides	94.3 (HCl 95.3 (KOH)	[39]
Exudate gum from *Dacryodes edulis*	Aluminum	Acid media	42	[40]
Gum arabic			80	
Exudate gum from *Raphia hookeri*	Aluminum	Acid media + halide	56.3	[41]
Exudate gum from *Pachylobus edulis*	Mild steel	Acid media	56	[42]
Bambusa arundinacea	Steel rebar	Chloride and nitrite	85	[43]
Parts of the kola tree (leaves, nuts, and bark) and tobacco	Steel rebar	NaCl	70–91	[44]
Artemisia pallens	Mild steel	Acid media	93–98	[45]
V. amygdalina	Steel rebar	NaCl	90.8	[46]
Chamaerops humilis L.	Steel rebar	Alkaline media	42.2	[47]
Mangrove tannin	Copper	Acid media	82.4	[48]
Chitosan	Copper	Acid media	93	[49]
Myrtus communis	Copper	Acid media	>85	[50]
Tagetes erecta	Copper	Acid media	98.07	[51]

Source of natural products	Metal/alloy studied	Agg. media studied	Efficiency (%)	References
Morinda lucida	Steel rebar	NaCl	92.8	[52]
Opuntia ficus indica	Carbon steel	Acid media	70–91	[53]
Alhagi maurorum plant extract	Copper	Acid media	33–83	[54]
Egyptian licorice extract	Copper	Acid media	89.55	[55]
Tridax procumbens Chromolaena odorata	Stainless steel	Oilfield environment	82.03 95.6	[56]
Corchorus olitorius	Mild steel	Acid media	93	[57]

Table 1. Natural products and their anticorrosive properties on metals or alloys in different aggressive.

value when different conditions are applied, according to conditions of study in each reference (temperature, concentration of inhibitors, concentration of aggressive media, conditions of study, and adsorption isotherm).

Figure 2 shows examples for the most used natural green corrosion inhibitors from past to present times as a real image.

Figure 2. Various sources for natural green corrosion inhibitors.

http://dx.doi.org/10.5772/intechopen.72753

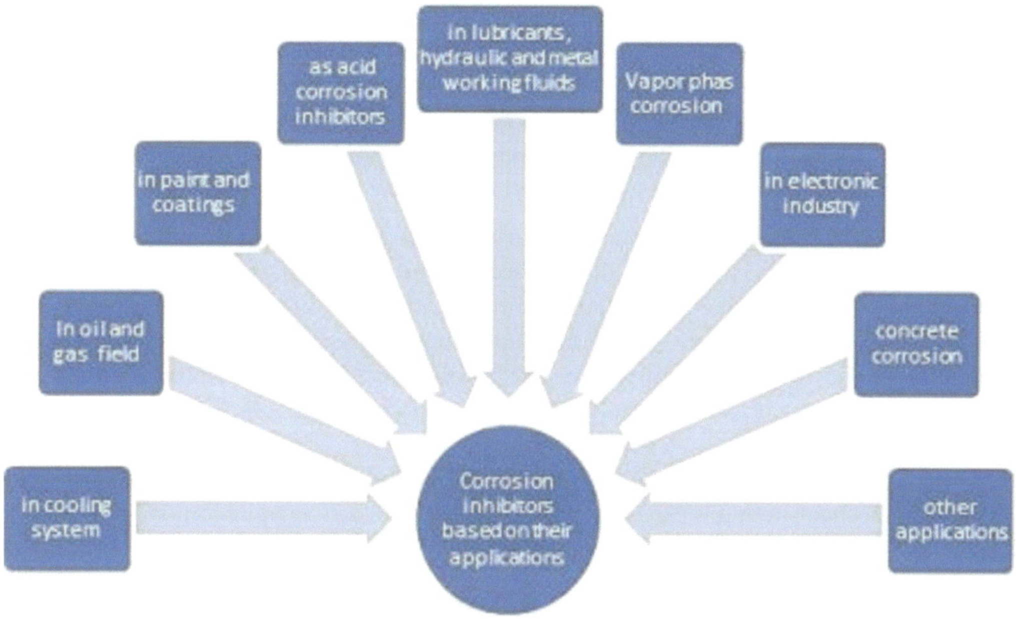

Figure 3. Usage of green corrosion inhibitors in various industries.

Figure 3 presents the importance of green corrosion inhibitors in large industrial applications. Most of these inhibitors contain functional groups and/or π-electrons in conjugation with triple or double bonds, which affect organic compounds by specific interaction between necessary elements such as nitrogen, sulfur, and oxygen through free lone pairs of electrons, which are adsorbed on the metal surface or by supplying electrons through π-orbitals [58].

2. Mechanism of green inhibitors as corrosion inhibitors

Corrosion is a spontaneous process; the relative rate of corrosion is related to the change in standard Gibbs free energy ($\Delta G°$). A more negative value of $\Delta G°$ is related to higher spontaneous reaction, i.e. higher corrosion rate [59]. Metals and alloys when exposed to the environment corrode to form stable corrosion products [60]. Utilization of additive corrosion inhibitors is necessary to mitigate corrosion rate. Corrosion products such as rust and scale can also act as corrosion inhibitors because they can accumulate on the surface and act as physical protective barriers; however, the relative rate of corrosion of any particular metal depends on the Pilling–Bedworth ratio [61, 62], which is used at high-temperature oxidation (corrosion); the ratio is defined as:

$$\frac{M * d}{n * m * D} \tag{5}$$

where m and d are the atomic weight and density of the metal, respectively, M and D are the molecular weight and density of scale (corrosion product) accumulated on the metallic surface, and n denotes the number of metallic atoms in the molecular formula of the corrosion product. The magnitude of the Pilling–Bedworth ratio can be used to explain the status of surface film, i.e. whether it will be protective or not. When the volume of corrosion product is smaller than the volume of metal from which it is formed, then *Md*/*nmD* < 1 and in this situation it is expected that the surface film of the corrosion product contains pores and cracks that would be relatively nonprotective. When the volume of the corrosion product is larger than the volume of metal for *Md*/*nmD* > 1, it is expected that the surface film of the corrosion product is relatively more compressed and compact than would result in a more protective film.

Adsorption is the first step in forming a corrosion protective film or coat in the presence of aggressive media that occurs on metallic surfaces on the active sites. Several factors affect the adsorption of inhibitor on the metallic surface and isolate it including adsorption mode, chemical and electronic characteristics of the inhibitor, temperature, type of electrolyte employed, steric effects, and the nature and surface charge of metals [63]. The Langmuir adsorption isotherm is the most favorable to clarify interaction between the inhibitor and metal surface [53]. Adsorption on the corroded surfaces approximates to a steady-state adsorption that may be physical adsorption (physisorption) or chemical adsorption (chemisorption), or a mixed adsorption mechanism that is considered ideal for effective corrosion inhibition.

Physical adsorption is related to the standard free energy of adsorption ΔG^o_{ads} in aqueous solution. If its value is −20 kJ.mol^{-1} or less negative it is associated with an electrostatic interaction between charged centers of molecules and the charged metal surface, which results in a dipole interaction of the molecules and metal surface [64]. However, chemical adsorption is a process that involves the transfer or sharing of electrons from the inhibitor to the metallic surface and results in the formation of a coordinate covalent bond. The bonding strength is much larger than physical adsorption where the value of ΔG^o_{ads} is around −40 kJ.mol^{-1} or more negative [65, 66]. Adsorption of green corrosion inhibitor retards corrosion by detraction of the active metallic surface area, leaving inactive sites on the surface exposed to corrosive media. Green inhibitors are more efficient at room temperature or low temperatures, while inhibition efficiency is decreased with an increase in temperature in most cases.

The action of green inhibitors depends on the structure of the active ingredient; many researchers have postulated numerous theories to explain the mechanism of their effect. The active ingredient derived from natural inhibitors changes from one plant species to another but their structures are closely related to their organic coordinate. As an example, garlic contains allyl propyl disulfide, mustard seeds contain an alkaloid berberine that has a long chain of aromatic rings and an N atom in the ring, carrot contains pyrrolidine, and castor seed contains the alkaloid ricinine. Eucalyptus oil contains monomtrene-1,8-cineole. Lawsonia extract contains 2-hydroxy-1,4-naphthoquinone resin and tannin, coumarin, gallic acid, and sterols. Gum exudate contains hexuronic acid, neutral sugar residues, volatile monoterpenes, canaric and related triterpene acids, and reducing and nonreducing sugars. Garcinia kola seed contains primary and secondary amines, unsaturated fatty acids, and bioflavonoids. Calyx extract contains ascorbic acid, amino acids, flavonoids, pigments, and carotene [67]. The corrosion inhibition activity of these plant extracts could be due to the presence of heterocyclic constituents like alkaloids, flavonoids, etc.

3. Green inhibitors for rebar embedded in concrete

Kundu et al. [68] reviewed green inhibitors' effect in rebar embedded into concrete under an alkaline environment of nearly pH 12–13 to protect rebar from premature deterioration. Its mechanism depends on the strong hydrophobic effect of *Bambusa arundinacea* that supports the formation of a product layer of two faced oxides (Fe_2O_3 and Fe_3O_4) adherent to the steel surface [69] or as spinel α-Fe_3O_4–γ-Fe_2O_3 solid solution, which is proposed to form a passive film on steel [43]. Moreover, efficiency of *B. arundinacea* was due to its strong adsorption parallel to the metal surface, which restricted the number of surface active sites available for chloride ion ingression. In another investigation, Eyu et al. [70] studied the effect of *Vernonia amygdalina* extract within 70 days of immersion on carbon steel reinforcement in concrete exposed to a chloride-laden environment to compare efficiencies of other inhibitors, while Loto et al. [44] studied plant extract by a potentiometric method to compare the percentage of efficiency between tobacco and extract of kola nut, leaves, and bark in 5% NaCl. Okeniyi et al. [71, 72] studied the bark extract of the *Rhizophora mangle* L. plant for rebar in concrete in an acidic environment.

4. Green corrosion inhibitor based on ionic liquids

Nowadays, worldwide growing ecological awareness and strict environmental protocols prevent synthesis and utilization of hazardous traditional volatile corrosion inhibitors. So, there is a vital need for improvement in synthetic and engineering chemistry by environmentally friendly materials. Use of alternative synthetic strategies toward green synthesis [73, 74] has become necessary as multicomponent reactions (MCRs) in combination with ultrasonic (sonochemical) and microwave irradiation. So, scientists have been directed to develop drugs from plant extracts that are characterized by their natural and biological origins and nontoxic nature [75] to act as green corrosion inhibitors. Pathak et al. [76] studied different classes of drugs in different media; their effectiveness depends on their chemical composition, molecular structure, and affinities for metallic surfaces. The most effective of them depend on heteroatoms like nitrogen, oxygen, and phosphorus.

However, extraction and purification of plant extracts is tedious, exhausting, and extremely expensive as well as time consuming, and requires relatively large amounts of organic solvents that may adversely affect the environment and living beings; high temperature can also decompose the active constituents and thereby decrease relative inhibition efficiency. For these reasons the use of drugs as corrosion inhibitors for ferrous and nonferrous metals is also limited [75]. Therefore, there is a need to develop green inhibitors by proper design of the synthesis, which can be achieved by using cheap and environmentally friendly products. This is onset of green chemistry as ionic liquids that are eco-friendly and sustainable solvents composed of ions that can displace a wide range of inorganic and organic compounds. Ionic liquids follow the principles of green chemistry proposed by Anastas and Warner [77]. Ionic liquids have a promising future in the field of green chemistry, and have attracted great attention because they are widely used in various applications [78, 79]. **Figure 4** shows important properties of ionic liquids and their applications as green corrosion inhibitors [80–82]. They show high performance to adhere to the surface of steel with the formation of thin films to protect steel [78], copper [83], zinc [84], titanium, and aluminum substrates [85].

Chitosan is a linear copolymer of (1-4)-2-amido-2-deoxy-d-glucan (glucosamine) and (1-4)-2-acetamido-deoxy-d-glucan (*N*-acetylglucosamine), which are considered sources for ionic liquids [86]. Chitosan is extracted from crab shell [87] based on aminopolysaccharides and is used to produce several new materials [88]. Their high ability for functionalization makes them applicable to apply in several industrial applications because of their better solubility

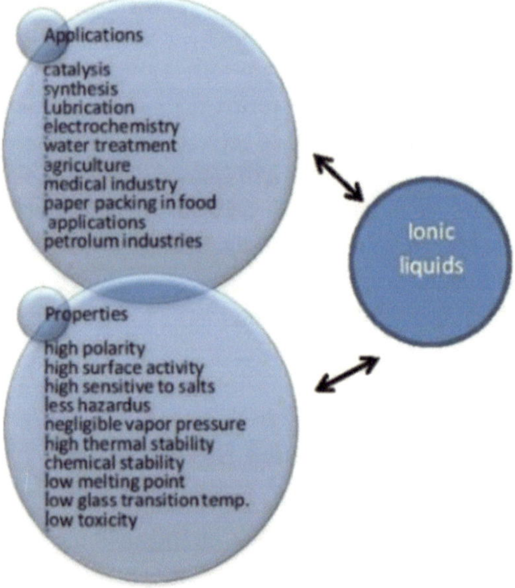

Figure 4. Important properties of ionic liquids and their impact as green corrosion inhibitors for many applications.

in organic solvents and water than chitosan itself. This chemical functionality is extremely relevant when chitosan is used as a corrosion inhibitor [89]. Chitosan can inhibit the corrosion of copper in acidic media because HCl is used as a mixed-type inhibitor, which follows a Langmuir isotherm [49]. Chitosan possesses unique physicochemical properties, namely, biocompatibility, antimicrobial activity, biodegradability, and excellent film-forming ability. Indeed, it can be used to form thin coatings to protect metallic surfaces.

Ionic liquids have been classified into 11 categories according to Hajipour and Refiee [90], namely, neutral ionic liquids, acid ionic liquids, basic ionic liquids, ionic liquids with amphoteric anions, functionalized ionic liquids, protic ionic liquids, chiral ionic liquids, supported ionic liquids, bio-ionic liquids, poly-ionic liquids, and energetic ionic liquids, and have described common features and properties of these ionic liquids. However, Angell et al. [91] classified ionic liquids in four classes, including aprotic, protic, inorganic, and solvate (chelate) ionic liquids. Their adsorption on metallic surfaces takes place via chemisorption, which obeys the Langmuir adsorption isotherm as reported by Verma et al. [92] in their recent review, which showed several ionic liquids and their properties as green corrosion inhibitors for different metals and alloys such as mild steel, aluminum, copper, zinc, and magnesium in several electrolytic media. A similar observation has been reported by other authors for different metals including copper, nickel, and stainless steel [93], while Shetty et al. [94] have reviewed adsorption of ionic liquids on an aluminum surface that obeyed the Temkin adsorption isotherm.

Inhibition of metallic corrosion in the presence of ionic liquids involves blocking of anodic oxidative metallic dissolution as well as cathodic hydrogen evolution reactions [95] described as follows:

$$M + X^{-} \leftrightarrow [(MX]^{-})\,ads \tag{6}$$

$$[(MX)^{-}]_{ads} + \text{ILs C}^{+} \leftrightarrow (\text{M } X^{-} \text{ Ils C}^{+})_{ads} \tag{7}$$

where M is the anodic dissolution of metals in aqueous corrosive solution, and $ILsC^+$ and X^- represent the cationic counterpart of the ionic liquid and anionic counterpart of the ionic liquid, respectively. Both anionic charged species attract positively charged cationic counterparts of the ionic liquids ($ILsC^+$) by electrostatic force of attraction (physisorption) and form a monomolecular layer as an insoluble complex on the metallic surface [95]. The adsorption of the $ILsC^+$ on the metallic surface causes a change in the surface polarity, which induces the adsorption of $ILsC^+$ and X^- ions again, which results in a multimolecular layer [95]. The multimolecular layers are stabilized by van der Waal's cohesion force acting between organic moieties of the ionic liquids, which causes a more closely adsorbed film at metal/electrolyte interfaces. Generally, the cationic part ($ILsC^+$) interacts with the metallic surface and forms the multimolecular layers, while the rest of the ionic liquids form hydrophobic hemimicelles, admicelles, and/or surface aggregation [95, 96]. The adsorbed multimolecular layers of the ionic liquids isolate the metal (M) from the corrosive environment and protect it from corrosive dissolution.

During cathodic hydrogen evolution reactions, hydronium ions are adsorbed on the metallic surface by the Volmer mechanism followed by discharge of hydrogen gas by the Heyrovsky and Tafel mechanism [97, 98], represented as follows:

$$M + H_2O^{+} + e^{-} \leftrightarrow MH_{ads} + H_2O \tag{8}$$

$$MH_{ads} + H_2O^{+} + e^{-} \leftrightarrow MH_{ads} + H_2O \tag{9}$$

$$MH_{ads} + MH_{ads} \leftrightarrow H_2 + 2M \tag{10}$$

$$M + ILsC^{+} + e^{-} \leftrightarrow M([ILsC])_{ads} \tag{11}$$

The cationic part of ionic liquids ($ILsC^+$) starts competing with hydrogen ions for electrons; however, $ILsC^+$ has a larger molecular size so replaces a greater number of water molecules from the metallic surface. After their adsorption the cationic part of the ionic liquids accepts electrons from the metal (M), which results in the formation of electrically neutral ionic liquids (inhibitors). The neutral species transfer (donation) their nonbonding (of heteroatoms) and π-electrons into the d-orbitals of the surface metallic atoms resulting in the formation of coordinate bonds between metal and ionic liquids (chemisorption) as reported for several organic conventional inhibitors [95]. However, metals are already electron-rich species; this type of donation causes interelectronic repulsion, which results in the transfer of electrons from d-orbitals of the surface metallic atoms to antibonding molecular orbitals of the ionic liquids (retrodonation). Both donation and retrodonation strengthen each other through synergism [99].

Development of corrosion inhibitors requires correlating the inhibition efficiency of the inhibitors with their molecular properties through study interactions between inhibitors and the metallic surface and describing the adsorption behavior of ionic liquids on the metallic surface. This is achieved by density functional theory (DFT). DFT is one of the most important methods used in theoretical chemistry for corrosion inhibition to accurately predict the inhibition efficiencies based on the calculations provided by mechanistic information about metal–inhibitor interactions and can be performed for inhibitor molecules even before their synthesis [100–105].

5. Vapor-phase inhibitors

Vapor-phase inhibitors (VPI) are volatile corrosion inhibitors. They are considered a revolution in green inhibitors and are required in many fields covering electronics, packaging, industrial processing, reinforced concrete, coatings, and metalworking fluids due to their nontoxic properties. They are free of nitrites, halogens, and phosphates, and are completely safe to handle. In addition, they also inhibit the corrosion of ferrous and nonferrous metals. In a review of the literature, Bastidas et al. [106] studied various factors influencing VPI performance and mechanisms. Nanovapor-phase inhibitors made from renewable agricultural by-products do not destabilize the natural balance of the environment. Their impact depends on their diffusion, filling all void spaces and active areas with protective vapor molecules that

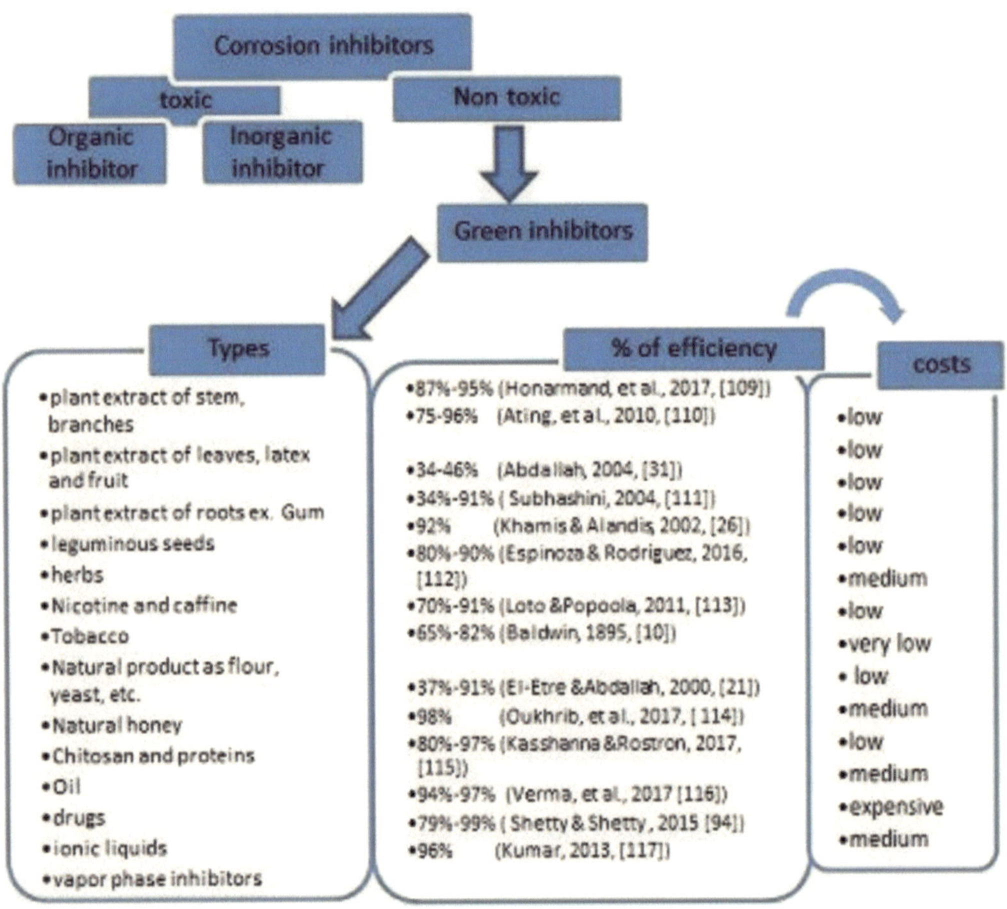

Figure 5. Accessibility of green corrosion inhibitors from nature with approximated cost of extraction and percentage of efficiency according to the cited literature [109–117].

are attracted to metallic surfaces and adsorbed physically and/or chemically onto the resulting formation of a nanoprotective barrier layer [107, 108].

Several factors must be considered to control the cost of extractions, which sometimes may be very high when the quantities needed are high. Also, the estimated percentage of an inhibitor's efficiency must be controlled, including temperature, concentration of inhibitors, type of metallic surface exposed to aggressive environments, etc. **Figure 5** summarizes the accessibility of green corrosion inhibitors from nature with approximate estimations of cost of extraction between low, medium, and expensive, and the percentage of efficiency according to the cited literature.

The objective of this chapter was to review natural compounds as effective green corrosion inhibitors because of their biodegradability, easy availability, and nontoxic nature. The literature revealed that natural plant extracts are effective green corrosion inhibitors against various metals and alloys. In addition, the chapter focused on why the efficiency of green and sustainable inhibitors of ionic liquids for the corrosion of metals and alloys is preferred compared to traditional corrosion inhibitors because of their advantageous physiochemical properties. Finally, attention was given to the adsorption behavior of ionic liquids generally, following the Langmuir isotherm, the Temkin adsorption isotherm, and in particular the adsorption behavior of ionic liquids on metallic surfaces using DFT-based quantum chemical calculations. Quantum chemical calculations (DFT) provide a good insight into the inhibition mechanism and experimental order of inhibition efficiency. The tendency to explore vapor-phase corrosion inhibitors and green nanoinhibitors is a new area for future research.

A lot of potential is still untapped, especially computational modeling of the major extract components of various metals and alloys. Further research should also be focused on plant extraction methods and their active constituents as well as scale-up experiments for industrial applications that are needed to commercialize these natural extracts to effectively replace conventional chemicals.

Author details

Omnia S. Shehata*, Lobna A. Korshed and Adel Attia

*Address all correspondence to: omniashehata@yahoo.com

Physical Chemistry Department, National Research Centre, Dokki, Cairo, Egypt

References

[1] Hefter GT, North NA, Tan SH. Organic corrosion inhibitors in neutral solutions; part 1—inhibition of steel, copper, and aluminum by straight chain carboxylates. Corrosion. 1997;**53**(8):657-667. DOI: 10.5006/1.3290298

[2] Bethencourt M, Botana FJ, Calvino JJ, Marcos M, Rodriguez-Chacon MA. Lanthanide compounds as environmentally-friendly corrosion inhibitors of aluminium alloys: A review. Corrosion Science. 1998;**40**(11):1803-1819. DOI: 10.1016/S0010-938X(98)00077-8

[3] Sabirneeza AAF, Geethanjali R, Subhashini S. Polymeric corrosion inhibitors for iron and its alloys: a review. Chemical Engineering Communications. 2015;**202**(22):232-244. DOI: 10.1080/00986445. 2014. 934448

[4] Branzoi F, Branzoi V, Licu C. Corrosion inhibition of carbon steel in cooling water systems by new organic polymers as green inhibitors. Materials and Corrosion—Werkstoffe und Korrosion. 2014;**65**(6):637-647. DOI: 10.1002/maco.201206579

[5] Krishnaveni K, Ravichandran J. Effect of aqueous extract of leaves of *Morinda tinctoria* on corrosion inhibition of aluminium surface in HCl medium. Transactions of Nonferrous Metals Society of China. 2014;**24**(8):2704-2712. DOI: 10.1016/S1003-6326(14)63401-4

[6] Singh P, Srivastava V, Quraishi MA. Novel quinoline derivatives as green corrosion inhibitors for mild steel in acidic medium: electrochemical, SEM, AFM, and XPS studies. Journal of Molecular Liquids. 2016;**216**(1):164-173. DOI: 10.1016/j.molliq.2015.12.086

[7] Oguzie EE. Inhibition of acid corrosion of mild steel by *Telfaria occidentalis* extract. Pigment & Resin Technology. 2005;**34**(6):321-326. DOI: 10.1108/03699420 510630336

[8] Sanatkumar BS, Nayak J, Shetty AN. Influence of 2-(4-chlorophenyl)-2-oxoethyl benzoate on the hydrogen evolution and corrosion inhibition of 18 Ni 250 grade weld aged maraging steel in 1.0 M sulfuric acid medium. International Journal of Hydrogen Energy. 2012;**37**(11):9431-9442. DOI: 10.1016/j.ijhydene.2012.02.165

[9] Manamela KM, Murulana LC, Kabanda MM, Ebenso EE. Adsorptive and DFT studies of some imidazolium based ionic liquids as corrosion inhibitors for zinc in acidic medium. International Journal of Electrochemical Science. 2014;**9**(6):3029-3046

[10] Baldwin J. British Patent. 1895:2327

[11] Saleh RM, El-Hosary AA. Corrosion inhibition by naturally occurring substances. Effect of Pomegranate Juice and the Aqueous Extracts of Pomegranate Fruit Shells, Tamarind Fruits and Tea Leaves on the Corrosion of Al. In: El-Hosary, editor. Electrochemistry; 13th ed.; Karaikudi. CECRI; 1972

[12] El-Hosary AA, Saleh RM, El Din AMS. Corrosion inhibition by naturally occurring substances: the effect of *Hibiscus subdariffa* (karkade) extract on the dissolution of Al and Zn. Corrosion Science. 1972;**12**(12):897-904. DOI: 10.1016/S0010-938X(72)80098-2

[13] Srivastav K, Srivastava P. Studies on plant materials as corrosion inhibitors. British Corrosion Journal. 1981;**16**(4):221-223. DOI: 10.1179/000705981798274788

[14] Saleh RM, Ismail AA, El Hosary AA. Corrosion inhibition by naturally occurring substances: VII. The effect of aqueous extracts of some leaves and fruit-peels on the corrosion of steel, Al, Zn and Cu in acids. British Corrosion Journal. 1982;**17**(3):131-135. DOI: 10.1179/000705982798274345

[15] Zucchi F, Omar IH. Plant extracts as corrosion inhibitors of mild steel in HCl solutions. Surface Technology. 1985;**24**(4):391-399. DOI: 10.1016/0376-4583(85)90057-3

[16] Zakvi SJ, Mehta GN. Acid corrosion of mild steel and its inhibition by Swertia angustifolia—study by electrochemical techniques. Transactions of the SAEST. 1988;**23**(4):407-410

[17] Ekpe UJ, Ebenso EE, Ibok UJ. Inhibitory action of *Azadirachta indica* leaves extract on the corrosion of mild steel in H_2SO_4. Journal of West African Journal of Biological and Applied Chemistry. 1994;**37**:13-30

[18] Quraishi MA, Farooqi IH, Saini PA. Investigation of some green compounds as corrosion and scale inhibitors for cooling systems. Corrosion. 1999;**55**(5):493-497. DOI: 10.5006/1.3284011

[19] Sakthivel P, Nirmala PV, Umamaheswari S, Antony AAA, Paruthimal Kalaignan G, Gopalan A, Vasudevan T. Corrosion inhibition of mild steel by extracts of *Pongamia glabra* and *Annona squamosa* in acidic media. Bulletin of Electrochemistry. 1999;**15**(2):83-86

[20] Verma SA, Mehta GN. Effect of acid extracts of *Acacia arabica* on acid corrosion of mild steel. Bulletin of Electrochemistry. 1999;**15**(2):67-70

[21] El-Etre AY, Abdallah M. Natural honey as corrosion inhibitor for metals and alloys. II. C-steel in high saline water. Corrosion Science. 2000;**42**(4):731-738. DOI: 10.1016/S0010-938X (99)00106-7

[22] Kliškic M, Radošević J, Gudic S, Katalinic V. Aqueous extract of Rosmarinus officinalis L. as inhibitor of Al–Mg alloy corrosion in chloride solution. Journal of Applied Electrochemistry. 2000;**30**(7):823-830

[23] Martinez S, Štern I. Inhibitory mechanism of low-carbon steel corrosion by mimosa tannin in sulphuric acid solutions. Journal of Applied Electrochemistry. 2001;**31**(9):973-978

[24] El-Etre AY. Inhibition of acid corrosion of aluminum using vaniline. Corrosion Science. 2001;**43**(6):1031-1039. DOI: 10.1016/S0010-938X(00)00127-X

[25] Quraishi MA, Jamal D. Approach to corrosion inhibition by some green inhibitors based on oleochemicals. Bulletin of Electrochemistry. 2002;**18**(7):289-294

[26] Khamis E, Alandis N. Herbs as new type of green inhibitors for acidic corrosion of steel. Material Wissenschaft und Werkstofftechnik. 2002;**33**(9):550-554. DOI: 10.1002/1521-4052

[27] Müller B. Corrosion inhibition of aluminium and zinc pigments by saccharides. Corrosion Science. 2002;**44**(7):1583-1591. DOI: 10.1016/S0010-938X(01)00170-6

[28] El-Etre AY. Inhibition of aluminum corrosion using *Opuntia* extract. Corrosion Science. 2003;**45**(11):2485-2495. DOI: 10.1016/S0010-938X(03)00066-0

[29] Avwiri GO, Igho FO. Inhibitive action of *Vernonia amygdalina* on the corrosion of aluminium alloys in acidic media. Material Letters. 2003;**57**(22):3705-3711. DOI: 10.1016/S0167-577X(03)00167-8

[30] Parikh KS, Joshi KJ. Natural compounds onion (*Allium cepa*), garlic (*Allium sativum*) and bitter gourd (*Momordica charantia*) as corrosion inhibitors for mild steel in hydrochloric acid. Transactions—Society for the Advancement of Electrochemical Science and Technology. 2004;**39**(1/2):29-35

[31] Abdallah M. Guar gum as corrosion inhibitor for carbon steel in sulfuric acid solutions. Portugaliae Electrochimica Acta. 2004;**22**(2):161-175

[32] Gunasekaran G, Chauhan LR. Eco friendly inhibitor for corrosion inhibition of mild steel in phosphoric acid medium. Electrochimica Acta. 2004;**49**(25):4387-4395. DOI: 10.1016/j.electacta.2004.04.030

[33] Orubite KO, Oforka NC. Inhibition of the corrosion of mild steel in hydrochloric acid solutions by the extracts of leaves of *Nypa fruticans* Wurmb. Materials Letters. 2004;**58**(11):1768-1772. DOI: 10.1016/j.matlet.2003.11.030

[34] Anthony N, Malarvizhi E, Maheshwari P, Rajendran S, Palaniswamy N. Corrosion inhibition of caffein–Mn^{2+} system. Indian Journal Chemical Technology. 2004;**11**(May):346-350

[35] Chaieb E, Bouyanzer A, Hammouti B, Benkaddour M. Inhibition of the corrosion of steel in 1 M HCl by eugenol derivatives. Applied Surface Science. 2005;**246**(1):199-206. DOI: 10.1016/j.apsusc.2004.11.011

[36] Sathiyanathan R, Maruthamuthu S, Selvanayagam M, Mohanan S, Palaniswamy N. Corrosion inhibition of mild steel by ethanolic extracts of *Ricinus communis* leaves. Indian Journal Chemical Technology. 2005;**12**(May):356-360

[37] Li Y, Zhao P, Liang Q, Hou B. Berberine as a natural source inhibitor for mild steel in 1 M H_2SO_4. Applied Surface Science. 2005;**252**(5):1245-1253. DOI: 10.1016/j.apsusc.2005.02.094

[38] Abdel-Gaber AM, Abd-El-Nabey BA, Sidahmed IM, El-Zayady AM, Saadawy M. Inhibitive action of some plant extracts on the corrosion of steel in acidic media. Corrosion Science. 2006;**48**(9):2765-2779. DOI: 10.1016/j.corsci.2005.09.017

[39] Oguzie EE. Corrosion inhibition of aluminium in acidic and alkaline media by Sansevieria trifasciata extract. Corrosion Science. 2007;**49**(3):1527-1539. DOI: 10.1016/j.corsci.2006.08.009

[40] Umoren SA, Obot IB, Ebenso EE, Obi-Egbedi N. Studies on the inhibitive effect of exudate gum from *Dacroydes edulis* on the acid corrosion of aluminium. Portugaliae Electrochimica Acta. 2008;**26**(2):199-209

[41] Umoren SA, Obot IB, Ebenso EE, Obi-Egbedi NO. The inhibition of aluminium corrosion in hydrochloric acid solution by exudate gum from *Raphia hookeri*. Desalination. 2009;**247**(1-3):561-572

[42] Umoren SA, Ekanem UF. Inhibition of mild steel corrosion in H_2SO_4 using exudate gum from Pachylobus edulis and synergistic potassium halide additives. Journal Chemical Engineering Communications. 2010;**197**(10):1339-1356

[43] Abdulrahman AS, Ismail M. Green plant extract as a passivation-promoting inhibitor for reinforced concrete. International Journal of Engineering Science and Technology. 2011;**3**(Aug.):6484-6489

[44] Loto CA, Loto RT, Popoola API. Electrode potential monitoring of effect of plants extracts addition on the electrochemical corrosion behaviour of mild steel reinforcement in concrete. International Journal of Electrochemical Science. 2011;**6**(Aug):3452-3465

[45] Garai S, Garai S, Jaisankar P, Singh JK, Elango A. A comprehensive study on crude methanolic extract of *Artemisia pallens* (Asteraceae) and its active component as effective corrosion inhibitors of mild steel in acid solution. Corrosion Science. 2012;**60**:193-204. DOI: 10.1016/j.corsci.2012.03.036

[46] Loto CA, Joseph OO, Loto RT, Popoola API. Inhibition effect of *Vernonia amygdalina* extract on the corrosion of mild steel reinforcement in concrete in 3.5M NaCl environment. International Journal of Electrochemical Science. 2013;**8**:11087-11100

[47] Left DB, Zertoubi M, Khoudali S, Benaissa M, Irhzo A, Azzi M. Effect of methanol extract of *Chamaerops humilis* L. leaves (MECHLL) on the protection performance of oxide film formed on reinforcement steel surface in concrete simulated pore solution. International Journal of Electrochemical Science. 2013;**8**(10):11768-11781

[48] Shah AM, Rahim AA, Hamid SA, Yahya S. Green inhibitors for copper corrosion by mangrove tannin. International Journal of Electrochemical Science. 2013;**8**:2140-2153

[49] El-Haddad MN. Chitosan as a green inhibitor for copper corrosion in acidic medium. International Journal of Biological Macromolecules. 2013;**55**:142-149. DOI: 10.1016/j.ijbiomac.2012.12.044

[50] Bozorg M, Shahrabi Farahani T, Neshati J, Chaghazardi Z, Mohammadi Ziarani G. Myrtus communis as green inhibitor of copper corrosion in sulfuric acid. Industrial & Engineering Chemistry Research. 2014;**53**(11):4295-4303. DOI: 10.1021/ie404056w

[51] Mourya P, Banerjee S, Singh MM. Corrosion inhibition of mild steel in acidic solution by Tagetes erecta (marigold flower) extract as a green inhibitor. Corrosion Science. 2014;**85**(Aug.):352-363. DOI: org/10.1016/j.corsci.2014.04.036

[52] Okeniyi JO, Loto CA, Popoola API. *Morinda lucida* effects on steel-reinforced concrete in 3.5% NaCl: implications for corrosion-protection of wind-energy structures in saline/marine environments. Energy Procedia. 2014;**50**:421-428. DOI: 10.1016/j.egypro.2014.06.051

[53] Flores-De los Ríos JP, Sánchez-Carrillo M, Nava-Dino CG, Chacón-Nava JG, González-Rodríguez, Huape-Padilla E, Neri-Flores MA, Martínez-Villafañe A. Opuntia ficus-indica extract as green corrosion inhibitor for carbon steel in 1 M HCl solution. Journal of Spectroscopy. 2015;**2015**:14692-14701/714692. DOI: 10.1155/2015/714692

[54] Abd-El-Nabey BA, El-Housseiny S, El-Naggar GA, Matter EA, Esmail G. Inhibitive action of Alhagi maurorum plant extract on the corrosion of copper in 0.5 M H_2SO_4. Physical Chemistry. 2015;**5**(3):49-62. DOI:10.5923/j.pc.20150503.01

[55] Deyab MA. Egyptian licorice extract as a green corrosion inhibitor for copper in hydrochloric acid solution. Journal of Industrial and Engineering Chemistry. 2015; **22**(Feb.):384-389. DOI: org/10.1016/j.jiec.2014.07.036

[56] Aribo S, Olusegun SJ, Ibhadiyi LJ, Oyetunji A, Folorunso DO. Green inhibitors for corrosion protection in acidizing oilfield environment. Journal of the Association of Arab Universities for Basic and Applied Sciences. 2016 (In Press). DOI: 10.1016/j.jaubas.2016.08.001

[57] Gobara M, Zaghloul B, Baraka A, Elsayed M, Zorainy M, Mokhtar Kotb M, Elnabarawy H. Green corrosion inhibition of mild steel to aqueous sulfuric acid by the extract of *Corchorus olitorius* stems. Materials Research Express. 2017;**4**(4):046504. DOI: 10.1088/2053-1591/aa664a

[58] Shehata OS, Khorshed LA, Mandour HS. Effect of acetamide derivative and its Mn-complex as corrosion inhibitor for mild steel in sulphuric acid. Egyptian Journal of Chemistry. 2017;**60**(2):243-259. DOI: 10.21608/ejchem.2017.674.1014

[59] Manimegalai S, Manjula P. Thermodynamic and adsorption studies for corrosion inhibition of mild steel in aqueous media by *Sargasam swartzii* (brown algae). Journal of Material and Environmental Science. 2015;**6**(6):1629-1637

[60] Roh Y, Lee SY, Elless MP. Characterization of corrosion products in the permeable reactive barriers. Environmental Geology. 2000;**40**(1-2):184-194. DOI: 10.1007/s002540000178

[61] Xu C, Gao W. Pilling-Bedworth ratio for oxidation of alloys. Materials Research Innovations. 2000;**3**(4):231-235. DOI: 10.1007/s100190050008

[62] Bedworth RE, Pilling NB. The oxidation of metals at high temperatures. Journal of the Institute of Metals. 1923;**29**(3):529-582

[63] Maayta AK, Al-Rawashdeh NAF. Inhibition of acidic corrosion of pure aluminum by some organic compounds. Corrosion Science. 2004;**46**(5):1129-1140. DOI: 10.1016/j.corsci.2003.09.009

[64] Damaskin BB, Petrii OA, Batrakov VV. Adsorption. New York: Plenum Press; 1971

[65] Roberge PR. Handbook of corrosion engineering. McGraw-Hill; 2000

[66] Martinez S, Štern I. Thermodynamic characterization of metal dissolution and inhibitor adsorption processes in the low carbon steel/mimosa tannin/sulfuric acid. Applied Surface Science. 2002;**199**(1):83-89. DOI: 10.1016/S0169-4332 (02) 00546-9

[67] Rani BE, Basu BB. Green inhibitors for corrosion protection of metals and alloys: an overview. International Journal of Corrosion. 2011;**2012**:15. DOI: 10.1155/2012/380217

[68] Kundu M, Prasad SK, Virendra K. A review article on green inhibitors of reinforcement concrete corrosion. International Journal of Emerging Research in Management & Technology. 2016;**5**:42-46

[69] Asipita SA, Ismail M, Majid MZA, Abdullah CS, Mirza J. Green *Bambusa arundinacea* leaves extract as a sustainable corrosion inhibitor in steel reinforced concrete. Journal of Cleaner Production. 2014;**67**:139-146. DOI: 10.1016/j.jclepro.2013.12.033

[70] Eyu DG, Esah H, Chukwuekezie C, Idris J, Mohammad I. Effect of green inhibitor on the corrosion behaviour of reinforced carbon steel in concrete. ARON Journal of Engineering and Applied Science. 2013;**8**(5):326-332

[71] Okeniyi JO, Loto CA, Popoola API. Corrosion inhibition performance of*Rhizophora mangle* L. bark-extract on concrete steel-reinforcement in industrial/microbial simulating-environment. International Journal Electrochemical Science. 2014;**9**(8):4205-4216

[72] Okeniyi JO, Loto CA, Popoola API. *Rhizophora mangle* L. effects on steel-reinforced concrete in 0.5 M H_2SO_4: implications for corrosion-degradation of wind-energy structures in industrial environments. Energy Procedia. 2014;**50**:429-436. DOI: 10.1016/j.egypro.2014.06.052

[73] Carp O, Visnescu D, Patrinou G, Tirsoaga A, Paraschiv C, Tudose M. Green synthetic strategies of oxide materials: Polysaccharides-assisted synthesis. Part 1. Polysaccharide Roles in Metal Oxides Synthesis. Revue Roumain de Chimie. 2010;**55**(10):705-709. DOI: 10.1002/chin.201139239

[74] Hulsbosch J, De Vos DE, Binnemans K, Ameloot R. Biobased ionic liquids: solvents for a green processing industry? ACS Sustainable Chemistry & Engineering. 2016;**4**(6):2917-2931. DOI: 10.1021/acssus chem eng.6 b00553

[75] Verma C, Chauhan DS, Quraishi MA. Drugs as environmentally benign corrosion inhibitors for ferrous and nonferrous materials in acid environment: an overview. Journal of Materials and Environmental Sciences (JMES). 2017;**8**(11):4040-4051

[76] Pathak RK, Mishra P. Drugs as corrosion inhibitors: a review. International Journal of Science and Research (IJSR). 2016;**5**(4):671-677

[77] Anastas PT, Warner JC. Green Chemistry: Theory and Practice. New York: Oxford University Press; 2000

[78] Kowsari E, Payami M, Amini R, Ramezanzadeh B, Javanbakht M. Task-specific ionic liquid as a new green inhibitor of mild steel corrosion. Applied Surface Science. 2014;**289**:478-486. DOI: 10.1016/j.apsusc.2013.11.017

[79] Scendo M, Uznanska J. The effect of ionic liquids on the corrosion inhibition of copper in acidic chloride solutions. International Journal of Corrosion. 2011;**2011**:3. DOI: 10.1155/2011/718626

[80] Crini G. Recent developments in polysaccharide-based materials used as adsorbents in wastewater treatment. Progress in Polymer Science. 2005;**30**(1):38-70. DOI: 10.1016/j.progpolymsci.2004.11.002

[81] Ng LT, Swami S. IPNs based on chitosan with NVP and NVP/HEMA synthesised through photoinitiator-free photopolymerisation technique for biomedical applications. Carbohydrate Polymers. 2005;**60**(4):523-528. DOI: 10.1016/j.carbpol.2005.03.009

[82] Ham-Pichavant F, Sèbe G, Pardon P, Coma V. Fat resistance properties of chitosan-based paper packaging for food applications. Carbohydrate Polymers. 2005;**61**(3):259-265. DOI: 10.1016/j.carbpol.2005.01.020

[83] Scendo M, Uznanska J. The effect of ionic liquids on the corrosion inhibition of copper in acidic chloride solutions. International Journal of Corrosion. 2011;**2011**:13. DOI: 10.1155/2011/718626

[84] Manamela KM, Murulana LC, Kabanda MM, Ebenso EE. Adsorptive and DFT studies of some imidazolium based ionic liquids as corrosion inhibitors for zinc in acidic medium. International Journal of Electrochemical Science. 2014;**9**(6):3029-3046

[85] Wang YC, Lee TC, Lin JY, Chang JK, Tseng CM. Corrosion properties of metals in dicyanamide-based ionic liquids. Corrosion Science. 2014;**78**:81-88. DOI: 10.1016/j.corsci 2013.09.002

[86] Shukla SK, Mishra AK, Arotiba OA, Mamba BB. Chitosan-based nanomaterials: a state-of-the-art review. International Journal of Biological Macromolecules. 2013;**59**:46-58. DOI: 10.1016/j.ijbiomac.2013.04.043

[87] Paul W, Sharma CP. Chitosan, a drug carrier for the 21st century: A review. STP Pharma Science. 2000;**10**(1):5-22

[88] Yao Z, Zhang C, Ping Q, Yu L. A series of novel chitosan derivatives: synthesis, characterization and micellar solubilization of paclitaxel. Carbohydrate Polymers. 2007;**68**(4):781-792. DOI: 10.1016/j.carbpol.2006.08.023

[89] Mima S, Miya M, Iwamoto R, Yoshikawa S. Highly deacetylated chitosan and its properties. Journal of Applied Polymer Science. 1983;**28**(6):1909-1917. DOI: 10.1002/app.1983.070280607

[90] Hajipour AR, Refiee F. Recent progress in ionic liquids and their applications in organic synthesis. Organic Preparations and Procedures International. 2015;**47**(4):249-308. DOI: 10.1080/00304948. 2015.1052317

[91] Angell CA, Ansari Y, Zhao Z. Ionic liquids: past, present and future. Faraday Discussions. 2012;**154**:9-27. DOI: 10.1039/C1FD00112D

[92] Verma C, Ebenso EE, Quraishi MA. Ionic liquids as green and sustainable corrosion inhibitors for metals and alloys: an overview. Journal of Molecular Liquids. 2017;**233**:403-414. DOI: 10.1016/j.molliq.2017.02.111

[93] Lin PC, Sun IW, Chang JK, Su CJ, Lin JC. Corrosion characteristics of nickel, copper, and stainless steel in a Lewis neutral chloroaluminate ionic liquid. Corrosion Science. 2011;**53**(12):4318-4323. DOI: 10.1016/j.corsci.2011.08.047

[94] Shetty SK, Shetty AN. Ionic liquid as an effective corrosion inhibitor on 6061 Al-15 Vol. Pct. SiC (p) composite in 0.1 M H_2SO_4 medium—an ecofriendly approach. Canadian Chemical Transactions. 2015;**3**:41-64. DOI: 10.13179/canchemtrans.2015.03.01.0160

[95] Likhanova NV, Domínguez-Aguilar MA, Olivares-Xometl O, Nava-Entzana N, Arce E, Dorantes H. The effect of ionic liquids with imidazolium and pyridinium cations on the corrosion inhibition of mild steel in acidic environment. Corrosion Science. 2010;**52**(6):2088-2097. DOI: 10.1016/j.corsci.2010.02.030

[96] Fuchs-Godec R. The adsorption, CMC determination and corrosion inhibition of some N-alkyl quaternary ammonium salts on carbon steel surface in 2M H_2SO_4. Colloids and Surfaces A: Physicochemical and Engineering Aspects. 2006;**280**(1):130-139. DOI: 10.1016/j.colsurfa.2006.01.046

[97] Gennero de Chialvo MR, Chialvo AC. Hydrogen evolution reaction: analysis of the Volmer-Heyrovsky-Tafel mechanism with a generalized adsorption model. Journal of Electroanalytical Chemistry. 1994;**372**(1-2):209-223. DOI: 10.1016/0022-0728(93)03043-O

[98] Gennero de Chialvo MR, Chialvo AC. The Tafel–Heyrovsky route in the kinetic mechanism of the hydrogen evolution reaction. Electrochemistry Communications. 1999;**1**(9):379-382

[99] Migahed MA. Electrochemical investigation of the corrosion behaviour of mild steel in 2 M HCl solution in presence of 1-dodecyl-4-methoxy pyridinium bromide. Materials Chemistry and Physics. 2005;**93**(1):48-53. DOI: 10.1016/j.matchemphys.2005.02.003

[100] Verma C, Olasunkanmi LO, Ebenso EE, Quraishi MA, Obot IB. Adsorption behavior of glucosamine-based, pyrimidine-fused heterocycles as green corrosion inhibitors for mild steel: experimental and theoretical studies. Journal of Physical Chemistry C. 2016;**120**(21):11598-11611. DOI: 10.1021/acs.jpcc.6b04429

[101] Verma C, Olasunkanmi LO, Obot IB, Ebenso EE, Quraishi MA. 2,4-Diamino-5-(phenylthio)-5H-chromeno [2,3-b] pyridine-3-carbonitriles as green and effective corrosion inhibitors: gravimetric, electrochemical, surface morphology and theoretical studies. RSC Advances. 2016;**6**(59):53933-53948. DOI: 10.1039/c6ra04900a

[102] Verma C, Olasunkanmi LO, Obot IB, Ebenso EE, Quraish MA. 5-Arylpyrimido-[4,5-b] quinoline-diones as new and sustainable corrosion inhibitors for mild steel in 1 M HCl: a combined experimental and theoretical approach. RSC Adv. 2016;**6**(19):15639-15654. DOI: 10.1039/C5RA27417F

[103] Suresh SJS. Recent advances in ionic liquids: green unconventional solvents of this century: part I. Green Chemistry Letters and Reviews. 2011;**4**(4):289-310. DOI: doi.org/10.1080/17518253.2011.572294310

[104] Murulana LC, Singh AK, Shukla SK, Kabanda MM, Ebenso EE. Experimental and quantum chemical studies of some bis(trifluoromethyl-sulfonyl) imide imidazolium-based ionic liquids as corrosion inhibitors for mild steel in hydrochloric acid solution. Industrial & Engineering Chemistry Research. 2012;**51**(40):13282-13299. DOI: 10.1021/ie300977d

[105] Yadav DK, Quraishi MA. Application of some condensed uracils as corrosion inhibitors for mild steel: gravimetric, electrochemical, surface morphological, UV–visible, and theoretical investigations. Industrial & Engineering Chemical Research. 2012;**51**(46):14966-14979. DOI: 10.1021/ie301840y

[106] Bastidas DM, Cano E, Mora EM. Volatile corrosion inhibitors: a review. Anti-Corrosion Methods and Materials. 2005;**52**(2):71-77. DOI: 10.1108/00035590510584771

[107] Subramanian A, Natesan M, Muralidharan VS, Balakrishnan K, Vasudevan T. An overview: vapor phase corrosion inhibitors. Corrosion. 2000;**56**(2):144-155. DOI: 10.5006/1.3280530

[108] Vuorinen E, Kálmán E, Focke W. Introduction to vapour phase corrosion inhibitors in metal packaging. Surface Engineering. 2004;**20**(4):281-284. DOI: 10.1179/026708404225016481

[109] Honarmand E, Mostaanzadeh H, Motaghedifard MH, Hadia M, Khayadkashanic M. Inhibition effect of opuntia stem extract on corrosion of mild steel: a quantum computational assisted electrochemical study to determine the most effective components in

inhibition. Protection of Metals and Physical Chemistry of Surfaces. 2017;**53**(3):560-572. DOI: 10.1134/S20702511703008X

[110] Ating EI, Umoren SA, Udousoro II, Ebenso EE, Udoh AP. Leaves extract of Ananas sativum as green corrosion inhibitor for aluminum in hydrochloric acid solution. Green Chemistry Letters and Reviews. 2010;**3**(2):61-68. DOI: 10.1080/17518250903505253

[111] Subhashini S. Corrosion inhibition study of mild steel in acid media by extract of some leguminous seeds as eco-friendly inhibitors. Journal Camel Practice Research. 2004;**11**(1):27-34

[112] Espinoza-Vázquez A, Rodríguez-Gómez FJ. Caffeine and nicotine in 3% NaCl solution with CO_2 as corrosion inhibitors for low carbon . RSC Advances. 2016;**6**(74):70226-70236. DOI: 10.1039/C6RA07673D

[113] Loto CA, Popoola API. Effect of tobacco and kola tree extracts on the corrosion inhibition of mild steel in acid chloride. International Journal of Electrochemical Science. 2011;**6**(Aug):3264-3276

[114] Oukhrib R, El Ibrahimi B, Bourzi H, El Mouaden K, Jmiai A, El Issami S, Bammou L, Bazzi L. Quantum chemical calculations and corrosion inhibition efficiency of biopolymer "chitosan" on copper surface in 3% NaCl. Journal of Material and Environmental Sciences. 2017;**8**(1):195-208

[115] Kasshanna S, Rostron P. Novel synthesis and characterization of vegetable oil derived corrosion inhibitors. Journal of Material and Environmental Sciences. 2017;**8**(12):4292-4300

[116] Verma C, Chauhan DS, Quraishi MA. Drugs as environmentally benign corrosion inhibitors for ferrous and nonferrous materials in acid environment: an overview. Journal of Material and Environmental Sciences. 2017;**8**(11):4040-4051

[117] Kumar H, Saini V, Yadav V. Study of vapour phase corrosion inhibitors for mild steel under different atmospheric conditions. International Journal of Engineering and Innovative Technology. 2013;**3**(3):206-211

Section 3

Recent Applications of Corrosion Inhibitors

Chapter 7

Opuntia ficus-indica (Nopal Extract) as Green Inhibitor for Corrosion Protection in Industrial Steels

Araceli Mandujano Ruiz,
Luis-Enrique Corona Almazán,
Héctor Herrera Hernández and
Jorge Morales Hernández

Additional information is available at the end of the chapter

http://dx.doi.org/10.5772/intechopen.72944

Abstract

Soluble extract from *Opuntia ficus-indica* (Nopal extract) has been proposed in this chapter as a green inhibitor due to its component called mucilage, which has the ability to retain water; for this reason, it has been used as metal corrosion protection in machinery pieces, tools and other metallic components that need to be stored for short periods. In this way, three industrial carbon steels (AISI 1018, 1045 and 4140) have been exposed in sulfuric acid (H_2SO_4) to evaluate the corrosion behavior with or without Nopal extract (NE). Some electrochemical techniques have been implemented to evaluate the corrosion inhibition efficiency (IE) such as DC linear polarization resistance (LPR) and AC electrochemical impedance spectroscopy (EIS). Results indicated a considerable superficial modification of steel in terms of dielectric constant and ion charge capacity. When the NE was added, the corrosion mechanism changed from localized to general attack, decreasing the corrosion rate in all cases. More susceptibility to fail by corrosion was observed in the 1045 carbon steel in comparison with the other two studied steels; these results were confirmed by the percentage of inhibitor's efficiency of about 95%.

Keywords: green inhibitor, corrosion protection, Nopal extract, industrial carbon steel, corrosion rate, electrochemical evaluation, electrochemical impedance spectroscopy

1. Introduction

Study of corrosion inhibitors has taken a new role in recent years, since new regulations restrict the use of several components in conventional corrosion inhibitors like nitrites, benzoates, chrome, lead, arsenic, among others, because they are toxic and dangerous to health and environment. Formulations of alternative compounds with more biocompatibility and low costs have led to the development of the ***green inhibitors*** also called ***eco-friendly inhibitors*** [1, 2]. Main sources from which these inhibitors are extracted come from plants, fruits, seeds, leaves and flowers whose components such as flavonoids, alkaloids and natural oils (pennyroyal oil, jojoba oil, etc.) are considered the inhibitor's active agent. Those extracts have been studied in low concentration and exposed to different aggressive media to protect metals (mainly carbon steels), obtaining good results at the superficial protection against corrosion during the first hours of exposure.

Figure 1 shows a general classification of the corrosion inhibitors according to its origin. Organic inhibitors are characterized by their high molecular weight structures and are polar molecules. Most organic inhibitors are adsorbed on the metal surface by displacing water molecules and forming a compact barrier. Inorganic inhibitors are salts of some metals, which have a passivation effect and reaction with the metal. Some synthetic compounds can reduce corrosion damage in carbon steels [3–5] because their oxygen, nitrogen, and sulfur heteroatoms react on the metal surface, blocking active sites where corrosion occurs.

Green corrosion inhibitors are biodegradable and do not contain heavy metals or toxic compounds. Mechanisms of the green inhibitors indicate that ions/molecules are adsorbed onto metal surface, interfering with the anodic and/or cathodic reactions and decreasing the diffusion rate for reactants to the metal surface. Usually, the electrical resistance of the metal surface is decreasing. Since their innovation, some researchers have proposed many plants as prospects to be green corrosion inhibitors. **Table 1** shows a review of several papers about green inhibitors collected from 2004 to 2017, where the carbon steels in acid media (HCl and H_2SO_4)

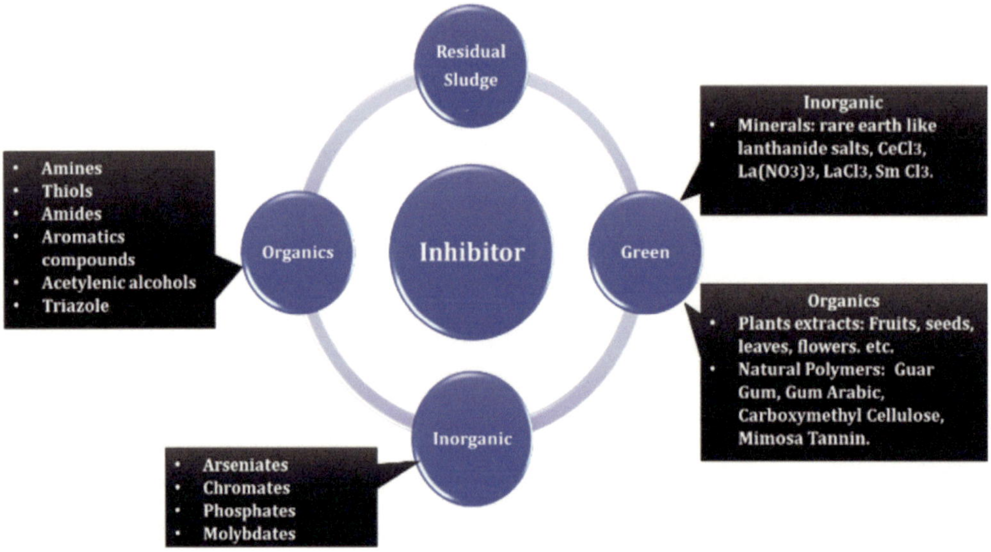

Figure 1. General classification of corrosion inhibitors.

Inhibitor	Metal	Medium	IE (%)	Year	References
Zanthoxylum armatum	Mild steel	H_3PO_4	> 90	2004	[8]
Simmondsia chinensis	Carbon steel	HCl	98	2004	[9]
Artemisia vulgaris	Carbon steel	HCl	85	2004	[10]
Datura metel	Carbon steel	HCl and H_2SO_4	84~97	2005	[11]
Mentha pulegium	Carbon steel	HCl	80	2006	[12]
Ammi visnaga	SX316	HCl	99.3	2006	[13]
Carica papaya	Mild steel	H_2SO_4	94	2007	[14]
Gongronema latifolium	Aluminum	HCl, NaOH	> 90	2007	[15]
Opuntia ficus-indica **(Cactus)**	Aluminum	HCl	94	2007, 2012	[16, 17]
Phyllanthus amarus	Mild steel	HCl and H_2SO_4	> 90	2008	[18]
Black pepper	Carbon steel	HCl and H_2SO_4	98	2008	[19]
Zallouh root	Carbon steel	H_2SO_4	90	2008	[20]
Justicia gendarussa	Mild steel	HCl	93	2009	[21]
Clove oil	Nickel, inconel 600, 690	HCl	88~92	2009	[22]
Lupinus albus	Carbon steel	HCl and H_2SO_4	86	2009	[23]
Gossypium hirsutum	Aluminum	NaOH	97	2009	[24]
Ananas sativum	Aluminum	HCl	96	2010	[25]
Murraya koenigii	Mild steel	HCl	~ 94	2011	[26]
Marine Alga *Caulerpa racemosa*	Mild steel	H_2SO_4	~85	2012	[27]
Spirulina Platensis	Mild steel	HCl and H_2SO_4	~80	2012	[28]
Aloe vera	Zinc, Galvanized iron	HCl, H_2SO_4	88.5	2014	[29]
Tagetes erecta	Mild steel	H_2SO_4	98.07	2014	[30]
Watermelon rind	Mild steel	HCl, H_2SO_4	~86	2014	[31, 32]
Mangifera indica **(Mango) and Orange**	Mild steel, Carbon steel	H_2SO_4, HCl	97	2014	[33]
Anise	Carbon steel	HCl	94	2014	[34]
Coconut coir	Aluminum, Mild Steel	HCl and H_2SO_4	80	2012,2014	[35–37]
Musa paradisiaca **(banana)**	Mild steel, SS304	H_2SO_4, NaCl	82.7	2011,2014	[38, 39]

Inhibitor	Metal	Medium	IE (%)	Year	References
Ficus hispana	Mild steel	HCl	90	2015	[40]
Nicotiana tabacum	Mild steel	H_2SO_4	94	2015	[41]
Litchi chinensis	Mild steel	H_2SO_4	97.8	2015	[42]
Capsicum annuum, Citrus aurantium, Moringa oleifera	Copper	HNO_3	60~80	2015	[43]
Valoniopsis pachynema	Brass	H_3PO_4	96.06	2015	[44]
Azadirachta indica	Al, Mild steel, Tin, SS, Carbon steel	HCl, H_2SO_4, HNO_3	> 85	2015	[45]
Petroselinum crispum **(Parsley),** *Eruca sativa* **(Arugula)** *Anethum graveolens* **(Dill)**	Carbon steel	HCl	92	2015	[46]
Morinda citrifolia	AISI 1045, AISI 8620	HCl	> 86	2015,2016	[47, 48]
Pomegranate	Carbon steel, Mild steel, α-Brass	HCl	> 90	2015,2013,2017	[49–51]
Gentiana olivieri	Mild steel	HCl	92	2016	[52]
Roasted coffee	Carbon steel	HCl	94	2016	[53]
Curcumin	Aluminum	**NaCl**	80	2016	[54]
Calotropis procera **leaves**	Mild steel	**NaCl**	> 90	2016	[55]
Ocimum basilicum	Carbon steel, Mild steel	NaCl + Na_2S, H_2SO_4	88.5	2016,2012	[56, 57]
Tilia	Mild steel	HCl	79	2017	[58]
Cedrus atlantica	Lead	Na_2CO_3	~ 70	2017	[59]

Table 1. Review of several papers related with green inhibitors from 2004 to 2017.

were the most studied, possibly because the total annual cost of metallic corrosion in a country is associated mainly with the corrosion of carbon steels of different industrial sectors. Most of these works were focused on demonstrating the efficiency of these extracts as well as their behavior with respect to temperature, identifying the absorption mechanism from the extract components [6, 7], however, other factors like cost and amount are not mentioned in detail.

Evaluation of corrosion inhibition efficiency was based on the implementation of electrochemical and gravimetric methods; the electrochemical techniques consist of linear polarization resistance (LPR) and electrochemical impedance spectroscopy (EIS). With LPR was obtained information about the inhibition reaction with the compound, classifying the inhibitor action as anodic, cathodic or mixed. EIS allowed to detect the resistance changes on the surface and how the active sites are reordered through changes at the interfacial capacitance of metal-electrolyte.

Corrosion inhibition efficiency can be calculated through the coverage degree "θ" on the metal surface by the following equations:

$$\theta = \frac{\frac{1}{R_{ct}}(blank) - \frac{1}{R_{ct}}(inhibitor)}{\frac{1}{R_{ct}}(blank)} \tag{1}$$

$$\theta = \frac{i_{corr(blank)} - i_{corr(inhibitor)}}{i_{corr_{(blank)}}} \tag{2}$$

$$\theta = \frac{V_{corr(blank)} - V_{corr(inhibitor)}}{V_{corr_{(blank)}}} \tag{3}$$

$$\theta = \frac{Weight\ loss\ in\ blank - Weight\ loss\ in\ inhibitor}{Weight\ loss\ in\ blank} \tag{4}$$

where R_{ct} is the charge transfer resistance, i_{corr} is the corrosion current and V_{corr} is the corrosion rate. Expressions (1), (2) and (3) are obtained from electrochemical techniques like LPR and EIS, and Eq. (4) is calculated using the gravimetric method. With any of those equations, the efficiency is obtained with:

$$EI(\%) = \theta \times 100 \tag{5}$$

In general, a good inhibitor must have an efficiency greater than 90%.

Efficiency of an organic compound as a corrosion inhibitor depends mainly on the anchor capacity by chemical or physical interactions on the metal surface, forming multilayers of adsorbed organic molecules that block the active sites of corrosion and retarding the anodic and/or cathodic reactions. Thermodynamic studies with adsorption isotherms (Temkin, Frumkin, Langmuir, Freundlich, among others) have allowed to estimate the stability of the layer adsorbed on the surface. These depend on the physicochemical properties of the active molecules related to the inhibitor functional groups like aromatic rings and amino acids with $-NH_2$ groups, which are readily protonated in acid media and adsorbed on the metal surface through acid anions (Cl^-, SO_4^{2-}) negatively charged. Electron density from the donor atoms, and the interaction between the orbital π from inhibitory and the orbital ***d*** of the iron, play an important role in the adsorption type [60–62].

Investment in the use of inhibitors for industrial scale to prevent corrosion represents a small percentage (1%) worldwide compared with the costs in the use of paints, surface treatments and materials selection, which represent 80% of the investment. The main sectors investing in inhibitors such as petrochemicals (production, synthesis, and refining), water treatment systems, and chemical and food processes, have reported an efficiency of up to 90% in the mitigation of corrosion, saving billions of dollars. However, despite its effectiveness, many of the commercial synthetic inhibitors that are currently used generate problems of toxicity and contamination to the environment. Environmental Protection Agency (EPA) regulations have determined that inorganic inhibitor compounds such as salts of chromates, phosphates and molybdates, as well as organic inhibitors like phosphates, amides or thiols, are dangerous. Due to this problem, green inhibitors open an opportunity for the replacement of these compounds (**Table 1**) where the synthesis process of natural extracts is of low cost and can be applied in closed systems as pipelines and packaging of pieces for transport. Nowadays, the limitation in the industrial staggering of these green inhibitors is in their preservation, since it has been shown that after 24 h, the natural extract begins to decompose which would imply the use of large volumes of extract to serve the industry [63].

1.1. *Opuntia ficus* plant

One plant that has caused interest about their potential as a green inhibitor is *Opuntia ficus-indica* (Nopal), which is a traditional vegetable in Mexico with a high soluble fiber content. Nopal is classified as a wild plant that survives in desert regions and does not require much water for cultivation. It is said that it has an important ecological performance, as it stops the degradation of the deforested soil, making the arid land productive because of its capacity to retain water. There are about 1600 species in 122 genera of the cactus family, from which comes the Nopal. It has fruits that are edible and are known with the name of Tunas. The soluble fiber of Nopal known as ***mucilage***, which has the capacity to retain up to 30 times its weight in water. For this particular compound, it has been promoted as a good candidate for the production of an ecological inhibitor for the protection of the steel against corrosion [64, 65]. Chemical analysis of mucilage indicates that it contains a large amount of pectin as well as minerals such as calcium, potassium and sodium whose quantities depend on the age of the cladodes (leaves). The presence of these minerals suggests that this plant may remain stable over long periods in alkaline media.

Mucilage is similar in consistency with Aloe vera (slimy) and has been used for the conservation of building materials. Chandra [66] and Torres Acosta [17] studied the effects of dehydrated Nopal and Aloe vera powder in the electrochemical performance of reinforcing steel in chloride contaminated concrete. Preliminary results suggest that adding these powders in small concentrations might be suitable for enhancing corrosion resistance of steel in concrete. Therefore, this investigation deals with the performance of Nopal mucilage as a corrosion inhibitor for industrial steels exposed in H_2SO_4.

2. Experimental details

2.1. Specimens preparation

Three types of carbon steels were evaluated in this research: AISI 1045, 1018 and 4140. The composition of each metal is shown in **Table 2**. Rods of each steel were cut mechanically to obtain specimens of 25 mm of diameter and 5 mm of thickness for the electrochemical evaluations. The surface preparation was grinding with SiC paper through numbers 80, 240, 320, 400, 600 and 1500 grits until a mirror finish. Finally, the specimens were washed with distilled water, degreasing with acetone and drying with hot air.

Element	Wt. %		
	1045	1018	4140
C	0.45	0.2	0.4
Mn	0.7	0.9	0.85
P	0.04	0.04	0.04
S	0.05	0.05	0.05
Fe	98.76	98.81	97.31
Mo	******	******	0.2
Cr	******	******	0.9
Si	******	******	0.25

Table 2. Metal composition with AISI classification.

2.2. Preparation of Nopal extract

Nopal extract was obtained through the selection of tender cladodes, a cleaning process and removal of thorns, cut and disinfected with 2 ml of chlorine +1 ml of Cu_2SO_4, assuring the elimination of microorganism and fungi that degrade the extract. Subsequently, it was crushed in an extractor where the pulp was removed from the juice; this last one was heated for 1 h at 78°C with a volume ratio of 2:1 in distilled water. Finally, the concentrated liquid was filtered through a fine sieve to ensure separation of solid particles larger than 180 μm, obtaining the final product as shown in **Figure 2**.

Figure 2. Nopal extract preparation process.

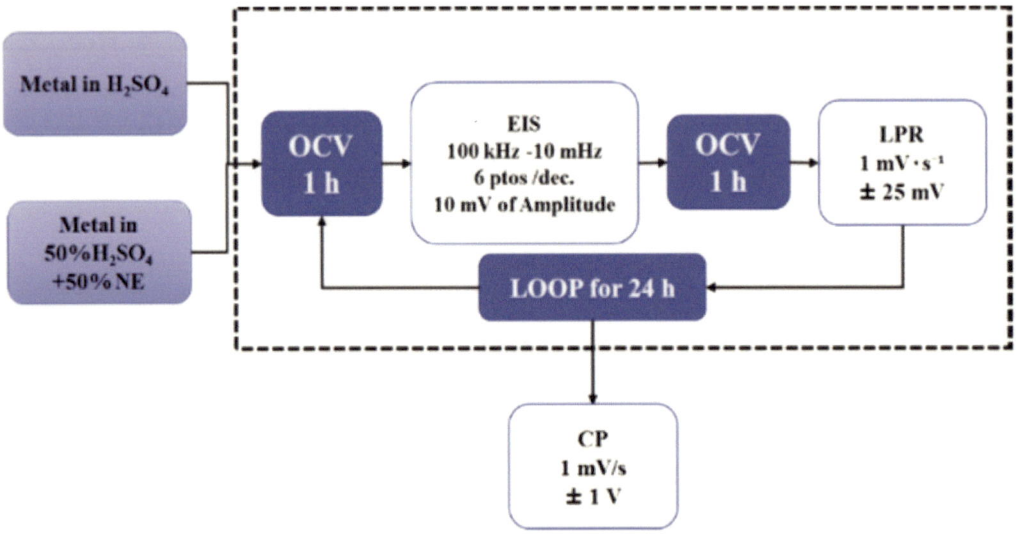

Figure 3. Experimental sequence.

2.3. Experimental sequence

Electrochemical characterization was carried out in an acrylic cell for flat probes [65], assembled with an arrangement of three electrodes. Steels 1018, 1045 and 4140 were placed as working electrode (WE), a grid of platinum was placed as a counter electrode (CE) and a calomelane electrode was used as reference electrode (RE). Solution of H_2SO_4 (0.6 mol·l^{-1}) was used as the test electrolyte, prepared with analytical grade reagents and deionized water. All experiments were performed under standard conditions of temperature and pressure. **Figure 3** shows the experimental sequence where the steels in H_2SO_4 with and without Nopal extract (NE) were evaluated. The relation of Nopal extract and H_2SO_4 ratio was 50:50 volume %. Open circuit potential (OCP) was monitored to ensure the system stability during 1 h, followed by the electrochemical impedance spectroscopy (EIS) technique, applying a sweep of frequencies from 100 kHz to 10 mHz with 10 mV amplitude. OCP measurement was run again for 1 h to continue with linear polarization resistance (LPR) with an overpotential range of ±25 mV and a sweep speed of 1 mV·s^{-1}. Cycle OCP-EIS-OCP-LPR was repeated for 24 h until record the final behavior with a polarization curve (CP) applying an overpotential of ±1 V at a scan rate of 1 mV·s^{-1}. Corrosion rates, capacitances and resistances for all steels were calculated.

3. Results and discussion

Corrosion potential results are shown in **Figure 4**, where only the steels exposed in acid medium reported positive potentials with a tendency to move toward more negative (active) potentials when the inhibitor is present, indicating a change in the surface activity of the metal. **Figure 5** shows the behavior of the resistance between the metal in the acid medium and the medium with inhibitor (H_2SO_4 + NE). A considerable growth of the Nyquist semicircle was observed with the addition of NE for the three steels, being the widest semicircle for the 4140 steel, which reaches a value of charge transfer resistance (Rct) of 1400 ohms at the beginning

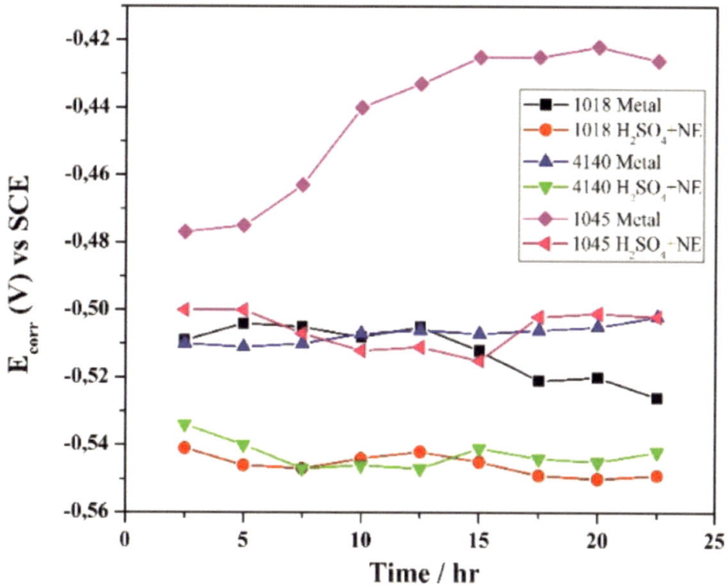

Figure 4. Potential corrosion behavior of metals for 24 h.

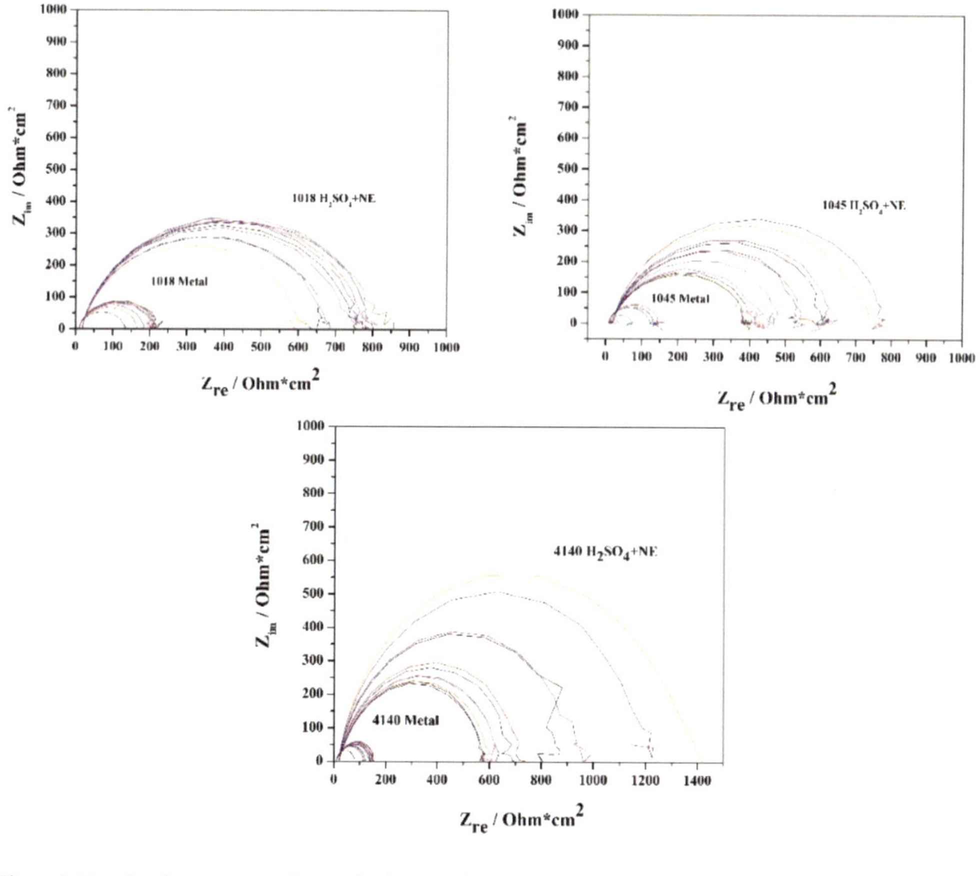

Figure 5. Nyquist plots corresponding to the three steels in H_2SO_4 and H_2SO_4 + Nopal extract.

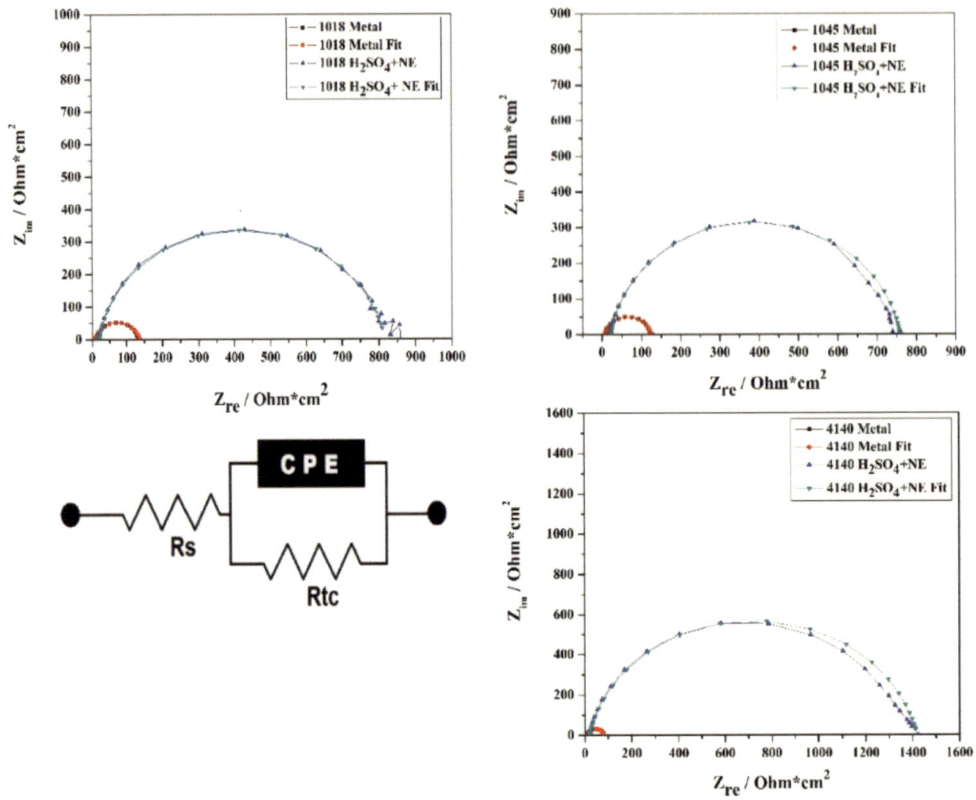

Figure 6. EEC and fit of Nyquist plots corresponding to metals in H_2SO_4 and H_2SO_4 + Nopal extract.

of the evaluation. As the exposure time increases, the semicircle shows a tendency to decrease slightly as a result of the alteration of the protection film formed by the inhibitor. At the end of the monitoring, the R_{ct} values from the inhibitor films do not approach the R_{ct} values of the blank, conserving some protection capacity with respect to the metal without inhibitor.

Semicircle adjustment was performed for each impedance cycles by taking an equivalent electrical circuit (EEC) as shown in **Figure 6**. Due to the shape of the Nyquist semicircle, the basic configuration called EEC Randles model was selected, which is formed by a resistance in series (solution resistance, R_s) coupled to a constant phase element (CPE) that is in parallel with another resistance attributed to the metal/electrolyte interface (R_{ct}).

To obtain the capacitance values (C_{dl}) from the constant phase element (CPE), it was necessary to obtain the maximum frequency of the Nyquist semicircle ($\omega_{\theta\ max}$) as well as the α exponent that was calculated during the CPE adjustment, since CPE = $F \cdot s^{(a-1)}$ where F and s indicate Faradios and seconds, respectively. Results in **Table 3** are shown in μF cm^{-2}. Equation (6) shows the calculation to C_{dl}:

$$C_{dl} = CPE^{*}\,\omega_{\theta\max}{}^{(a-1)} \tag{6}$$

Metal	Medium	LPR			EIS	
		Jo (A/cm²)	E_{corr} vs. SCE (V)	R_p (Ohm ·cm²)	R_{ct} (Ohm ·cm²)	C_{dl} (µF·cm²)
1018	H_2SO_4	2.92E-05	−0.511	268.182	187.538	**179.36**
	H_2SO_4 + NE	7.85E-06	−0.545	1015.013	717.162	**155.96**
4140	H_2SO_4	4.74E-05	−0.506	179.379	121.76	**1202.42**
	H_2SO_4 + NE	7.71E-06	−0.543	1085.28	815.98	**185.14**
1045	H_2SO_4	3.34E-04	−0.439	64.865	50.78	**492.49**
	H_2SO_4 + NE	1.15E-05	−0.505	729.44	514.16	**120.41**

Table 3. Electrochemical results.

Polarization curves of carbon steel samples exposed in H_2SO_4 with inhibitor are shown in **Figure 7(A)**, where a decrease in current corrosion of the three steels studied is observed.

Anodic slope shows a change near the 400 mV, associated with a change of their activation by the presence of a coating from the inhibitor's organic molecules, limiting the ionic interchange between the metal and electrolyte.

Figure 7(A) shows the behavior of the polarization in which there is a decrease in the flow of current on the three steels exposed to the inhibitor. **Figure 7(B)** shows the corrosion behavior for all the steels; these values were obtained by extrapolation of Tafel, where the steels show a lower corrosion rate to 1018, 4140 and 1045, respectively; these values remain stable throughout the 24 h of exposure.

Table 3 shows the values obtained with the electrochemical techniques LPR and EIS; a correspondence was observed between the values of resistance for both techniques; however,

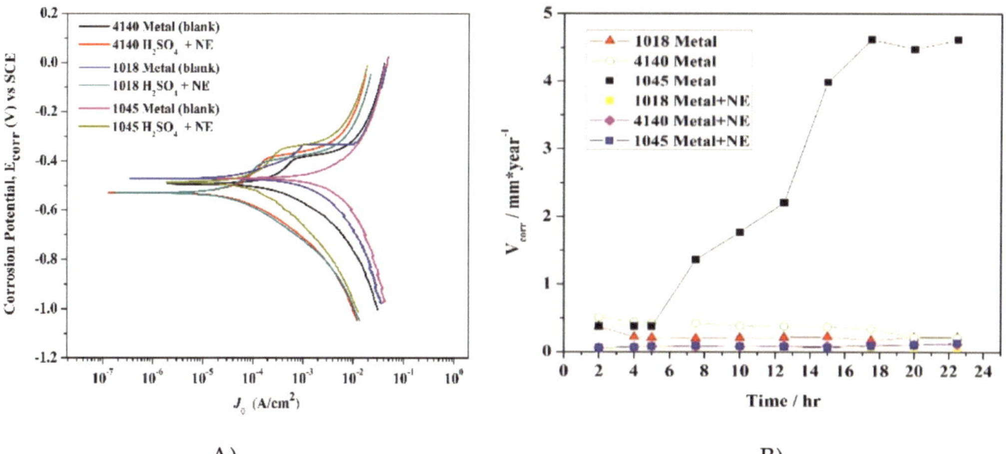

Figure 7. (A) Polarization curve for steels in H_2SO_4 and H_2SO_4 + NE medium and (B) corrosion rate of metals for 24 h.

the resistance values by the impedance technique are lower because they are depurated due to the contribution of the resistance to the solution that has been separated with the adjustment of the EEC. The current indicates that the 1018 and 4140 steels have low corrosion rate. Capacitance values indicate a decrease with the presence of the inhibitor and are lower for 1045 and 1018 steels; these values tend to decrease due to a difficulty of ionic exchange.

Figure 8 shows the metallography obtained for all steels on a macroscopic scale, where for (a)–(c), the area exposed to H_2SO_4 shows that the main mechanism for the three metals is a combination between pitting and generalized corrosion, being more notable the pitting attack for the 1018 and 1045 steels before the NE addition.

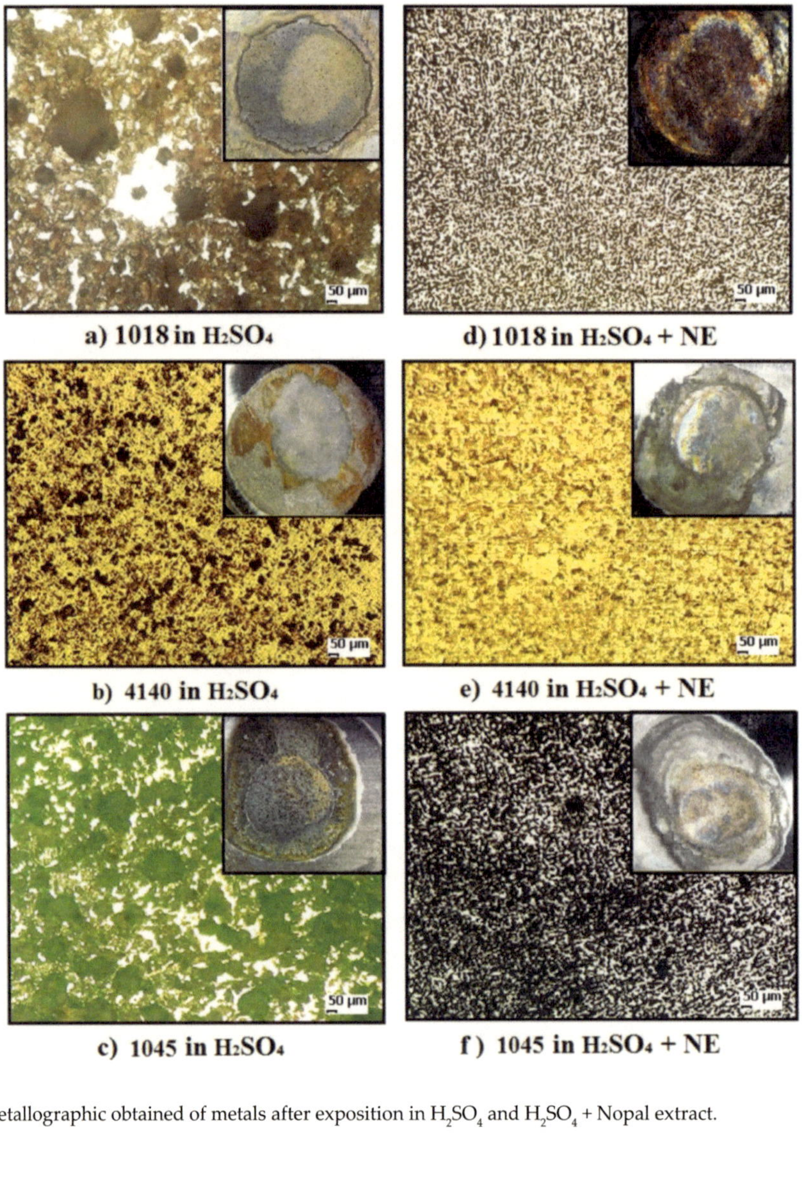

Figure 8. Metallographic obtained of metals after exposition in H_2SO_4 and H_2SO_4 + Nopal extract.

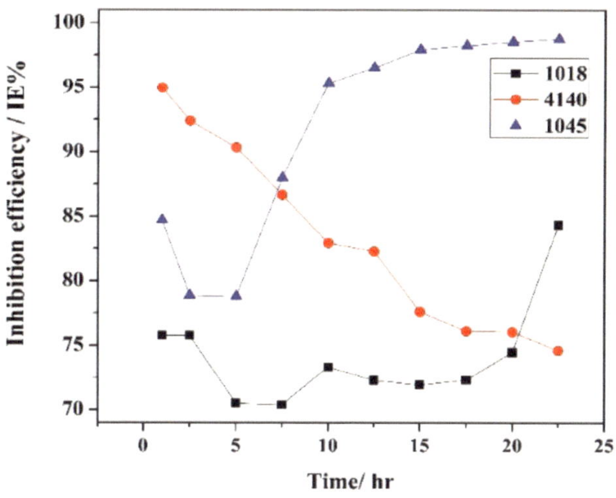

Figure 9. Inhibition efficiency for the three metals in H_2SO_4 + Nopal extract after 24 h.

Figure 8(d)–(f) shows the behavior of the steels after exposed to the acid mixture plus the Nopal extract. Change of the corrosion mechanism was observed with the addition of the NE from a localized attack (pitting) to a generalized attack due to the redistribution of the active sites by the presence of the organic molecules from the inhibitor, having better behavior in the steel 1045 where the corrosion was considerably diminished.

Figure 9 shows the efficiency calculated for the three steels, observing the performance of the Nopal extract during the 24 h of exposure. It was observed that steel 1045 obtains the highest percentage of efficiency, which is associated with the inhibitor coverage, and remains constant from the 10th hour of exposure, while the other two steels (1018 and 4140) do not achieve sufficient efficiency values to guarantee their protection.

4. Conclusions

Opuntia ficus-indica (Nopal extract) was used as a green inhibitor to evaluate the corrosion behavior in the protection of three industrial steels (AISI 1018, 1045 and 4140), exposed in sulfuric acid (H_2SO_4) plus the addition of NE in the same proportion (50/50 volume %). Corrosion potential showed a tendency to be more negative when the NE was added, forming a compact barrier on the metal surface that increased the surface resistance, according to the growth of the Nyquist semicircles and with respect to the steel substrates.

Charge transfer resistance (R_{ct}) was reduced as the exposition time increased due to the deterioration of the protective film. A 4140 steel reported less corrosion rate with respect to the 1018 and 1045 steels, respectively. Polarization resistance (R_p) showed the same behavior with the best behavior for the steel 4140 after 24 h.

Change of the corrosion mechanism was observed with the addition of the NE from a localized attack (pitting) to a generalized attack due to the redistribution of the active sites by the presence of the organic molecules from the inhibitor, having better behavior in the steel 1045 where the corrosion was considerably diminished.

The NE addition changed the corrosion mechanism in the steels from a localized attack (pitting) to a generalized attack due to the adsorbed inhibitor molecules on the metal surface, reducing the anodic and/or cathodic reactions and as a consequence the corrosion rate.

Alloyed steel (4140) showed a decrease in the inhibitor efficiency with the exposition time, showing better tendency (above 90%) for the 1045 steel. This is possible because the chemical composition of the steel is interacting with the adsorbed organic molecules, increasing or reducing the actives sites for corrosion. Difference in corrosion rate for the 4140 steel with the others was the combination of the green inhibitor and their alloyed elements (Mo and Cr).

Acknowledgements

The authors wish to thank our institution (CIDETEQ) and (UAEM) for the facilities and the National Council of Science and Technology (CONACYT) for financial support. We appreciate the participation of the student Luis E. Almazán Corona, who actively participated in the experimental development through the program UAEM-SIyEA (Secretary of Investigation and Advance Studies) 3817/2014/CID.

Author details

Araceli Mandujano Ruiz[1]*, Luis-Enrique Corona Almazán[2], Héctor Herrera Hernández[2] and Jorge Morales Hernández[1]

*Address all correspondence to: amandujano@cideteq.mx

1 Centro de Investigación y Desarrollo Tecnológico en Electroquímica (CIDETEQ), Querétaro, México

2 Universidad Autónoma del Estado de México, Atizapán de Zaragoza, Estado de México, México

References

[1] Lim HL. Deeper understanding of green inhibitors for corrosion of reinforcing steel in concrete. In: Handbook of Research on Research Development in Materials Science and Corrosion Engineering Education, E.S. Reference: EUA. 2015

http://dx.doi.org/10.5772/intechopen.72944

[2] Bockris JO, Singh WP. Toxicity issues of organic corrosion inhibitors; applications of QSAR Model. In: Proceedings of Corrosion. Houston: NACE, International; 1996

[3] Espinoza VA, Negrón SGE, González OR, Angeles BD, Herrera HH, Romero RM. Mild steel corrosion inhibithion in HCl by di-alkyl and di-1,2,3, triazole derivatives of uracil and thyamine. Materials Chemistry and Physics. 2014:407-417

[4] Espinoza VA, Negrón SGE, Angeles BD, Palomar PME, Romero RMA, Herrera HH. Pantoprazol as inhibitors of API 5L X52 steel corrosion in HCl 1M. ECS Transactions. 2011:207-2016

[5] Palomar PM, Romero RM, Herrera HH, Abfeu QMA, Likhanova N, Uruchurtu J, et. al. Influence of the alkyl chain length of 2 amino 5 alkyl 1,3,4 thidiazole compounds on the corrosion inhibithion of steel immersed in sulfuric acid solutions. Corrosion Science. 2012:231-243

[6] Majdouline L et al. An investigation of mild carbon steel corrosion inhibition in hydrochloric acid medium by environment friendly green inhibitors. Journal of Coatings Technology and Research. 2011;**9**:15-26

[7] Ibrahim TH, Zour MA. Corrosion inhibition of mild steel using fig leaves extract in hydrochloric acid solution. International Journal of Electrochemical Science. 2011;**6**:6442-6455

[8] Gunasekaran G, Chauchan LR. Eco friendly inhibitor for corrosion inhibition of mild steel in phosphoric acid medium. Electrochimica Acta. 2004;**49**:4387-4395

[9] Chetouani A, Hammouti B, Benkaddour M. Corrosion inhibition of iron in hydrochloric acid solution by jojoba oil. Pigment & Resin Technology. 2004;**33**:26-31

[10] Bouyanzer A, Hammouti B, Chetouani A. A study of anti-corrosive effects of Artemisia oil on steel. Pigment & Resin Technology. 2004;**33**:287-292

[11] Sethuraman MG, Bothi RP. Corrosion inhibition of mild steel by Datura metel. Pigment & Resin Technology. 2005;**34**:327-331

[12] Bouyanzer A, Hammouti B, Majidi L. Pennyroayl oil from *Mentha pulegium* as corrosion inhibitor for steel in 1 M HCl. Materials Letters. 2006;**60**:2840-2843

[13] El-Etre AY. Khilah extract as inhibitor for acid corrosion of SX 316 steel. Applied Surface Science. 2006;**252**:8521-8525

[14] Okafor PC, Ebenso EE. Inhibitive action of Carica papaya extracts on the corrosion of mild steel in acidic media and their adsorption characteristics. Pigment & Resin Technology. 2007;**36**:134-140

[15] Oguzie EE, Onuoha GN, Ejike EE. Effect of Gongronema latifolium extract on aluminium corrosion in acidic and alkaline media. Pigment & Resin Technology. 2007;**36**:44-49

[16] Torres Acosta AA. *Opuntia-ficus-indica* (Nopal) mucilage as a steel corrosion inhibitor in alkaline media. Journal Application of Electrochemical. 2007;**38**:835-841

[17] Torres-Acosta AA, Molina MW, Guzmán AEM. State of the art on cactus additions in alkaline media as corrosion inhibitors. International Journal of Corrosion. 2012:1-9

[18] Okafor PC et al. Inhibitory action of *Phyllanthus amarus* extracts on the corrosion of mild steel in acidic media. Corrosion Science. 2008;**50**:2310-2317

[19] Raja PB, Sethuraman MG. Inhibitive effect of black pepper extract on the sulphuric acid corrosion of mild steel. Materials Letters. 2008;**62**:2977-2979

[20] El- Etre AY. Inhibition of C-steel corrosion in acidic solution using the aqueous extract of zallouh root. Materials Chemistry and Physics. 2008;**108**:278-282

[21] Satapathy AK et al. Corrosion inhibition by *Justicia gendarussa* plant extract in hydrochloric acid solution. Corrosion Science. 2009;**51**:2848-2856

[22] Abdullah M, Al-Karanee SO, Abdel FAA. Inhibition of the corrosion of nickel and its alloys by Natural Clove Oil. Chemical Engineering Communications. 2009;**196**:1406-1416

[23] Abdel- Gaber AM, Adb- El NBA. The role of acid anion on the inhibition of the acidic corrosion of steel by lupine extract. Corrosion Science. 2009;**51**:1038-1042

[24] Abiola OK, Otaigbe JOE, Kio OJ. *Gossipium hirsutum* L. extracts as green corrosion inhibitor for aluminum in NaOH solution. Corrosion Science. 2009;**51**:1879-1881

[25] Atinga EI et al. Leaves extract of Ananas sativum as green corrosion inhibitor for aluminium in hydrochloric acid solutions. Green Chemistry Letters and Reviews. 2010;**3**:61-68

[26] Beenakumari KS. Inhibitory effects of *Murraya koenigii* (curry leaf) leaf extract on the corrosion of mild steel in 1 M HCl. Green Chemistry Letters and Reviews. 2011;**4**:117-120

[27] Chennappan K, Mathur GS. Caulerpin—A bis-indole alkaloid as a green inhibitor for the corrosion of mild steel in 1 M HCl solution from the marine alga *Caulerpa racemosa*. Industrial and Engineering Chemistry Research. 2012:10399-10407. DOI: 10.1021/ie3010379

[28] Kamal C, Sethuraman MG. Spirulina platensis—A novel green inhibitor for acid corrosion of mild steel. Arabian Journal of Chemistry. 2012;**5**:155-161

[29] Shah P, Agarwal S. Aloe-Vera: A green corrosion inhibitor. International Journal for Research in Applied Science and Engineering Technology. 2014;**2**:14-17

[30] Mourya P, Banerjee S, Singh MM. Corrosion inhibition of mild steel in acidic solution by *Tagetes erecta* (Marigold flower) extract as a green inhibitor. Corrosion Science. 2014

[31] Odewunmi NA, Umoren SA, Gasem ZM. Utilization of watermelon rind extract as a green corrosion inhibitor for mild steel in acidic media. Journal of Industrial and Engineering Chemistry. 2014:1-9

[32] Odewunmi NA, Umoren SA, Gasem ZM. Watermelon waste products as green corrosion inhibitors for mild steel 2 in HCl solution. Journal of Environmental Chemical Engineering. 2014:1-11

[33] Da Rocha CJ, Da Cunha PG, D'Eliab E. Aqueous extracts of mango and orange peel as green inhibitors for carbon steel in hydrochloric acid solution. Materials Research. 2014;**17**:1581-1587

[34] Fouda AS et al. Anise extract as green corrosion inhibitor for carbon steel in hydrochloric acid solutions. International Journal of Innovative Research in Science Engineering and Technology. 2014;**3**:11210-11228

[35] Umoren SA et al. Coconut coir dust extract: A novel eco-friendly corrosion inhibitor for Al in HCl solutions. Green Chemistry Letters and Reviews. 2012;**5**:303-313

[36] Umoren SA, Solomon MM, Eduok UM, Obot IB, Israel AU. Inhibition of mild steel corrosion in H_2SO_4 solution by coconut coir dust extract obtained from different solvent systems and synergistic effect of iodide ions: Ethanol and acetone extracts. Journal of Environmental Chemical Engineering. 2014. DOI: 10.1016/j.jece.2014.03.024

[37] Umoren SA. et al. Inhibition of mild steel corrosion in H_2SO_4 solution by coconut coir dust extract obtained from different solvent systems and synergistic effect of iodide ions: Ethanol and acetone extracts. Journal of Environmental Chemical Engineering. 2014:1-55

[38] Mayanglambam RS, Sharma V, Singh G. Musa Paradisiaca extract as a green inhibitor for corrosion of mild steel in 0.5 M sulphuric acid solution. Portugaliae Electrochimica Acta. 2011;**29**:405-417

[39] Ismail A, Tajuddin MAM. Banana peel as green corrosion inhibitor for stainless steel 304. International Conference on X-Rays & Related Techniques in Research & Industry. 2014:109-110

[40] Muthukrishnan P et al. Stigmasterol extracted from *Ficus hispida* leaves as a green inhibitor for the mild steel corrosion in 1 M HCl solution. Arabian Journal of Chemistry. 2015:1-12

[41] Bhawsar J, Jain PK, Jain P. Experimental and computational studies of *Nicotiana tabacum* leaves extract as green corrosion inhibitor for mild steel in acidic medium. Alexandria Engineering Journal. 2015;**54**:769-775

[42] Singh MR, Gupta P, Gupta K. The litchi (*Litchi chinensis*) peels extract as a potential green inhibitor in prevention of corrosion of mild steel in 0.5 M H_2SO_4 solution. Arabian Journal of Chemistry. 2015:1-7

[43] Savita et al. Fruit extract as a green inhibitor for copper corrosion in nitric acid solution. International Journal of Innovative Research in Science Engineering and Technology. 2015;**4**:4545-4553

[44] Kumar RS, Chandrasekaran V. Valoniopsis pachynema extract as a green inhibitor for corrosion of brass in 0.1 N phosphoric acid solution. Metallurgical and Materials Transactions. 2015:1-8

[45] Sharma SK, Peter A, Obot IB. Potential of Azadirachta indica as a green corrosion inhibitor against mild steel, aluminum, and tin: A review. Journal of Analytical Science and Technology. 2015;**6**:1-16

[46] Al-Senani GM. The use of green leafy vegetables extracts as corrosion inhibitors for carbon steel in acidic medium. World Applied Sciences Journal. 2015;**33**(10):1659-1666

[47] Herrera HH, et al. Morinda Citrifolia como inhibidor natural de la corrosión de aceros en medio ácido. In: Congreso Internacional de Metalurgia y Materiales 2015. Concepción-Chile. 2015

[48] Herrera HH, et al. Morinda Citrifolia como Inhibidor Natural de la Corrosión de Aceros en Medio Ácido. In: Memorias del Congreso Internacional de Investigación Academica Journals Tabasco 2016; Tabasco; 2016

[49] Abboud Y et al. Corrosion inhibition of carbon steel in hydrochloric acid solution using pomegranate leave extracts. Corrosion Engineering, Science and Technology. 2015:1-9

[50] Abboud Y et al. Punica granatum leave extract as green corrosion inhibitor for mild steel in Hydrochloric acid. MATEC Web of Conferences. 2013

[51] Cheyad MS, Salman TA. Pomegranate peel extract as green corrosion inhibitor for α-brass in 2M HCl solution. Journal of Materials Science & Surface Engineering. 2017;**5**:597-601

[52] Baran E, Cakir A, Yazici B. Inhibitory effect of *Gentiana olivieri* extracts on the corrosion of mild steel in 0.5 M HCl: Electrochemical and phytochemical evaluation. Arabian Journal of Chemistry. 2016:1-17

[53] Souzaa ECDCA et al. Roasted coffee extracts as corrosion inhibitors for mild steel in HCl solution. Materials Research. 2016;**19**:1276-1285

[54] Juhaiman LAA. Curcumin extract as a green inhibitor of leaching from aluminum cookware at quasi-cooking conditions. Green and Sustainable Chemistry. 2016;**6**:57-70

[55] Taleb I, Elron G, Ime BO, Mustafa K, Mohamed AZ. Corrosion inhibition of mild steel by *Calotropis procera* leaves extract in a CO_2 saturated sodium chloride solution. Journal of Adhesion Science and Technology. 2016;**30**(23):2523-2543

[56] SFA E-A et al. The role of aqueous *Ocimum basilicum* extract for monitoring the corrosion inhibition of carbon steel used in sanitation plants and its effect on *Escherichia coli*. Zastita Materijala. 2016;**57**:583-596

[57] Mayanglambam RS, Sharma V, Singh G. *Ocimum basilicum* extract as a potential green inhibitor for corrosion of mild steel in 0.5M H_2SO_4 solution. Asian Journal of Chemistry. 2012;**24**:1773-1778

[58] SFA E-A, SAA E-M, Mostafa MH. Tilia leafs as eco-friendly corrosion inhibitor for mild steel in aqueous solutions. Zastita Materijala. 2017;**58**:217-282

[59] Elmiziani I et al. Lead corrosion inhibition by *Cedrus atlantica* as a green inhibitor in 0.1M Na_2CO_3 solution. International Journal of Advanced Chemistry. 2017;**5**:1-7

[60] Khaled KF. Experimental, density function theory calculations and molecular dynamics simulations to investigate the adsorption of some thiourea derivatives on iron surface in nitric acid solutions. Applied Surface Science. 2010;**256**:6753-6763

[61] Sanyal B. Organic compounds as corrosion inhibitors in different environments—A review. Progress in Organic Coatings. 1981;**9**:165-236

[62] Ibrahim TH, Chehade Y, Zour MA. Corrosion inhibition of mild steel using potato peel extract in 2M HCL solution. International Journal of Electrochemical Science. 2011;**6**: 6542-6555

[63] Sastri VS. Challenges in corrosion, costs, causes, consequences and control. Wiley series in corrosion. 2015

[64] De los Ríos JP, Sánchez CM, Nava DC, Chacón NJG, González RJG, Huape PE, Neri FA, Martínez VA. *Opuntia ficus-indica* extract as green corrosion inhibithor for carbon steel in 1M HCl solution. Journal of Spectroscopy. 2015:1-9

[65] Herrera HH, Franco TMI, Miranda HJG, Herrera SE, Espinoza VA, Fajardo G. Gel de Aloe-Vera como Potencial inhibidor de la corrosión del acero de Refuerzo Estructural. Avances en Ciencias e Ingeniería. 2015;**6**:9-23

[66] Chandra S, Eklud L, Villareal RR. Use of cactus in mortars and concrete. Cement and Concrete Research. 1998;**28**:41-51

Chapter 8

Vapor Inhibitors for Corrosion Protection in Humid and Saline, Natural, and Industrial Environments

Nelson Cheng, Benjamin Valdez Salas,
Michael Schorr Wiener and
Juan Ricardo Salinas Martinez

Additional information is available at the end of the chapter

http://dx.doi.org/10.5772/intechopen.72815

Abstract

Corrosion is a pernicious phenomenon that appears in engineering materials, infrastructure assets, and industrial equipment exposed to the atmosphere, water, and soil. The aim of this chapter is to produce a body of knowledge on the theory and practice of corrosion inhibition so it can be used in the selection of corrosion inhibitors, to prevent and/or minimize corrosion in natural environments and industrial facilities. Corrosion inhibitors are employed in different forms such as emitters, pellets, powders, films, paints, aerosols, and aqueous and solvent solutions, depending on their chemical composition: organic or inorganic. They are impregnated on papers or plastics; closed in pouches and sachets; or added to coatings and paints to form a barrier against the attack of corrosive agents. They are employed in many industries: automotive, offshore/onshore, water, military, marine, manufacture, oil and gas, electronics, and concrete structures. A special group of corrosion inhibitors are extracted from vegetables or derived from industrial vegetable wastes, which are related to the advanced field of "Green Chemistry." The use of vapor corrosion inhibitors (VCI) has expanded rapidly in the last decades, and is frequently mentioned in the relevant literature, in particular in professional congresses and conferences, organized by international and national corrosion associations.

Keywords: corrosion, corrosion inhibitors, natural environments, industrial facilities, impregnated papers

1. Introduction

The global potable water scarcity, clean efficient energy, and environmental quality are recognized as central disciplines in modern science, engineering, and technology. Today, they are linked to the critical problems of climate change, worldwide warming, and greenhouse gas emissions, all interrelated phenomena [1, 2].

In recent years, advances, discoveries, and developments related to corrosion management, including theory and practice, have been made by universities, research institutes, international and national laboratories, and private industrial enterprises [3–5]. The importance and relevance of corrosion is evident by the periodical publication of 20 scientific and technical journals, in many languages, devoted to corrosion, materials, coatings, paints, structures, fouling, and scaling. Moreover, the economic and social value of the use of VCI for corrosion control in many critical industrial fields is graphically demonstrated in **Figure 1**.

The VCI triangle shows the harmful action of corrosive agents encountered in water bodies, atmosphere and land environments, and in industry facilities. The VCI are actively implemented to prevent, mitigate, and control corrosion in numerous industries and natural locations.

A special electronic newsletter is published weekly by NACE Corrosion Press describing corrosion events, corrosion problems and their solutions, novel instruments for corrosion detection and measurements, application of nanoparticles in corrosion devices, and repair and rehabilitation of bridges and infrastructure assets.

The influence of VCI technology is expressed by many international and national patents gathered in recently published reviews [6, 7].

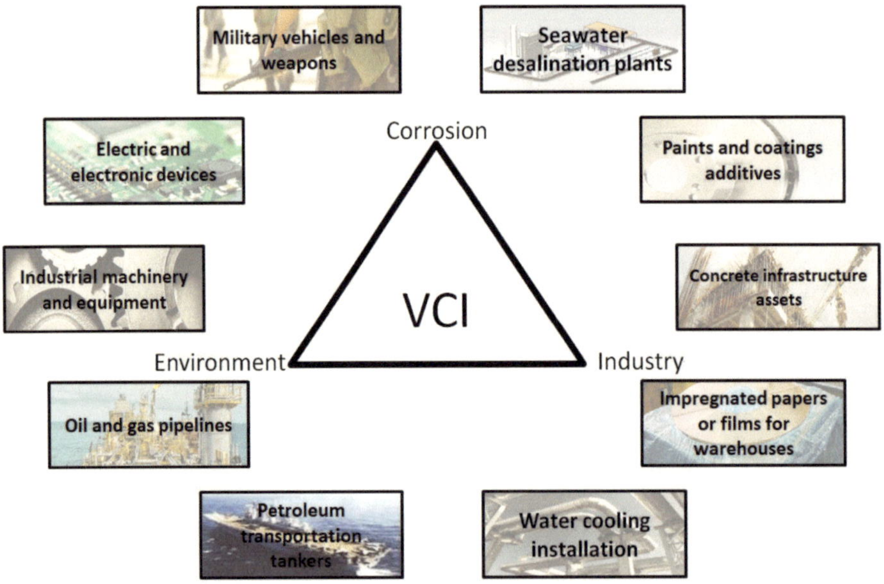

Figure 1. Schematic illustration of the great variety of VCI applications.

http://dx.doi.org/10.5772/intechopen.72815

2. Corrosion and pollution

Corrosion and deterioration of industrial equipment occurs by an interaction between the surface of the material and the corrosive factors of its environment. Localized corrosion happens on parts of these equipment with a geometry prone to corrosion. Prevention and mitigation of these forms of corrosion require adequate methods and techniques, including VCI. Corrosion and pollution are harmful processes that are strongly intertwined since many pollutants, e.g., SO_X, NO_X, H_2S, NH_3, CH_4, CO_X, O_3, and acids, accelerate corrosion, and corrosion products such as rust, oxides, and soluble salts pollute water bodies such as lakes and rivers. Both are noxious phenomena that destroy the environment, impair the productivity of industry, and affect the durability of the materials and infrastructure assets [1–3].

The surface of a metallic piece of a machine or a vehicle reacts with the corrosive factors of its environment such as humidity and salinity, regulated by anodic and cathodic electrochemical reactions, as follows:

$$\text{Anodic} \quad 2\text{Fe} \rightarrow 2\,\text{Fe}^{+2} + 4\,\text{e}^- \tag{1}$$

$$\text{Cathodic} \quad \text{O}_2 + 2\,\text{H}_2\text{O} + 4\,\text{e}^- \rightarrow 4\,\text{OH}^- \tag{2}$$

The total corrosion process is

$$2\text{Fe} + \text{O}_2 + 2\,\text{H}_2\text{O} \rightarrow 2\text{Fe}\,(\text{OH})_2 \tag{3}$$

Moreover, a posterior oxidation transforms the ferrous hydroxide into ferric oxide $Fe_2O_3xH_2O$, which covers the steel piece as rust.

Corrosion may appear in several forms: as general, uniform, covering the whole surface or as localize corrosion in special spots and sometimes producing perforations called pits.

3. VAPPRO corrosion inhibitors

Magna International, with headquarters in Singapore, consists of several enterprises located in Australia, Canada, Indonesia, China, and South Korea. The company is one of the world's leaders devoted to the critical field of corrosion-preventive technology. One of its main products is Vapor Corrosion Inhibitor, known as VAPPRO, which stands for Vapor-Phase-Protection, suitable for environmental, industrial defense, and commercial fields.

VAPPRO products conform to U.S. Military specifications, especially for the protection of weapons and vehicles. NACE International, the Worldwide Corrosion Authority partners with the U.S. Department of Defense (DOD) to launch CorrDefense, and online e-magazine highlighting corrosion control efforts of the DOD, the Coast Guard and the National Aeronautical and Space Administration (NASA) as well as projects the agency shares with laboratories, universities, research institutions, and private companies.

Magna VAPPRO products are accessible through the North Atlantic Treaty Organization and its Master Cross Reference List (NATO MCRL). These VCI are assigned NATO Stock Numbers, to facilitate their quoting.

The VAPPRO products conform to the requirements of the European Union (EU) Directive on Restriction of Hazardous Substances (RoHS), which control the use of harmful chemicals that may be noxious to the environment and consumer's well-being.

VAPPRO VCI have been tested and certified by the Bureau Veritas, a well-known organization dealing in quality, health, safety, and environment (QHSE) for the benefit of the general public.

Magna International cooperates with NACE International, with headquarters in Houston, U.S. and having a membership of about 30,000 experts on corrosion management and with the World Corrosion Organization (WCO) which associates many national corrosion organizations, with its central office in China.

Magna International maintains a cooperation relation with the American Chemical Society (ACS), U.S. using the correct chemical definitions and terms for the chemical molecules and reactions.

The North American Free Trade Agreement (NAFTA), which associates the United States, Mexico, and Canada, is familiar with the VAPPRO VCI, being widely applied in natural environments, industrial facilities, and military assets. The NAFTA trilateral trade deal is now being renegotiated an experts believe that an upgrade to this deal is possible, including new rules on digital trade and e-commerce.

Magna worldwide operations are based on the application of the International Standards Organization-ISO 9001-2000, ISO 14001-2004, and ISO 9001-2008 that satisfy the requirements of the VAPPRO consumers. Magna provides VCI and service in many countries including the NATO countries, NAFTA region, the Asia Pacific, and Europe zones, maintaining a high level of science, technology, and innovation, for the creation of novel products in the demanding twenty-first century. Magna International marketing executives are conducting an expansion of their organization and enhancing their knowledge and strategy. All these multiple activities generate a drive towards intense work and useful achievements.

Magna International, as a global enterprise, is committed to social and environmental concerns, to promote greater awareness that minimizes or eliminates harmful corrosion and pollution.

4. Corrosion inhibitors

Practical methods that minimize or eliminate corrosion in the chemical, mineral, fertilizer, electrons, food and beverages, energy, and manufacture industries include:

- Selection of corrosion resistant construction materials, metallic, plastic, ceramic, and composites for fabrication of the plants equipment, machinery, and structures.

http://dx.doi.org/10.5772/intechopen.72815

- Application of paints, coating, and linings for protection of the industrial plants and energy facilities.
- Cathodic protection (CP) applying sacrificial anodic metals such as aluminum and magnesium and their alloys. An additional CP system employs impressed direct electrical current.
- Corrosion inhibition is a technique widely accepted, the vast industry of corrosion inhibitors (CI) is fast spreading worldwide for various technological and industrial applications such as petroleum oil and natural gas facilities, structures built from steel reinforced concrete, pickling and cleaning steel vehicles during production, protected transportation, and storage of electronic and military equipment; for cooling water systems; as additives for paints, coatings, elastomers, and plastics, to avoid corrosion in natural gas pipelines, to control corrosion in desalination plants [8–12].

To prevent atmospheric corrosion, in particular in humid environments, military equipment, vehicles, and weapons are covered, during long times, with papers, canvas, and plastic sheets fully impregnated with VCI.

Three types of CI, with different inhibition mechanisms, are known:

a. CI that are adsorbed on the metal surface, avoiding metal deterioration. They comprehend organic chemicals such as waxes, oils, and amines.

b. CI that hinder the anodic reaction of metal dissolution and, frequently, for a passive layer.

c. The cathodic CI controls corrosion, for example, by the reduction of dissolved oxygen (O_2).

In the last decade, a new family of inhibitors has emerged, called green corrosion inhibitors, which are relevant in this crucial time of energy problems and economic havoc since they will extend the life of the water infrastructure and save large expenses in materials, equipment, and structures. They belong to the advanced field of green chemistry, also known as sustainable chemistry, which involves the design of chemical products and processes that reduce or eliminate the use or generation of hazardous substance [13, 14].

Types of VCI include nitrite of amines, amine carboxylates, heterocyclic compounds (thiazole, triazole, pyrrole, mercaptans, imidazoline, etc.), carboxylic acid esters, amines, acetylenic alcohols, mixtures, and reaction products of these substances. **Figure 2** shows four molecular structures of synthetic chemicals, developed in modern laboratories, that are the base of the well-known VCI.

In this time of energy crisis, with petroleum prices up and down, and economic global turmoil, it is necessary to develop and implement safe "green" VCI. The oil inventories in many industrialized countries are weighing on oil prices. But frequent filling and emptying of these huge storage tanks, permitting access to humid air, can lead to harmful corrosion events, which, logically, should promote the application of useful corrosion inhibitors.

4.1. Natural corrosion inhibitors

In the context of this section on CI it is worthwhile to present a particular group of CI, called natural CI, which are obtained applying the chemical technology of extraction from vegetables

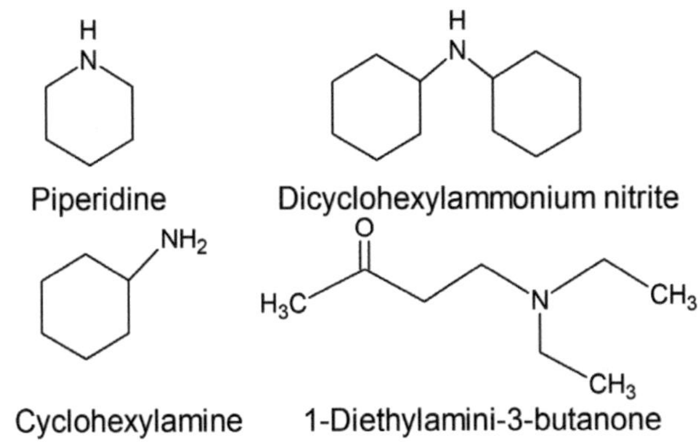

Figure 2. Molecular structures used for production vapor corrosion inhibitors.

using water or organic solvents as extractors. The ancient human communities located around forests and jungles, especially in tropical and arid areas, discovered and extracted medicines from parts of plants, shrubs, and trees such as roots, leaves, flowers, fruits, and seeds. These people took them orally or applied them as oil, cream, or ointments, to prevent and cure diseases, to relieve pain, and to improve their quality of life—in short, to attain the highest possible level of health.

Some natural CI are under experimental development stage and others are applied to prevent, diminish, or control corrosion damage [15–29].

5. Protection mechanisms

VAPPRO VCI protects ferrous and nonferrous metals, including their alloys and multimetals, machinery, structures, and devices. VCI molecules penetrate in recessed areas, voids, crevices, and cavities exposed to saline water and to humid atmospheres.

The protection mechanism of VCI depends on two basic chemical processes described in the following examples.

Ionization is the formation of ions as a result of chemical breakdown of molecules:

$$\underset{\text{Molecule}}{NaCl} \rightleftarrows \underset{\text{Cation}}{Na^{+}} + \underset{\text{Anion}}{Cl^{-}} \tag{4}$$

Hydrolysis is a chemical decomposition involving the splitting of water in hydrogen cation and the hydroxide anion:

$$NaCl + H_2O \rightleftarrows Na^{+} + Cl^{-} + H^{+} + OH^{-} \tag{5}$$

Figure 3 describes how the VCI molecule ionizes and hydrolyzes in the presence of moisture, forming a protective layer on the metal surface.

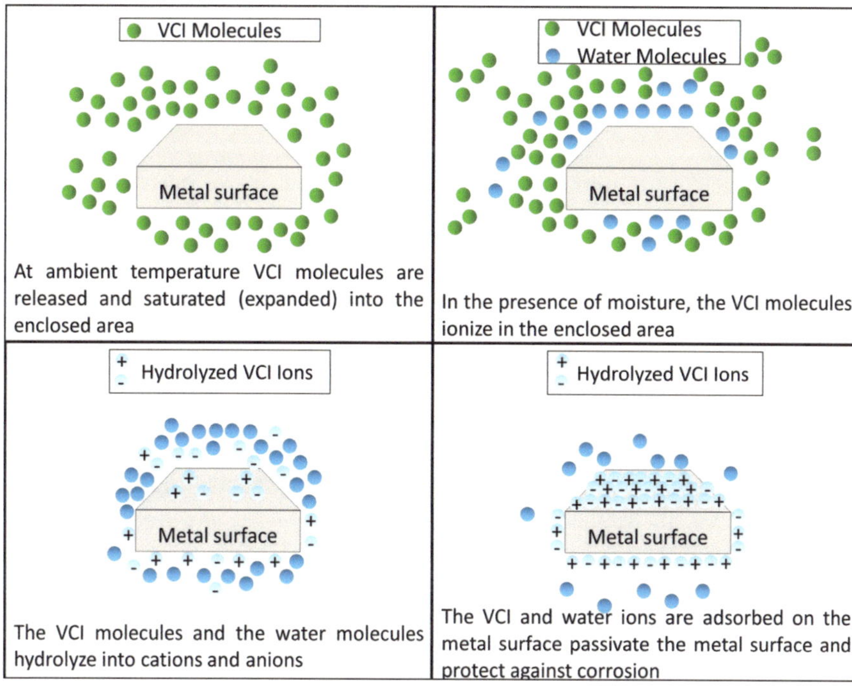

Figure 3. VCI molecule ionizes and hydrolyzes in the presence of moisture, forming a protective layer on the metal surface.

6. Practical applications

VCI products in solid and liquid forms are widely applied to protect components and parts of many industry products: automotive, oil, and gas, offshore and onshore, water supply and cooling, marine including oil tankers, military vehicles, and weapons, manufacture machines and tools, electronic and electrical devices, and concrete structures. These VCI are presented, packaged, and distributed in accordance to the form, size, and characteristics of the system being protected against corrosion.

6.1. Emitters and foams

Emitters consist of a piece of plastic foam impregnated with VCI, particularly suitable for the protection of enclosures holding electrical gear, relays, contactors, switches, and connectors. (**Figure 4**) The vapor does not increase the electrical resistance of these devices. Diverse foam types are used to produce emitters such as breathable polyurethane which is utilized as a tape with a self-adhesive backing. A VCI foam pad combines the resiliency and light weight of the inhibitor impregnated foam.

6.2. Pouches

Pouches are made from DuPont Tyvek, plastic material, to meet packaging applications for electronic devices, military parts of weapons, and protection against atmospheric corrosion factors. Other pouches are prepared from spun-bonded polyethylene fibers, which are located in the enclosed space holding the metallic parts to be protected (**Figure 5**).

6.3. Films

Thin antistatic polyethylene film containing VCI to protect metal parts from rust, tarnish, stains, water spots, and sulfide attack. They appear as custom size bags, sheets, tubing, sleeves, and roll stock. They are appropriate for protection during shipment and storage. Their UV stabilization extends the film life making it useful for outdoors, corrosive environment [30].

6.4. Impregnated papers and canvas

Military assets require implementation of corrosion control and monitoring techniques, in particular during long periods of storage, under the open skies, sometimes with falling rains and snow. Special papers such as kraft and stone, plastic sheets, and textile canvas, well impregnated with VCI, are applied to cover equipment, vehicles, and weapons [31].

The U.S. armed forces under the guidance of the U.S. Department of Defense (DOD) and its Office of Corrosion Policy and Oversight (CPO) maintain a Website (www.CorrDefense.org) and

Figure 4. Foam emitter device with VCI.

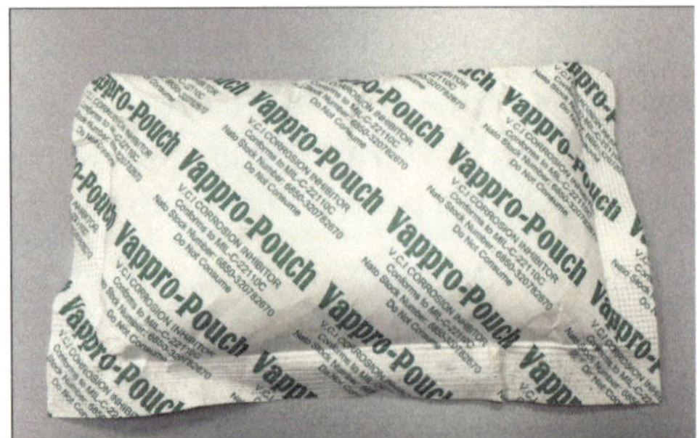

Figure 5. Breathable pouches containing vapor corrosion inhibitor.

publish an electronic magazine that features content on corrosion and corrosion control of military facilities. The diverse VCI impregnated sheets have proven to work for long periods [32].

Delicate machinery parts are kept in warehouse in closed VCI impregnated paper envelopes, cardboard boxes, or wrapped with textiles.

Most of the military equipment and arms are continually kept in reserve, so it is important to protect them while maintaining their original properties such as reliability and readiness for warlike operations. Heavy cannons on wheels, military hardware on wood pallets, and large weapons are wrapped with distinct fabrics, all impregnated with VCI. These covers are easy to install and are resistant to UV radiation, moistures, rain, snow, dust, and mold and are durable (**Figure 6**).

6.5. Coatings and paints

VCI are used as additives to improve their function as a barrier that impedes to reach the metal surfaces by the corrosive parameters and damage the machines and structures.

A wax coating, containing VCI, is particularly useful in marine and industrial rust proofing application in ferrous and nonferrous metals. A paint made from an acrylic polymer with an added nontoxic, organic VCI offers corrosion protection in outdoor environments. Magtan is an unpigmented primer that converts the rust into a more stable iron compound. Pro-Disc is a water-based coating that combines VCI with a nonslip, safe material. It was developed to protect brake disks of military vehicles during short storage or long mothballing without affecting the braking efficiency of the vehicle. Electrospray provides protection for circuit boards, relays, and electrical equipment without changing electrical properties of installations (**Figure 7**).

6.6. Powders

VCI in the form of powder are water soluble, do not contain phosphates and chromates, are used in low dosage for closed loop cooling system, hydrotesting, and marine fields.

These polymolecular VCI are added to seawater as a powder; then it converts into a colloidal suspension with nanoparticles dissolved in the water. These particles are adsorbed on the steel wall, forming a thin protective film. Their performance depends on physical, biological, and chemical factors.

Figure 6. Steel combustible tanks, ready for covering with a VCI protective cloth.

Figure 7. Printed circuit board protected by VCI electrospray.

Petroleum tankers (**Figure 8**) are ships specially designed and constructed at American and Asian shipyards for the bulk transportation of crude oil from extraction sites to the refineries. In its way back the tanker holds are full of seawater to provide it adequate stability. VCI are added to this ballast water to provide protection against corrosion. According to U.S. federal mandate all tankers in U.S. waters should have double hulls to diminish oil spill and damage to the environment, in case of hull rupture accident. Pipes, fire-water storage tanks, and hydrotest water are dosed withe these powder VCI [33].

6.7. Gun lubricants

Weapons, small and personal such as pistols and rifles, and big, installed and operated on land, sea and air vehicles should be kept clean and well lubricated for an efficient action. Arm-Kleen, which includes VCI, was specially developed for the cleaning of small arms and for automatic weapons mounted in all kinds of vehicles. The weapons should be cleaned before and after firing, learning a temporary protective coating for a period of several weeks. Once applied, the removal of VCI is unnecessary [34].

Such anticorrosive gun lubricant fuses itself into the metal gun barrel, smoothening out the metal surface of the asperities, and greatly reducing the coefficient of friction. The lubricant

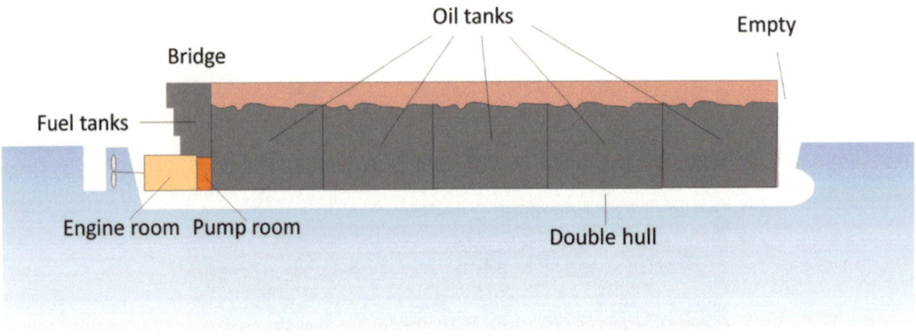

Figure 8. Petroleum transportation tanker showing holds containing ballast water with VCI powder.

that cleans also displaces moisture of the weapon's internal parts. These products conform to military standards: MIL-SPEC:MIL-PRF-634600 and MIL-C-372.

6.8. Aerosols

These VAPPRO aerosols, fortified with VCI, that protect electrical equipment, are presented in a variety of forms. Electrospray avoids corrosion against humid environments. A light oily liquid gives off an invisible vapor to protect machine tools, hand tools, and weapons subject to rusting and atmospheric corrosion. A penetrant dissolved in a solvent provides reliability in locating surface flaws; it is easily removed at the end of testing. It conforms to MIL-I-25135-E.

6.9. Master batch additives

The master batch VCI, identified as MB, are often dissolved in many media: water, oil, solvent, and kerosene, in accordance to their intended application, e.g., metal working fluids, inks, rust inhibitors and removers, gears, and engine and machine oils. Some are tacky, emulsifiable, viscous liquids with film forming properties.

VCI MB can be produced for special objectives, e.g., a low density polyethylene (LDPE), a resin to protect equipment functioning in aggressive, corrosive environments such as petroleum refineries and chemical plants.

6.10. Concrete rebar inhibitors (CRI)

Concrete is a composite material, useful for the construction of structures in land, water, and seas. It is made of mixture of water, Portland cement, sand, and mineral aggregates. Steel reinforced concrete is generally very durable; nevertheless concrete infrastructure, in particular in marine environments, can undergo visible damage due to penetration of seawater, reaching the steel reinforcement [35, 36].

Concrete and steel are the primordial engineering materials for the construction of the infrastructure of a nation, including bridges, buildings, highways, tunnels, parking garages, breakwaters, port piers, dam walls, offshore oil platforms, sport stadia, and electric transmission tower foundation. The principal cause of degradation of the steel reinforced structure is the corrosion attack to the rebar embedded in the concrete by CO_2 dissolved in seawater, increasing its pH to about 8.0, in conformance with the following equilibria:

$$CO_2 + H_2O \rightleftarrows HCO_3^- + H^+ \tag{6}$$

$$CaCO_3 + H_2O \rightleftarrows Ca^{+2} + HCO_3^- + OH^- \tag{7}$$

CO_2 and chlorides enter into the concrete body, reach the steel rebar, leading to the breakdown of passivity, a process called carbonation. The rebar loses thickness and, consequently, its strength, being the primary cause of damage and the collapse of the concrete structure. CRI-VCI have the ability to prevent and/or mitigate this event (**Figure 9**).

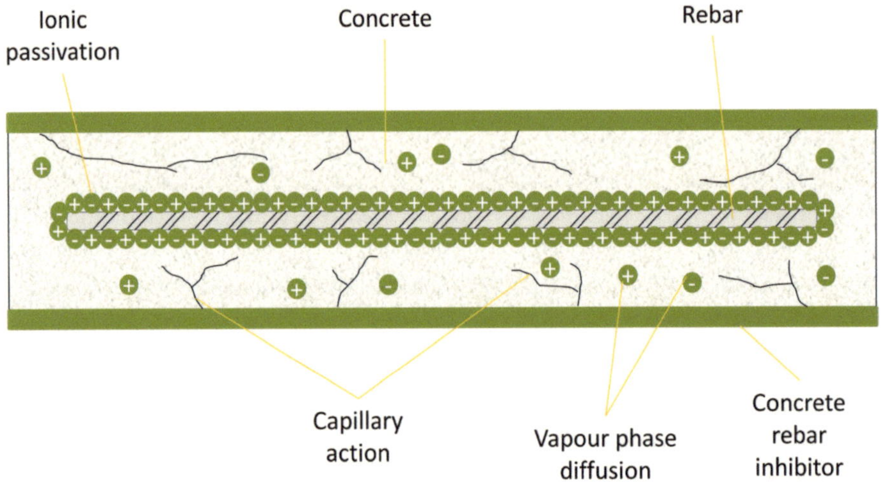

Figure 9. Protection mechanism of the concrete steel rebar.

VAPPRO concrete rebar inhibitors (CRI) are produced in special physical form and chemical composition, in accordance with their intended application, as follows:

- Concrete surface treatment (CST), to assure an impervious concrete surface, to avoid penetration of noxious agents present in the environment.
- VAPPRO rebar coating (VRC), to impart a protective coating on the rebar, that will be stable in contact with humid and dry concrete.
- Concrete repair products, to rehabilitate damaged concrete structure, covering them with fresh, alkaline concrete, providing conditions of its stable curing.
- Concrete additive treatment, adding to fresh concrete VCI soluble in solvents, specially designed for corrosion control in high humidity and high salinity environments.

7. Corrosion testing

The determination of the protection properties of the VCI and their corrosion avoidance performance are done by laboratory tests, simulating the relevant environmental and industrial conditions. ASTM and NACE standards were applied based on gravimetric, electrochemical, and surface examination methods and instruments, as recorded in the following.

ASTM standards detailed in:

- G1, preparing, cleaning, and evaluating corrosion test specimens.
- G3, electrochemical measurements in corrosion testing.
- G4, conducting corrosion test in field applications.

http://dx.doi.org/10.5772/intechopen.72815

- G5, potentiostatic and potentiodynamic anodic polarization measurements.
- G31, laboratory immersion corrosion testing of metals.
- G59, conducting potentiodynamic polarization resistance measurements.

NACE standards:

RP 0300	Corrosion of metals and alloys
TM 0169	Laboratory corrosion testing of metals
SP 0308	Inspection methods for corrosion evaluation of conventionally reinforced concrete structures
TM 0208	Laboratory test to evaluate the vapor-inhibiting ability of volatile corrosion inhibitor materials for temporary protection of ferrous metal surfaces

8. Conclusions

The VAPPRO corrosion inhibitors are designed and produced, taking into account the metallic materials and the structures to be protected and the physical, chemical, biological, and thermal conditions of the natural or industrial environment in which they operate.

VCI are utilized in diverse forms such as emitter devices, pellets, powders, additives to coatings and paints, aerosols, gun lubricants, master batch additives, impregnated papers and textiles, for concrete rebar protection, aqueous and solvent solutions, and in pouches and sachets.

The proper use of VCI in combination with a system of automated corrosion sensors is an effective means of preserving machinery, equipment, vehicles, and weapons, during long-term storage, with savings in maintenance costs.

The selection of the most appropriate VCI for a corrosion-prone system is based on its cost, ease of availability, workers safety, chemical stability, and saving in maintenance. The choice should be made in accordance with the corrosion expert knowledge and experience, and it is a compromise between technological and economic criteria.

Magna executives, designers, organizers, and employees know that to progress the key is to keep learning and improving and to increase knowledge and enrich experience to constantly renew themselves.

Acknowledgements

The authors are grateful to Magna International for offering their basic knowledge and practical experience in the development and implementation of VCI. In particular, the authors appreciate the contribution of Chemical Engineer Patrick Moe, Magna International Singapore.

Author details

Nelson Cheng[1], Benjamin Valdez Salas[2], Michael Schorr Wiener[2*] and Juan Ricardo Salinas Martinez[2]

*Address all correspondence to: mschorr2000@yahoo.com

1 Magna International Pte Ltd., Singapore

2 Departamento de Materiales, Minerales y Corrosión, Instituto de Ingeniería, Universidad Autónoma de Baja California, Mexicali, Baja California, Mexico

References

[1] Valdez B, Schorr M, et al. Effect of climate change on the durability of engineering materials in hydraulic infrastructure: An overview. Corrosion Engineering, Science and Technology. 2010;**45**(**1**):34-41

[2] Raichev R, Valdez B, Veleva L. Corrosion de metales y degradacion de materiales. Mexico: Universidad Autonoma de Baja California; 2009. pp. 281-284

[3] Hammel R. Alternative Futures for Corrosion and Degradation Research. Arlington: Potomac Institute Press; 2014. pp. 2-13

[4] Roberge PA. Handbook of Corrosion Engineering. USA: McGraw-Hill; 2000

[5] Roberge PR. Handbook of Corrosion Engineering. New York: McGraw-Hill; 1999. pp. 833-862

[6] Bastidas DM, Cano E, Mora EM. Volatile corrosion inhibitors: A review. Anti-Corrosion Methods and Materials. 2005;**52**(**2**):71-77

[7] Inzunza RG, Valdez B, Schorr M. Corrosion inhibitors patents for industrial applications: A review. Corrosion Science. 2013;**3**(**2**):71-78

[8] Al-Sayed T, Eid A, Al-Marzooqi M, Jason U. Protection of offshore platform caisson legs with a vapor corrosion inhibitor: A case study. In: NACE Corrosion Conference; Texas: NACE; March 2014. Paper No. 4200

[9] Garcia R, Valdez B, Schorr M, Eliezer A. Green corrosion inhibitors for water systems. Materials Performance. 2013;**52**(**6**):48-51

[10] Reams B. NACE International Maritime Program. Materials Performance. 2017;**4**:67-69

[11] Schorr M, Valdez B, Quintero M. Effect of H_2S in polluted waters: A review. Corrosion Engineering, Science and Technology. 2016;**41**:221-227

[12] Schorr M, Valdez B. Corrosion of the marine infrastructure in polluted seaports. Corrosion Engineering, Science and Technology. 2010;**45**(**1**):34-41

[13] Sastri VS. Green Corrosion Inhibitors: Theory and Practice. John Wiley & Sons; 2011

[14] Sharma SK. Green Corrosion Chemistry and Engineering. Wiley-VCH Verlag GmbH; 2012

[15] Barbosa HM, Espinoza A, Rodriguez FJ. Fenil-cumarine glucoside as a natural inhibitor in the protection of low carbon steel against corrosion. In: XXVI International Materials and Research Congress Symposium D.6, NACE: Corrosion and Metallurgy; Cancun. August 2017

[16] Zambrano GE, Carvo F, Garcia E, Chavez JC, Brito JR, Potenciano LG, Canche JE. Corrosion inhibitory effects of *Annona muricata* leaves extract on copper surfaces in acidic media. In: XXVI International Materials and Research Congress Symposium D.6, NACE: Corrosion and Metallurgy; Cancun. August 2017

[17] Torres PJ. Rodriguez JG, Martinez H. Electrochemical study of a coffee oil amide as a copper corrosion inhibitor. In: XXVI International Materials and Research Congress Symposium D.6, NACE: Corrosion and Metallurgy; Cancun. August 2017

[18] Gomez NB, Porcayo J, Gonzalez JG, Martinez L. Performance of inhibitors derived from spent coffee ground under dynamic conditions. In: XXVI International Materials and Research Congress Symposium D.6, NACE: Corrosion and Metallurgy; Cancun. August 2017

[19] Sotelo O, Valdez S, Porcayo J, Martinez L. Exploitation of avocado waste as green corrosion inhibitor. In: XXVI International Materials and Research Congress Symposium D.6, NACE: Corrosion and Metallurgy; Cancun. August 2017

[20] Salinas G, Martinez L, Porcayo J. Development and evaluation of a green corrosion inhibitor based on rice bran oil as agro-industrial waste. In: XXVI International Materials and Research Congress Symposium D.6, NACE: Corrosion and Metallurgy; Cancun. August 2017

[21] Rojas V, Martinez ME, Posadas A. Electrochemical characterization of corrosion by *Pseudomonas aeruginosa* in API X52 pipe for gas production. In: XXVI International Materials and Research Congress Symposium D.6, NACE: Corrosion and Metallurgy; Cancun. August 2017

[22] Diaz A, Torres A, Martiez W, Guzman EA. Steel electrochemical characterization in cement based mortar with Nopal slime. In: XXVI International Materials and Research Congress Symposium D.6, NACE: Corrosion and Metallurgy; Cancun. August 2017

[23] Garcia R, Valdez B, Kharshan R, Schorr M. Interesting behavior of *Pachycormus* discolor leaves, ethanol extract as a corrosion inhibitor of carbon steel in 1 M HCl: A preliminary study. International Journal of Corrosion. 2012;**2012**:1-8

[24] Garcia R, Valdez B, Schorr M, Ramos R. Aqueous extract of creosote bush (*Larrea tridentata*) leaves as green inhibitors for carbon steel hydrochloric acid solution. International Journal of Electrochemical Science. 2013;**8**:6433-6448

[25] Inzunza RG. Steel corrosion inhibitors of natural extracts for acid environments, PhD thesis, University of Baja California, Mexico. 2014 (Spanish)

[26] Choudary YK, Sabhpondit A, Ranganathan D. Inulin as corrosion inhibitor, U.S. Patent 20120238479A1. 2012

[27] Von FJ. Coatings including tobacco products as corrosion inhibitors WO Patent 151028 A3. 2008

[28] Vaszilesin CG, Dan ML, Duca DA, Labosel MA. Resveratrol as corrosion inhibitor for metals in wine and food industries. In: 22nd International Symposium on Analytical and Environmental Problems. 2017

[29] Oguzzie EE, Adindu CB, Enenebeaku CK, Ogukwe CE, Chidiebere MA, Oguzie KL. Natural products for materials protection: Mechanism of corrosion inhibition of mild steel by acid extracts of *Piper guineense*. Journal of Physical Chemistry C. 2012;**116**:13603-13615

[30] Garcia I. Development of a polymeric coating containing corrosion inhibitors for corrosion control in marine environments, PhD thesis, University of Baja California, Mexico. 2015 (Spanish)

[31] Cheng N, Valdez B, Schorr M, Salinas R. Corrosion inhibitors for prolonged protection of military equipment and vehicles. Materials Performance. 2016;**55**(1):54-57

[32] Corrdefense. Corrdefense-DOD news about preserving military assests [Internet]. 2017. Available from: http://corrdefense.nace.org/ [Accessed: 2017-10-25]

[33] Cheng N, Cheng J, Valdez B, Schorr M, Bastidas JM. Inhibition of seawater steel corrosion via colloid formation. Materials Performance. 2016;**55**(4):48-51

[34] Valdez B, Schorr M. Conference on lubrication and corrosion inhibitors, Latin America Area, Magna International, Singapore and Canada. Materials Performance. 2016;**55**(1)

[35] Salinas R, Schorr M, Valdez B. Corrosion inhibitors for concrete road deicing operations. In: Latincorr Conference; October 2016; Mexico. Report in Materials Performance. 2017. pp. 176-177

[36] Sanchez AN, Taylor CD. Corrosion management of sustainable concrete infrastructure. Materials Performance. 2016;**55**:41-45

Chapter 9

Dimethyl Sulfoxide-Treated Starch of *Dioescorea hispida* as a Green Corrosion Inhibitor for Low Carbon Steel in Sodium Chloride Medium

Norinsan Kamil Othman, Emee Marina Salleh,
Zulhusni Dasuki and Azwan Mat Lazim

Additional information is available at the end of the chapter

http://dx.doi.org/10.5772/intechopen.73552

Abstract

This research was focused on discovering the utilization of *Dioscorea hispida* (*gadong* tuber) starch as a potential green corrosion inhibitor for SAE 1045 carbon steel in 0.6 M NaCl media. Raw starch from *gadong* tubers was obtained through the extraction and precipitation process prior to drying in producing fresh starch powder. The insoluble starch powder was then dissolved in 90% dimethyl sulfoxide (DMSO) solvent and was ready for corrosion inhibitor testing. The extraction of starch powder was analyzed using Fourier Transform Infrared Spectroscopy (FTIR) to identify the presence of functional groups that may contribute to the inhibition of the carbon steel sample. The inhibitive performance of *gadong* tuber starch (GTS) as a green corrosion inhibitor was studied by weight loss measurement and potentiodynamic polarization in a corrosive medium using 0.6 M sodium chloride (NaCl) solution. Effect of GTS concentration on corrosion behavior was investigated. Based on the findings, 1500 ppm of GTS demonstrated the highest corrosion inhibition performance as the corrosion rate of carbon steel immersed in plain 0.6 M NaCl was reduced from 0.131 to 0.018 mm/year, which is up to 86.3 IE% of inhibition efficiency in the presence of the green corrosion inhibitor. According to potentiodynamic polarization analysis, corrosion current density (I_{corr}) was also decreased from 1.13×10^{-5} to 1.55×10^{-6} mA/cm^{-2}.

Keywords: green corrosion inhibitor, DMSO-treated starch, *gadong* tubers, sodium chloride, electrochemical

1. Introduction

Carbon steel is the most common form of steels and because of its low cost, it is the chief material of construction. Moreover, carbon steel is extensively used in many industries such as automotive, pipelines and chemical industries [1]. However, the metal corrodes when it comes in contact with the environment containing chlorides [2, 3], such as when the metals are exposed to seawater. Corrosion study of carbon steel is a fundamental academic and industrial concern that has acquired a considerable amount of interest [4]. Since carbon steel is one of the widely used alloys, with most heavy industries such as construction materials for pipe work in the oil and gas production and engineering materials in the ocean that use low-alloy steels for its piping system in order to transport corrosive media to the storage tank. Note that the corrosion potential of carbon steel in seawater is as low as −0.8 V [5], which means that carbon steel is vulnerable to corrosion attack when in contact with seawater due to the presence of aggressive chloride ions (Cl^-). Over a certain period of time, it attacks a wide range of scales [6].

The application of corrosion inhibitor is extensively known for its capability to prevent metal alloy structure from corrosion. Commonly, corrosion inhibitor is applied mostly in the oil and gas field, piping system and paint coating. [7, 8]. Basically, the ability of the corrosion inhibitor in retarding corrosion is due to the existence of one or more heteroatoms such as sulfur (S), nitrogen (N), oxygen (O) and phosphorus (P). These compounds usually manage to facilitate adsorption on the metal surface and are believed to form a protective thin film on the carbon steel surface. Normally, the efficiency factors for the adsorption properties are related to the nature and the state of substrate surfaces, the type of corrosive medium and chemical structure of the inhibitor [4].

By and large, there are two types of corrosion inhibitors: inorganic and organic corrosion inhibitors. Inorganic compounds such as chromate, dichromate, nitrite and nitrate are commonly used as corrosion inhibitors in several media and for different metals and alloys [5]. However, the biotoxicity of these products, especially chromate, is well documented [6], as well as their non-environmental-friendly characteristics [7], which limit their application. Currently, research in corrosion is oriented to the development of "green corrosion inhibitors", compounds with good inhibition efficiency and low risk of environmental pollution and hazard [8].

Efforts have been paid in the pursuit to replace synthetic corrosion inhibitors with inhibitors obtained from eco-friendly substances. This interest has increased considerably due to the fact that naturally sourced corrosion inhibitor products contain mixtures of compounds having oxygen; sulfuric and nitrogen are eco-friendly, cheap, renewable and ecologically anti-corrosive substances, biodegradable and sourced from renewable materials [9, 10]. The successful uses of naturally occurring substances to inhibit the corrosion of metals in various environments have been reported by several researchers [11, 12, 13, 14].

The choice of optimal inhibitor should be based on its structure, ease of synthesis, cost of raw material and its toxicity in the environment must be negligible. The naturally eco-friendly corrosion inhibitors are indistinguishable due to several drawbacks such as: plant extracts are rich valuable sources of natural compounds that could be extracted simply without economic cost and most of the natural products are easily biodegradable, have no toxicity and have continuous feed sources of materials [11, 13]. Many plant extracts have been used as effective corrosion inhibitors of low carbon steel in corrosive media [15]. Carbohydrate polymers, which represent a set of chemically stable, biodegradable and eco-friendly macromolecules, have been widely used as corrosion inhibitors [16]. The use of biopolymers is gaining ground as inhibitor formulations for various field applications. A number of green corrosion inhibitors of biopolymeric base have successfully been produced (**Table 1**).

Starch is one of the most abundant natural polymer which can be obtained at low cost as only a simple extraction process is involved [17]. It is regarded as an inexpensive, renewable and readily available alternative with essential and active ingredients responsible for the corrosion inhibition. Moreover, starch also falls under the biodegradable source which is a suitable candidate for developing environmentally friendly corrosion inhibitors [18]. This organic corrosion inhibitor would benefit several applications in industries involving seawater, such as for cooling purposes, ship cleaning and desalination plants. Apart from that, starch from *gadong* tuber (*Dioscorea hispida sp.) hauls* huge potential as an organic corrosion inhibitor replacing synthetic inorganic corrosion inhibitors in the market and is harmless to the environment and does not require enormous processing scale.

Corrosion inhibitor	**Finding**	**Inhibition efficiency**	**Authors**
Xanthium strumarium leaves (XSL)	Adsorption of *XSL* extracts obeyed Langmuir adsorption isotherm model which acted as a mixed mode of physical and chemical adsorption, as the free energy of adsorption was more than −20 kJ/mol.	94.82% at mL/L of XSL	Khadom et al. [11]
Gum *Arabic Acacia* (GA)	GA acted as a mixed-typed inhibitor which suppressed both the anodic and cathodic process by physical and chemical adsorption on the steel surface following Langmuir adsorption model.	97% at 1.0 gL^{-1} of GA	Azzaoui et al. [12]
Polysaccharide from *Plantago (AX)*	Inhibition efficiency increased with increasing AX concentration and solution's temperature, which indicated a major role of chemical adsorption. The adsorption of AX on carbon steel surface obeys Langmuir adsorption isotherm. The adsorption process is mixed adsorption and endothermic accompanied by a decrease in entropy.	93.54% at 1000 ppm AX	Mobin and Rizvi [13]
Rollinia occidentalis (RO) extract	*RO* extract might be considered a mixed-type inhibitor since it affected both the anodic and cathodic reactions. Moreover, this behavior can be understood in terms of an inhibition mode resulting from the geometric blocking of the metallic surface by the adsorbed inhibitive species.	85.7% at 1.0 gL^{-1}	Patricia et al. [14]

Table 1. Comparative summary of various natural corrosion inhibitors used in corrosive media containing Cl^- on low carbon steel substrates.

Gadong tuber is a monocotile plant, which is an exotic tuber in our country, Malaysia (as shown in **Figure 1**). In previous years, it had been one of the sources of carbohydrates, but now the tubers are forgotten. Because it contains cyanide components in its alkaloid group, these tubers need to be soaked and cleaned first to remove the poison (*discorina*) before being eaten or used as a medicine. Interestingly, the extract from this tuber has been used as one of the ingredients in the production of poisons for the traditional Malay community in order to fish and control pests for rice cultivation.

There are about 1137 species of *Dioscorea* found throughout the world and scientific studies have begun since 1953 by scientists in the United States. The distribution of this species is widespread, from northern India encompassing South China to New Guinea. It grows extensively in areas such as secondary forests, bushes and sometimes planted in the yard and able to survive in various weather conditions and can survive for years and does not perish easily. There are several trade names for the *Dioscorea hispida* species around the world including *Mitsuda dokoro* (Japan), *Podava* (India) and *Sikapa* (Makassar). *Gadong* tubers grow in soil such as cassava and sweet potato. In the hole or base of trees, there are some round tubers that can grow up to several kilograms, hard skinned and with a fibrous root around them. Its skin is brownish yellow or slightly darker depending on the type of soil. When the *gadong* tuber has ripened in the soils, it turns greenish indicating that it contains high quantities of toxins. Planting methods usually use tubular cuttings and rarely use seed or stem cuttings. **Figure 2** shows the structure of the *Dioscorea hispida sp. plant*.

Since these tubers are no longer used as food source, our research group has exploited and modified the yields of these tubers to a more sophisticated material. In a series of *gadong* tuber studies, as an introduction, our research group introduces the benefits that can be derived from these toxic tubers. This chapter focuses on the development of green corrosion inhibitor from a natural source which is extracted from starch of *gadong* tuber (*Dioscorea hispida sp.*). *the* performance of starch from *gadong* tuber has been tested on a carbon steel sample in 0.6 M NaCl stimulating seawater. Corrosion evaluation was performed by weight loss measurement and electrochemical analysis. The experimental flow is shown in **Figure 3**.

Figure 1. (a) *Gadong* tuber plant and (b) *gadong* tuber.

Figure 2. Parts of *Dioscorea hispida sp.*, 1. Stalks, 2. Tuber, 3. Stem, 4. Branch, 5. Male flowers and 6. Fruit (capsules).

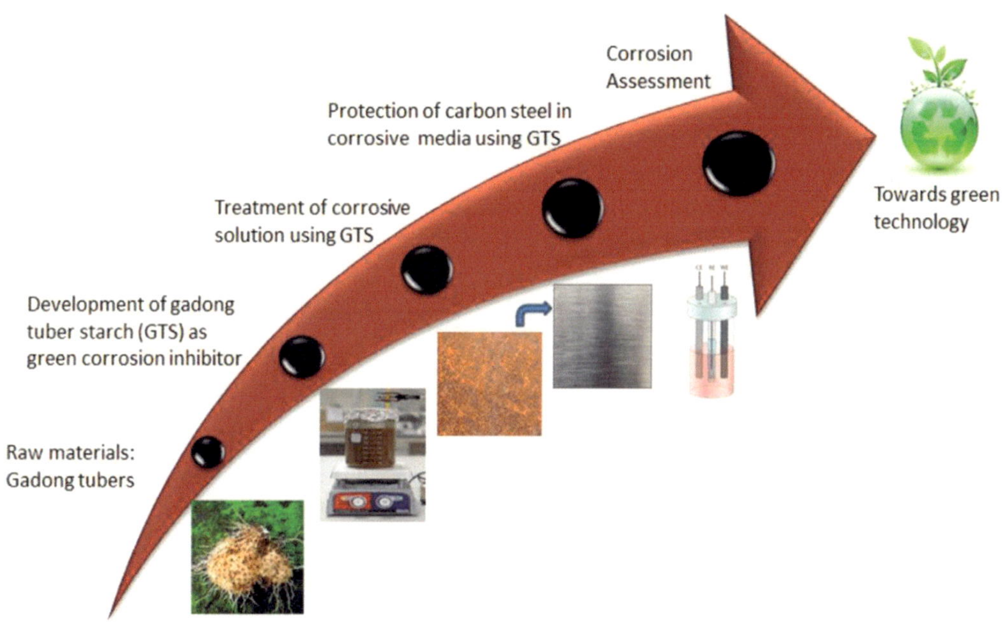

Figure 3. Development of starch from *gadong* tubers as a green corrosion inhibitor.

As for the future, the availability of abundant raw materials from natural sources can be the prime alternative to solve many industrial problems, especially corrosion. By consuming the naturally safe and economic resource, that is, *gadong* tubers, as corrosion inhibitors, the tubers can produce efficient, harmless and inexpensive functional starch. Apart from that, the uses of the starch can be numerous and advance in the future. The three major objectives of this research are the following:

1. To synthesize green corrosion inhibitor based on starch from the *gadong* tuber extract (*Dioscorea hispida sp.*).
2. To determine inhibition efficiency of the natural starch based on corrosion inhibitor in 0.6 M NaCl.
3. To investigate corrosion behavior and surface morphology of low carbon steel exposed in treated simulated seawater using *gadong* tuber extract.

2. Experimental procedure

2.1. Separation of starch

Starch separation method was conducted according to [19, 20] with a slight modification whereby, the freshly obtained tuber was washed with minimum distilled water, peeled and chopped into small pieces using a knife. Then, the chopped tuber was suspended in sufficient amount of distilled water prior to shattering using a blender (Panasonic MX-900) for 5 minutes. Then, the slurry suspension was filtered using a cheese cloth into a container before the filtrate was left for at least 5 hours for the starch to settle at the bottom of the container. The suspension was left to settle for 12 hours at normal room temperature in order to prevent the suspension from spoiling. Next, the supernatant was removed and the starch slurry was dried at 50°C for 2 days to remove all the water content. Finally, the starch powder was collected and stored for further usage. Subsequently, a method established by [21] was followed by dispersing 10% w/w of starch with 90% dimethyl sulfoxide (DMSO)-10% water. Then, the mixture was homogenized by a high-speed homogenizer for 15 minutes to achieve a homogenous brown-yellowish slurry before being diluted to the desired concentration.

2.2. Characterization of corrosion inhibitor

A mobile phase of the liquid chromatography was pumped through 0.5 mm filter into an injector fitted with a 100 ml injection loop to the chromatography column to a photometer with a helium-neon laser source ($\lambda = 632.8$ nm) and a K-5 flow cell. Meanwhile, the frit porosity and particle packing sizes for the column sets were 10 and 20 mm, respectively. The columns were placed in a column heater which was kept at 55°C. The mobile phase was 50 mM and the flow rate was 0.4 ml/min. The functional group of the treated and untreated starch with DMSO was identified by FTIR Spectrum, Perkin Elmer at a range of 400–1200 Hz. The untreated starch was characterized in powder form; meanwhile, the treated starch was in an aqueous form at 3000 ppm. Both the resultant spectra were obtained in order to compare the differences in the spectrum before and after being treated with DMSO.

2.3. Chemical composition of carbon steel used and sample preparation prior to corrosion measurement

The carbon steel used as a test sample was SAE 1045 with a composition of 98.63% Fe, 0.26% C, 0.47% Mn, 0.01% S, 0.01% P, 0.04% Cr, 0.43% Si, 0.02% Ni and 0.13% Al. Before performing morphological observation and corrosion measurement, the samples were ground by abrasive silicon carbide (SiC paper) from 120 to 1200 grit and polished using a lubricant and 1 μm diamond spray on polishing cloth.

2.4. Corrosion assessments

Gravimetric measurement was conducted according to ASTM G-31 Immersion Corrosion Testing of Metals standard. The corrosion behavior of carbon steel specimens with dimensions of 16.0 mm diameter and 3.0 mm thickness was weighed and measured, before being immersed in 240 ml solution containing 0.6 M NaCl. Different concentrations of the corrosion inhibitor ranging from 500 to 2500 ppm were studied during the 7 days of the immersion test. Then, the specimens were taken out and cleaned with cleaning solution as stated in ASTM G1–90 before being reweighed and measured to ascertain weight loss.

The electrochemical measurements were obtained using potentiostat-galvanostat GAMRY 3.2 which was connected to a computer with CMS 105 Gamry Framework Ver. 3.20 software. The three electrode cells were set up using a graphite rod as counter electrode (CE) and a saturated calomel electrode (SCE) as reference electrode. The working electrode (WE) was a carbon steel coupon with an exposed surface area of 2.01 cm^2 immersed in 150 ml of the respective solution. After the WE was immersed in the solution, open circuit potential was established for 30 minutes in order to achieve steady state potential before performing the electrochemical measurement. The current-potential Tafel slope were plotted by changing the electrode potential automatically from −0.25 to +0.25 V with the scanning rate of 0.1667 mV/s. The corrosion potential (E_{corr}) and corrosion density (I_{corr}) were assessed from the interception of the anodic and cathodic section of attained Tafel slopes.

2.5. Morphological study

The morphological study of the carbon steel sample surface was investigated using Ultra High Resolution field emission scanning electron microscopy (FESEM), Merlin. The sample surface was examined after immersion test in corrosive medium using 0.6 M NaCl solution and with the presence of corrosion inhibitor.

3. Results and discussion

3.1. FTIR analysis

Figure 4 shows the FTIR spectrum of extracted starch powder from the *gadong* tubers before and after treatment with DMSO. Through rough observations on the peak, (A) raw starch powder and (B) raw starch powder treated with DMSO solvent, both spectra show similarities at the range of 3700–1500 cm^{-1} bands. They also possessed the same extremely broad bands at 3700 cm^{-1}. These spectra were typically referred to the mixture of two molecular identities of polysaccharides of starch [22].

On the other hand, the raw starch powder extracted from *gadong* tubers (A) showed a large number of adsorption bands in the region of ~1200 cm^{-1}, in which it was referring to the minor component of proteins and lipids in the GTS [23]. Moreover, at the fingerprint region of bands between 900 and 1500 cm^{-1}, bands at 997, 1077, 1149 and 1340 cm^{-1} were identified as -C-O-C- bonds in the anhydroglucose unit. The anhydroglucose unit is the most important chemical building block of polysaccharides of starch without containing water [24].

FTIR spectra of extracted GTS (A) before and after treatment with DMSO (B) displayed that both spectra exhibited a transmittance peak at 3288 cm^{-1} with broader band of spectrum B. This phenomenon was attributed to the complex vibrational stretches associated with free inter- and intra-molecular hydroxyl group (-R-OH). Spectrum B possessed a broader band of hydroxyl group as a result of a high amount of water content during dissolution process with DMSO [11]. Moreover, spectrum B had a base line slope at 1700 cm^{-1} and below which was due to DMSO water content during the dissolution of starch that made the starch granules swell, and almost all the semi-crystalline characteristics of the granules vanished. As the swelling enlarged, the starch granules burst releasing the starch into the solution and solubilizing the starch [25].

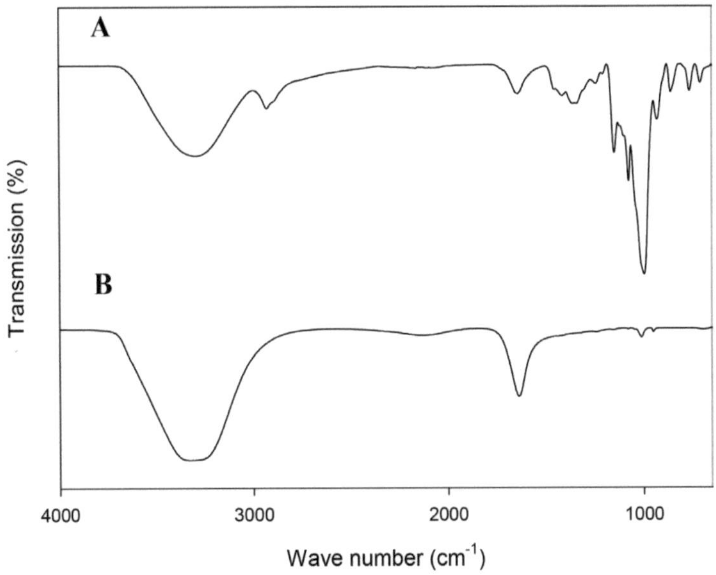

Figure 4. FTIR graph of extracted starch powder (A) before and after being treated with DMSO (B).

3.2. Corrosion measurement: gravimetric analysis

The corrosion behavior of SAE 1045 carbon steel in corrosive medium using 0.6 M NaCl was studied by the gravimetric method for 7 days of immersion, containing the concentrations of GTS ranging from 0 to 2500 ppm as a corrosion inhibitor. The calculated value of corrosion rate (mm/year) immersion test in NaCl medium is tabulated in **Table 2** and **Figure 5**. As displayed in **Figure 2**, corrosion rate of carbon steel in 0.6 M NaCl solution decreased with increasing concentrations of GTS up to 2000 ppm with lowest rate of corrosion at 0.048 mm/year. However, a further addition of GTS (using 2500 ppm) resulted in a slight increase in corrosion rate of 0.051 mm/year. This phenomenon indicated that the GTS might potentially act as corrosion inhibitor by enhancing the corrosion resistance of SAE 1045 carbon steel at a certain concentration of GTS.

Since inhibition efficiency is significantly dependent on the surface degree coverage of the GTS molecule on the surface of the carbon steel surface, the degree of the coverage can be calculated using the following formula [26]:

$$\theta = \frac{C_{R^\circ} - C_R}{C_{R^\circ}} \tag{1}$$

where C_{R° and C_R are the corrosion rates in the absence and presence of GTS as a corrosion inhibitor, respectively. **Table 2** shows the degree of the surface coverage (θ) that increased as a function of GTS concentration. The highest surface coverage of 0.59 was achieved by incorporating 2000 ppm of GTS. The phenomenon explained the improvement of corrosion protection by GTS which was due to the adsorption coverage of the corrosion inhibitor compound on the surface of the metal substrate. The adsorbed compound managed to impede the surface of the substrate from the surrounding corrosive media and thus stopped the corrosion process from taking place [22]. Moreover, the GTS molecules that adsorbed on the surface of the metal substrate interact between the lone pairs of electrons of the oxygen atom and the carbon steel surface, which supported the adsorption process. So, the presence of d π-vacant orbital of low energy in the iron atom also contributed to the adsorption process [27].

Concentration of GTS (ppm)	Corrosion rate (mm/year)	Coverage (θ)
0	0.116	0.00
500	0.108	0.06
1000	0.091	0.21
1500	0.050	0.57
2000	0.048	0.59
2500	0.051	0.56

Table 2. Calculated value of corrosion rate (mm/year), coverage area (θ) and inhibition efficiency (%) of carbon steel SAE 1045 after immersion in 0.6 M NaCl for 7 days.

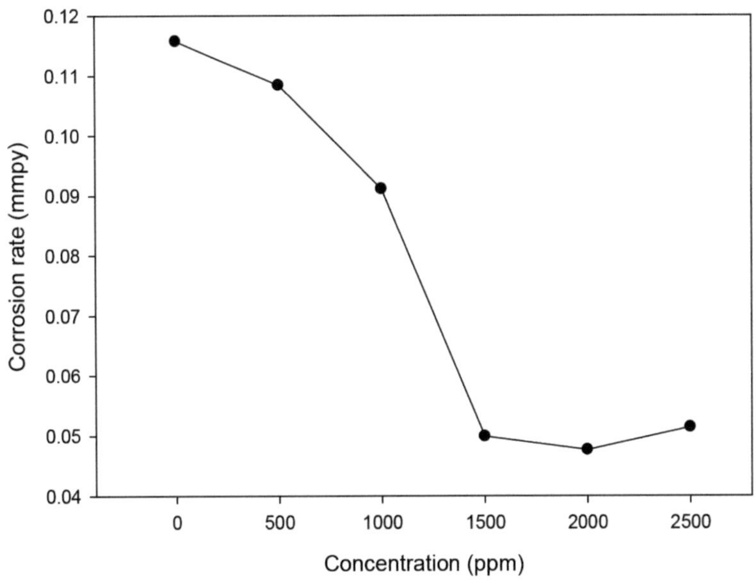

Figure 5. Gravimetric analysis of corrosion rate for various concentrations of *gadong* starch.

Due to the exposure to the aqueous solution, the inhibitor molecules which consisted of polymer compounds started to hydrolyze, inducing the bulk anions and cations in the media to increase and disturb the inhibition process. The hydrolysis of NaCl media caused the NaCl compounds to decompose into Na^+ cations and Cl^- anions. Some of these ions have reacted to water molecules and inhibitor molecules. The Na^+ ions that reacted with OH^- and Cl^- which combined with H^+ cations form a neutral solution or salt as corrosion products.

Apart from that, an increase in oxygen and carbon dioxide concentration in the inhibitor solution as the experiment was done in an open space for a 7-day period of immersion affected the inhibition and corrosion activity of the carbon steel samples. As reported by previous researchers, certain concentrations of inhibitors were affected by the concentration of oxygen and carbon dioxide contained in the inhibitor solution [28, 29]. In this current finding, the optimal concentration of the GTS was 2000 ppm for the 7 days of immersion testing. Further increasing of the starch to 2500 ppm was not significant in inhibting corrosion activity as the corrosion rate increased to 0.051 mm/year.

3.3. Potentiodynamic polarization analysis

The potentiodynamic polarization measurement was used to capture the behavior of corrosion inhibition of GTS on the SAE 1045 carbon steel samples. The transition changes in Tafel slopes could be used to determine whether the GTS acts an anodic, cathodic or mixed type of inhibition process. The polarization curves of the carbon steel in 0.6 M NaCl corrosive medium with various concentrations of GTS are shown in **Figure 6**. According to the Tafel plot, corrosion potential (E_{corr}) of all concentrations of GTS added were mostly shifted towards

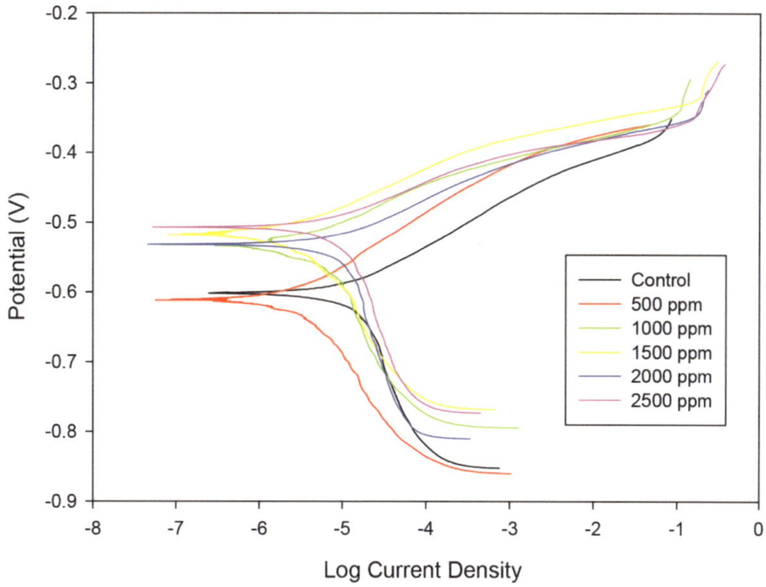

Figure 6. Potentiodynamic polarization curves of SAE 1045 carbon steel samples for various concentrations of GTS in 0.6 M NaCl solution.

a more positive position compared to the blank (control) sample. The E_{corr} shifting to a more positive position indicated that the GTS possessed a potential role in improving corrosion resistance of the carbon steel in treated NaCl solution by reducing the chemical reaction during immersion time [23, 28].

As shown in **Table 3**, E_{corr} of the immersed sample in plain 0.6 M NaCl was −601.4 mV. The E_{corr} was increased to −517.1 mV with an addition of 1500 ppm GTS. However, further increase of GTS to 2000 ppm decreased the E_{corr} which was −531.9 mV but then increased to 507.1 mV with an addition of and 2500 ppm. This fluctuating E_{corr} signified that the corrosion improvement was only achieved at a certain concentration of GTS. In general, maximum E_{corr} shift which is <85 mV demonstrated that the inhibitor is classified as a mixture inhibitor while for E_{corr} shift >85 mV, the inhibitor is a type of actual cathodic or anodic inhibitor [21]. From this finding, by incorporating the GTS into NaCl solution up to 2000 ppm provided maximum E_{corr} shift of 69.5 mV. Thus, it can be proposed that at these concentrations, the GTS is a type of mixture inhibitor as the maximum E_{corr} shift was less than 85 mV [19, 30].

Apparently, the incorporation of GTS offered an inclination to inhibit the corrosion activity by anode as the E_{corr} was shifted towards positive position. However, it was found that with addition of 2500 ppm GTS, this corrosion inhibitor potentially works as an actual anodic and cathodic type as the maximum E_{corr} is 94.3 mV, slightly larger than 85 mV. Due to the shifting plot towards positive position, the incorporation of GTS at this concentration was suggested to act as an anodic inhibitor where the corrosion activity was dominated by anodic reaction rather than cathodic reaction [20].

Concentration of GTS (ppm)	Corrosion potential, E_{corr} (mV)	Corrosion density, I_{corr} ($mAcm^{-2}$)	βa (mV)	βc (mV)	Corrosion rate, (mm/year)	Inhibition efficiency (IE%)	Polarization resistance, R_p
0	−601.4	1.13×10^{-5}	66.9	232.2	0.131	0.00	1993
500	−612.0	2.52×10^{-6}	75.5	146.1	0.029	77.86	8587
1000	−532.8	2.24×10^{-6}	48.0	149.6	0.026	80.15	7036
1500	−517.1	1.55×10^{-6}	48.2	131.6	0.018	86.26	9912
2000	−531.9	6.66×10^{-6}	50.3	259.9	0.077	41.22	2750
2500	−507.1	5.93×10^{-6}	47.0	220.3	0.069	47.33	2836

Table 3. Electrochemical parameters obtained from potentiodynamic polarization analysis.

I_{corr} was reduced from 1.13×10^{-5} of plain 0.6 M NaCl to 1.55×10^{-6} mA/cm^2 in the presence of 1500 ppm GTS. This situation was in agreement with the corrosion rates of the tested materials, where the lowest corrosion rate (0.018 mm/year) was provided by treating the corrosive medium with 1500 ppm GTS. Increasing GTS concentration slightly increased the I_{corr}, resulting in an increase in the corrosion rate. However, the I_{corr} of the samples immersed in GTS-treated NaCl was considerably lower than the sample immersed in non-treated solution. The reduction of I_{corr} shows that the corrosion inhibitor adsorbed on the surfaces expressively protects the surfaces from further corrosion attack; thus, the corrosion rate was potentially reduced.

In addition, there was no consistent change of βa and βc for the tested samples immersed in various concentrations of the incorporated GTS. This phenomenon signified that the corrosion inhibitor does not stop the corrosion activity but significantly retards the chemical reaction on the exposed surfaces. The inhibition reaction that occurred was due to a restriction mechanism towards active site of chemical reaction on the coupon's surface [24]. A maximum inhibition efficiency was achieved at 86.26 IE% with the addition of 1500 ppm GTS. This result proved that GTS significantly improved the corrosion resistance of SAE 1045 carbon steel in 0.6 M NaCl corrosive media.

Chemically, the carbon steel (Fe^{2+}) surrounded by the presence of NaCl begins to lose electrons, thus becoming a cation in solution:

Oxidation reaction: Fe is oxidized (loses electrons).

$$Fe_{(S)} \rightarrow Fe^{2+}{}_{(aq)} + 2\,e^{-} \quad (2)$$

The lost electrons are now used to decrease the chlorides dissolved in the simulated seawater (the water will contain some Na^{+} and Cl^{-} ions):

$$NaCl \rightarrow Na^{+} + Cl^{-} \quad (3)$$

Reduction reaction: Cl is decreased (gains electrons).

The Fe^{2+} ions then chemically react and bond with the OH^- ions in water to create iron hydroxide, which eventually dries up to form rust:

$$Fe^{2+}_{(aq)} + 2Cl^-_{(aq)} \rightarrow FeCl_{2\,(s)}\ (salt) \tag{4}$$

The highest inhibition efficiency of 86.26 IE% was provided by incorporating 1500 ppm GTS into 0.6 M NaCl medium. This phenomenon is attributed to the optimum adsorption of natural starch corrosion inhibitor's molecules that managed to cover the surface of carbon steel, in which it makes a split-up between the surface of carbon steel samples and the corrosive media. In addition, the presence of Na^+ cations and Cl^- anions also has fortified the corrosion inhibition mechanism of carbon steel samples in seawater [30, 31]. This scenario was supported by the formation of stable $FeCl_2$ salt as a corrosion product.

Figure 7 shows the SEM images and EDX characterization on carbon steel samples after immersion in 0.6 M NaCl for 7 days. The carbon steel immersed in plain NaCl shows a rough and thick layer of corrosion product. A thick deposition layer that covered most of the carbon steel surface was identified as ferrum (III) oxide (Fe_2O_3), which would easily fall off the surface, exposing a black layer (magnetite) underneath, which was due to corrosion process of steel [32]. This phenomenon was confirmed by EDX analysis that ferrum and oxygen elements were present on the surface of the carbon steel as a result of the reduction reaction of oxygen that originated from aqueous NaCl solution. The reduction reactions are as follows [33, 34]:

$$O^-_{2\,(aq)} + 2H_2O_{(aq)} + 4e^- \rightarrow 4OH^- \tag{5}$$

The Fe^{2+} ions then chemically react and bond with the OH^- ions in water to create iron hydroxide, which eventually dries up to form rust: (Iron (III) hydroxide).

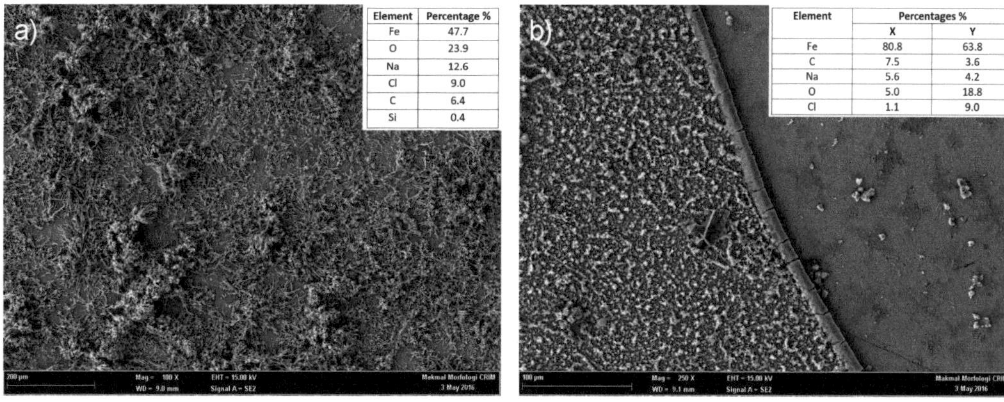

Figure 7. SEM images and EDX analysis of SAE 1045 carbon steels samples (a) without the presence of *gadong* starch (control) and (b) with the presence of 1500 ppm *gadong* starch as corrosion inhibitor after being immersed in 0.6 M NaCl after 7 days.

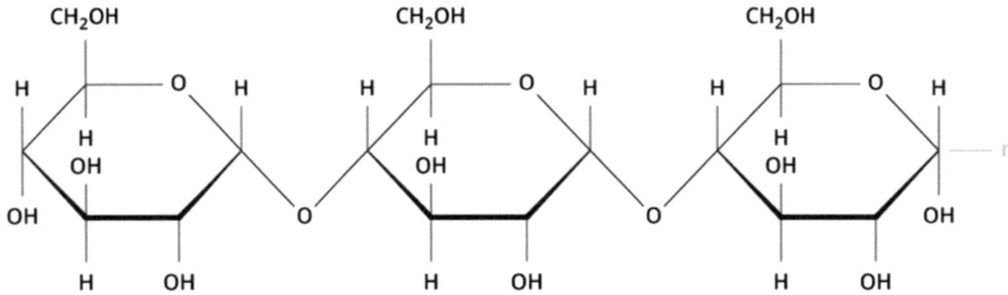

Figure 8. Basic chemical structure of polymeric starch.

$$Fe^{2+}_{(aq)} + 2OH^{-}_{(aq)} \rightarrow Fe\,(OH)_{2\,(aq)} \rightarrow Fe_2O_{3\,(s)}\ \text{(rust)} \quad (6)$$

The factors that will speed up the rate of corrosion include an increase in the amount of oxygen, temperature, chemical salts and humidity. However, as illustrated in **Figure 4(b)**, the carbon steel immersed in NaCl solution containing GTS as the corrosion inhibitor at 1500 ppm concentration shows most of the carbon steel surface with smooth terrain with small area of rough surface of corrosion product. This scenario signified that the corrosion resistance of carbon steel was remarkably enhanced with the presence of GTS as a corrosion inhibitor, with respect to its smoothness and less-corrosion product formed, indicating the reduction of corrosion rate.

The polysaccharides of starch were amylose, a long linear molecule, and amylopection, a highly branched version of amylose molecule [13]. The basic chemical formula of the starch molecule is $(C_6H_{10}O_5)_n$. Starch is a polysaccharide comprising glucose monomers joined in α 1,4 linkages (**Figure 8**). The corrosion inhibition activity of GTS might be due to the presence of various organic compounds in its composition including anhydroglucose unit of polysaccharides (-C-O-C- bonds) and hydroxyl group (-R-OH). These organic compounds contain the most important chemical building block of polysaccharides of starch and complex vibrational stretches associated with free inter- and intra-molecular hydroxyl groups which are the major adsorption centers [33]. Consequently, we suggest that the protective films may be formed on the metal surface through the adsorption process to inhibit the corrosion of mild carbon steel in artificial seawater [34].

The study showed that the inhibition of GTS molecules on the carbon steel surface in artificial seawater is a mixed inhibition, which might result either by electrostatic interaction of the polymer with positively charged carbon steel surface or through the lone pairs of electrons on oxygen atoms [21]. In the beginning, the protonated GTS molecules electrostatically interacted with the positively charged carbon steel. Then, the adsorbed GTS molecules began to participate with H^+ ions for electrons on the carbon steel surface [9]. This scenario resulted in the release of H_2 gas (due to aqueous state of NaCl solution) and GTS molecules returned to their neutral form. The carbon steel surface was reduced to be more negative which attributed to accumulation of electrons. Accordingly, an electron from the d-orbital of Fe might be transferred to a vacant π^* (anti-bonding) orbital of the GTS molecules to relieve the steel from extra negative charge and hence strengthen the corrosion inhibition of GTS via adsorption on the steel surface [28].

Natural products	Metal exposed	Test media	Concentration	IE%	References
Gadong tuber	SAE 1045 steel	0.6 M NaCl	1.5 g L^{-1}	86.26	Current study
Saffron	Copper	Natural seawater	0.1 g L^{-1}	67.00	[35]
Tobacco rob	Q235 steel	Artificial seawater	0.1 gL^{-1}	83.90	[36]
Santolina chamaecyparissus	304 stainless steel	3.5% NaCl	1.0 gL^{-1}	86.90	[37]
Lupine seed	Aluminum	2 M NaOH +0.5 M NaCl	0.16 gL^{-1}	45.90	[38]
Burley tobacco	Steel/Al	3.5% NaCl	0.1 gL^{-1}	~90.0	[39]
Tobacco	Aluminum	3.5% NaCl	1.0 wt%	96.60	[40]
Olive leaf	Copper	0.5 M NaCl	2.42 mmol L^{-1}	90.0	[41]

Table 4. Comparison of the inhibition efficiency of GTS with other natural products extracts.

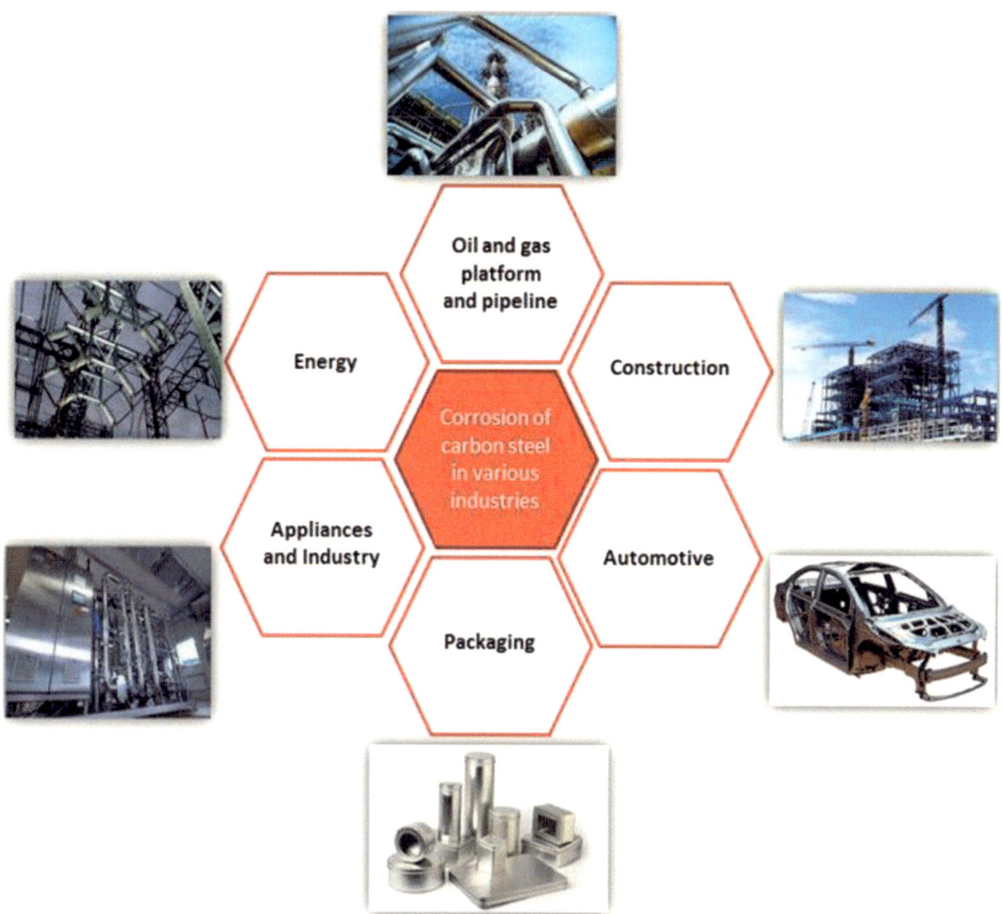

Figure 9. Schematic illustration of potential uses of green GTS in various industries.

Table 4 shows a comparison of the inhibition efficiency of currently produced GTS with other natural product extracts. The inhibition efficiency, IE% of GTS is 86.26 IE% by incorporating 1.5 gL^{-1} GTS on low carbon steel in artificial seawater was considerably comparable to other natural products. However, the inhibition efficiency of Saffron (0.10 g L^{-1}) was only 67.0% in natural seawater. The inhibition efficiency of Tobacco rob (0.1 gL^{-1}) on medium carbon steel, *Santolina chamaecyparissus* (1.0 g L^{-1}) on stainless steel, Lupine seed (0.16 g L^{-1}) on aluminum, Burley tobacco (0.1 gL^{-1}) on steel/Al, tobacco (1.0 wt%) on aluminum and olive leaf (2.42 $mmolL^{-1}$) on copper were 83.90, 86.90, 45.90, ~90.00, 96.60 and 90.0 IE%, respectively. A comparison of these data with the results of previous investigation indicated that GTS has good inhibition performance in artificial seawater. This finding proposed that the green GTS can be used in numerous industries suffering from corrosion attacks as shown in **Figure 9**.

4. Conclusion

Starch extracted from *Dioescorea hispida* sp. is an exceptional corrosion inhibitor for SAE 1045 carbon steel in 0.6 M NaCl solution. Through potentiodynamic polarization analysis, the GTS as corrosion inhibitor was identified as mixed type of corrosion inhibitor which was more prone towards anodic inhibition process. A maximum inhibition efficiency of 86.26 IE% was achieved with incorporation of 1500 ppm of GTS. Lower IE% at higher concentrations of GTS was due to excess amount of the starch molecules that failed to inhibit and alter the corrosive media from corroding the carbon steel sample. SEM images of the carbon steel samples upon immersion in NaCl media showed less-corrosion products formed on the steel surface upon the addition of the GTS.

Author details

Norinsan Kamil Othman*, Emee Marina Salleh, Zulhusni Dasuki and Azwan Mat Lazim

*Address all correspondence to: insan@ukm.edu.my

School of Applied Physics, Faculty of Science and Technology, Universiti Kebangsaan Malayasia (The National University of Malaysia), Bangi, Selangor, Malaysia

References

[1] Bentiss F, Lagrenee M, Traisnel M. 2,5-Bis(n-Pyridyl)-1,3,4-Oxadiazoles as corrosion inhibitors for mild steel in acidic media. Corrosion. 2000;**56**(7):733-742

[2] Yu D et al. Anticorrosion behavior of Superhydrophobic composite coating on carbon steel in seawater. Corrosion. 2013;**70**(4):329-336

[3] Idris MN, Daud AR, Othman NK. Electrochemical impedance spectroscopy study on corrosion inhibition of benzyltriethylammonium chloride. In: THE 2013 UKM FST POSTGRADUATE COLLOQUIUM: Proceedings of the Universiti Kebangsaan Malaysia, Faculty of Science and Technology 2013 Postgraduate Colloquium. 2013. AIP Publishing

[4] Cruz J et al. Experimental and theoretical study of 1-(2-ethylamino)-2-methylimidazoline as an inhibitor of carbon steel corrosion in acid media. Journal of Electroanalytical Chemistry. 2004;**566**(1):111-121

[5] Raja PB, Sethuraman MG. Natural products as corrosion inhibitor for metals in corrosive media — A review. Materials Letters. 2008;**62**(1):113-116

[6] Bello M et al. Modified cassava starches as corrosion inhibitors of carbon steel: An electrochemical and morphological approach. Carbohydrate Polymers. 2010;**82**(3):561-568

[7] Olusegun KA, James AO. The effects of Aloe Vera extract on corrosion and kinetics of corrosion process of zinc in HCl solution. Corrosion Science. 2010;**52**:661-664

[8] Alves A, Schwanninger M, Pereira H. Analytical pyrolysis as a direct method to determine the lignin content in wood part 1: Comparison of pyrolysis lignin with Klason lignin. Journal of Analytical and Applied Pyrolysis. 2006;**7**:209-213

[9] Al-Saade KA, Abas HA. Corrosion inhibition of hexamine for galvanized steel in hydrochloric acid solution. International Interdisciplinary Research Journal. 2013;**3**:79-90

[10] Amin. A newly synthesized glycine derivative to control uniform and pitting corrosion processes of al induced by SCN-anions-chemical, electrochemical and morphological studies. Corrosion Science. 2010;**52**:3243-3257

[11] Khadom AA, Abd AN, Ahmed NA. Xanthium Strumarium leaves extracts as a friendly corrosion inhibitor of low carbon steel in hydrochloric acid: Kinetics and mathematical studies. South African Journal of Chemical Engineering. June 2018;**25**:13-21. DOI: 10.1016/j.sajce.2017.11.002

[12] Azzaoui K, Mejdoubi E, Jodeh S, Lamhamdi A, Rodriguez-Castell'on E, Algarra M, Zarrouk A, Errich A, Salghi R, Lgaz H. Eco friendly green inhibitor gum arabic (GA) for the corrosion control of mild steel in hydrochloric acid medium. Corrosion Science. December 2017;**129**:70-81. doi.org/10.1016/j.corsci.2017.09.027

[13] Mobin M, Rizvi M. Polysaccharide from Plantago as a Green Corrosion Inhibitor for Carbon Steel in 1m Hcl Solution. 2016. DOI: 10.1016/j.carbpol.2016.12.056

[14] Patricia E, Alvarez M, Victoria FB, Adriana N, Silvia AB, Claudio AG. Rollinia occidentalis extract as green corrosion inhibitor for carbon steel in HCl solution. December 2017;**129**: 70-81. doi.org/10.1016/j.jiec.2017.09.012

[15] Anejjar A, Salghi R, Zarrouk A, Benali O, Zarrok H, Hammouti B, Ebenso EE. Inhibition of carbon steel corrosion in 1 M HCl medium by potassium thiocyanate. Journal of the Association of Arab Universities Basic and Applied Sciences. 2014;**15**:21-27

[16] Sakakibara I, Katsuhara T, Ikeya Y, Hayashi K, Mitsuhashi H. Cannabisin a, an arylnaphthalene lignanamide from fruits of *Cannabis sativa*. Phytochemistry. 1991;**30**:3013-3016

[17] Vanquez G, Antorrena G, Gonzalez J, Freire S. FTIR, H-1 and C-13 NMR characterization of acetosolv-soloubilized pine and eucalyptus lignins. Holzforschung. 1997;**51**:158-166

[18] Cyran MR, Saulnier L. Association and structural diversity of hemicelluloses in the cell walls of rye outer layers: Comparison between two ryes with opposite breadmaking quality. Journal of Agricultural and Food Chemistry. 2007;**55**:2329-2341

[19] Nand AV et al. Isolation and properties of starch from some local cultivars of cassava and taro in Fiji. The South Pacific Journal of Natural and Applied Sciences. 2008;**26**(1):45-48

[20] Ashri A et al. Physicochemical characterization of starch extracted from Malaysian wild yam (*Dioscorea hispida* Dennst.). Emirates Journal of Food and Agriculture. 2014;**26**(8):652

[21] Shon K-J, Lim S-T, Yoo B. Rheological properties of Rice starch dispersions in dimethyl Sulfoxide. Starch-Starke. 2005;**57**(8):363-369

[22] Rosliza R, Nik WW. Improvement of corrosion resistance of AA6061 alloy by tapioca starch in seawater. Current Applied Physics. 2010;**10**(1):221-229

[23] Wu Y et al. Effect of agar on the microstructure and performance of potato starch film. Carbohydrate Polymers. 2009;**76**(2):299-304

[24] Ochoa N et al. Modified cassava starches as potential corrosion inhibitors for sustainable development. Materials Research. 2013;**16**:1209-1219

[25] Dona A et al. A new NMR method for directly monitoring and quantifying the dissolution kinetics of starch in DMSO. Carbohydrate Research. 2007;**342**(17):2604-2610

[26] Khaled K. Molecular simulation, quantum chemical calculations and electrochemical studies for inhibition of mild steel by triazoles. Electrochimica Acta. 2008;**53**(9):3484-3492

[27] Mobin M, Khan M, Parveen M. Inhibition of mild steel corrosion in acidic medium using starch and surfactants additives. Journal of Applied Polymer Science. 2011;**121**(3): 1558-1565

[28] Eliyan FF, Alfantazi A. Mechanisms of corrosion and electrochemical significance of metallurgy and environment with corrosion of iron and steel in bicarbonate and carbonate solutions—A review. Corrosion. 2014;**70**(9):880-898

[29] Ghali E, Sastri VS, Elboujdaini M. Corrosion Prevention and Protection: Practical Solutions. John Wiley & Sons.; 2007

[30] Moller H, Boshoff E, Froneman H. The corrosion behaviour of a low carbon steel in natural and synthetic seawaters. Journal of the Southern African Institute of Mining and Metallurgy. 2006;**106**(8):585

[31] Hortling B, Tamminen T, Kentta E. Determination of carboxyl and non-conjugated carbonyl groups in dissolved and residual lignins by IR spectroscopy. Holzforschung. 1997;**51**: 405-410

[32] Abiola OK, Oforka NC. Corrosion Inhibition effect of Cocos Nuciferal juice on mild steel in 5% hydrochloric acid solution. Sci. Africa. 2003;**2**:82

[33] Danaee I, Khomami MN, Attar A. Corrosion behavior of AISI 4130 steel alloy in ethylene glycol–water mixture in presence of molybdate. Materials Chemistry and Physics. 2012;**135**:658-667

[34] Adapa PK, Karunakaran C, Tabil LG, Schoenau GJ. Potential applications of infrared and raman spectromicroscopy for agricultural biomass. Agriculture Engineering International. 2009;**1081**:1-25

[35] Oukhrib R, El Issami S, Chaouay A, El Mouaden K, Jmiai A, Ibrahimi B, Bazzi L, Bammou L, Hilali M. The inhibitory effect of saffron extract (Crocus sativus., L) on copper corrosion in seawater. Chemical Science Review Letters. 2015;**4**:241-251

[36] Wang H, Gao M, Guo Y, Yang Y, Hu R. A natural extract of tobacco rob as scale and corrosion inhibitor in artificial seawater. Desalination. 2016;**398**:198-207

[37] Shabani-Nooshabadi M, Ghandchi MS. *Santolina chamaecyparissus* extract as a natural source inhibitor for 304 stainless steel corrosion in 3.5% NaCl. Journal of Industrial and Engineering Chemistry. 2015. doi.org/doi:10.1016/j.jiec.2015.06.028

[38] Abdel-Gaber A, Khamis E, Abo-Eldahab H, Adeel S. Novel package for inhibition of aluminium corrosion in alkaline solutions. Materials Chemistry and Physics. 2010;**124**:773-779

[39] Fouda A, Elewady G, Shalabi K, Habouba S. Tobacco plant extracts as save corrosion inhibitor for carbon steel in hydrochloric acid solutions. International Journal of Advanced Research. 2014;**2**:817-832

[40] Von Fraunhofer JA. Coatings including Tobacco Products as Corrosion Inhibitors. U.S. 2008. Patent No. 20080295728A1

[41] Rahal C, Masmoudi M, Sabot RAR, Jeannin M, Bouaziz M, Refait P. Olive leaf extract as natural corrosion inhibitor for pure copper in 0.5 M NaCl solution: A study by voltammetry around OCP. J. Electroanal. Chem. 2016. DOI: 10.1016/j.jelechem.2016.03.010

Chapter 10

Theobroma cacao Peel Extract as the Eco-Friendly Corrosion Inhibitor for Mild Steel

Yuli Yetri, Gunawarman, Emriadi and
Novesar Jamarun

Additional information is available at the end of the chapter

http://dx.doi.org/10.5772/intechopen.73263

Abstract

Corrosion control for metals has technical, economical, environmental, and aesthetical importance. The use of corrosion inhibitors from natural product is one of the best ways to slow the rate of corrosion of metals and alloys. The use of extract inhibitors from natural product is the best alternative because it does not contain heavy metals and is nontoxic, environmentally friendly, biodegradable, easy to obtain, inexpensive, and available in large quantities. The organic compounds contained in the extracts of natural products such as tannins, amino acids, phenolics, and alkaloids contain heteroatomic groups, which can inhibit corrosion rate and are interesting to study. With the development of science and technology, many methods are used to protect the corrosion, such as electroplating and coating, and organic inhibitor shows real promise such as *Theobroma cacao* peel extract. Although substantial research has been devoted to corrosion inhibition by plant extracts, reports on the detailed mechanisms of the adsorption process and identification of the active ingredient are still scarce. The addition of polar extract of cacao peel into a solution of HCl 1.5 M is very effective to reduce the attack surface corrosion on mild steel and can maintain its mechanical properties. This chapter consciously restricts itself mainly to plant materials, especially *Theobroma cacao* peel extract, as green corrosion inhibitors.

Keywords: *Theobroma cacao*, corrosion inhibitor, plant product, coating

1. Introduction

Nowadays, steels are applied in a variety of needs, like constructions, buildings, bridges, cars, industry tools and installation such as water treatment, oil and waste. The steel is chosen as it is accessible and easily fabricated with high tensile strength. Nevertheless, the most weaknesses of this steel are easily undergone corrosion in corrosive medium [1], like in acid environmental condition during the acid cleaning process, acid solution transportation, storage, de-scaling, and pickling process. The frequently used acid solution in pickling process by industry is hydrochloride acid [2].

Corrosion problem is a major and serious one as it is related with the working safety, economics, environmental damage, and health. Generally, industry takes the above-average steel, which is made from mild steel. The lost caused by this corrosion in Indonesia is huge, reaching 1.5 billion rupiah per year. Some estimate that this loss is up to 2–3.5% of the GNP. This cost included the maintenance expense, the replacement for the material, working hours, and the lost benefits due to the stop of production, customer dissatisfaction, administration fee, physical lost, and health treatment [3].

In order to reduce the loss because of the corrosion on the steel, the protection for the steel should be done. Many attempts have been made to slow the corrosion rate such as coating, material selection, changing the environmental condition, design selection, painting, anodic and cathodic protection, and inhibitor substance addition into corrosive medium [4]. A frequently used way in preventing the corrosion is taking a particular inhibitor. This corrosion inhibitor derives from organic and inorganic substances which consist of groups of free electron or substances with heteroatom group of P, N, S, O like nitrate, chromate, phosphate, urea, phenylalanine, imidazoline, and amine compounds for inorganic one. The functional group within this compound can be adsorbed on the steel surface, and then, it is bonded to one another, forming the film layers, which can inhibit corrosion reaction. In fact, synthetic material that is used as inhibitors is dangerous to living organisms, relatively high in price, and environmentally harmful [5]. Organic compound like polyphenol derives from natural extract or natural products. They are an easy-to-use natural material which can be used as an alternate inhibitor as it is safe, readily available, biodegradable, low cost, and eco-friendly [2, 6].

Recently, there have been many researches that focus on finding a new source for corrosion inhibitor particularly on plants. Parts of the plants that can be taken for the research are leaves, barks, fruits, heartwoods, or roots. Several extracts of natural product that have been studied as eco-friendly corrosion inhibitors can be seen in **Table 1**.

On another side, Indonesia is well known as the largest cacao producer in the world, certainly resulting in the largest solid waste as well. The production of this waste reaches 60–70% of the total fruit production or six times of the dried cacao peels production [24]. In 2013, cacao production was 800,000 ton with 600,000 ton waste. By having the National Cacao Movement in the area of 1.6 million acres, it is estimated that in 2014, the production increases into 1.1 million ton. The raise in production is in the synergy with the raise of dried cacao peels into 825,000 ton [25]. As the government program becomes priority to recycle the waste into useful products, the use of this cacao peel waste becomes the corrosion inhibitor, which is one of the alternatives to reduce the pollution. Cacao peel extract contains several chemical compounds such as flavonoid or condensed tannin like antosianin, tannin, catechin, and leucoanthocyanidin. Tannin is a complex organic compound which consists of polyphenol and has a number of hydroxy groups where sometimes it occurs in the form of glycosides. Tannin is known as tannic acid or galatanat acid. As the inhibitor, it can replace the use of lead oxide or chromate, which is frequently added into paint for protecting steel corrosion [23]. Not many has paid attention on the use of cacao peel extract which acts as the inhibitor on mild steel, and it is documented, whereas its waste is readily available in great number. Thus, in order to balance

No.	Metal	Inhibitor source	Media corrosive	Inhibition efficiency (%)	Reference
1	Mild steel	*Lawsonia inermis*	HCl	92.06	[2]
2	Steel	*Azadirachta indica* leaves	HCl	60.40	[7]
3	Mild steel	*Palicourea guianensis*	HCl	89.00	[8]
4	Mild steel	*Murraya koenigii leaves*	HCl, H_2SO_4	94.50	[9]
5	Steel	*Phyllanthus amarus*	H_2SO_4	98.00	[10]
6	Carbon steel	*Pachycormus* leaves	HCl	93.50	[11]
7	Mild steel	*Emblica officinalis* leaves	HCl	88.60	[12]
8	Mild steel	*Euphorbia hirta* leaves	H_2SO_4	90.90	[13]
9	Mild steel	*Ananas comosus* L	HCl	91.02	[14]
11	Steel	*Prosopis cineraria*	HCl	86.67	[15]
12	Mild steel	*Uncaria gambir*	HCl	86.45	[16]
13	Mild steel	*Acacia Senegal*	HCl	90.32	[17]
14	Mild steel	*Argan hulls*	HCl	97.30	[18]
15	Mild steel	*Centella asiatica*	HCl	95.08	[19]
16	Mild steel	*Laurus nobilis*	H_2SO_4	87.60	[20]
17	Mild steel	*Musa acuminate*	HCl	96.00	[21]
18	Mild steel	*Dodonaea viscose*	HCl	95.40	[22]
19	Mild steel	*Theobroma cacao peels*	HCl	96.26	[23]

Table 1. Green inhibitors used for corrosion inhibition of metal.

and preserve the nature, the use of this waste should be applied more in order to be useful, like to be the eco-friendly inhibitor.

The object of this study is mild steel. It can be found on public needs for example railings, concrete bailout, and others; however, it easily undergoes corrosion. The corrosive medium used in this study is hydrochloric acid, and cacao peel extract acts as the inhibitor which slows down the corrosion rate.

Inhibition efficiency on cacao peel extract in chloride acid medium is set by applying the weight loss method, whereas its inhibition mechanism was learned based on Gibbs adsorption free energy analysis (ΔG_{ads}), the value of activation energy (Ea), SEM, EDX, FT-IR corrosion product spectrum analysis, and electrochemical parameters which derive from potentiodynamic polarization method and electrochemical impedance spectroscopy (EIS). However, for its inhibition effect on mechanical property, the hardness and tensile tests are applied.

2. Corrosion inhibitor

Corrosion inhibitor is the solution which is added in inconsiderable quantity to slow down or to prevent the reaction between the steel with its medium. The inhibitor has a role to form the barrier from one or several molecular layers on the acid attack, salt, and other corrosive environment [26]. Sulfur and nitrogen containing heterocyclic with their varied substituent are effective as the corrosion inhibitor. Inorganic compounds such as phosphate, chromate, dichromate, silicate, borate, molybdate, and arsenate compound have been found effective as the steel corrosion inhibitor. Pyrrole and its derivatives are shown as the best protection to corrosion in acidic medium. The major weakness of this inorganic inhibitor is its toxicity, inorganic products, difficulty to degrade, and containing heavy metal. The safest alternative in reducing the corrosion rate is by using inorganic compound which consists of polar function group like nitrogen, sulfur, and oxygen, where they can be used as an effective inhibitor [27]. This corrosion inhibitor works effectively to prevent the corrosion rate in various ways, and the inhibitor adsorption on the steel surface is one of the mechanisms for anodic corrosion inhibitor. The adsorption that occurred is physisorption and chemisorption or chemisorption, which is proceeded by physisorption. However, the adsorption occurred is up to the inhibitor structure itself, the type of the steel, and the media condition (pH, temperature, and inhibitor's number) [28].

The plant extract contains heteroatom organic compound that can slow down the corrosion rate; meanwhile, O, N, S, and P are the active center of adsorption process on steel layer [27, 29, 30]. The presence of free pair of electron in inhibitor molecule causes the electron transfer from inhibitor toward the steel surface as the coordination of covalent bond is formed. The characteristics of this compound inhibition are derived from the molecular adsorption ability, and polar group acts as the reaction center on adsorption process. The resulted layers of adsorption separate the steel from the corrosion media [31, 32].

Nevertheless, the nutrition in the cacao peels is considered high enough, and the waste of cacao production also contains alkaloid theobromine compounds (3,7 dimethyl xanthine), flavonoid or condensed tannin such as antosianin, tannin, katekin, and leucoanthocyanidin [33–38]. Tannin, which is present in cacao peels, contains active group functioned as corrosion inhibitor [36].

Tannin compound is able to form complex with steel(II) and steel(III). Steel(II) complex tannin is colorless, dissolved, and oxidized easily. In the presence of oxygen, it turns this complex into steel(III) complex tannin which is called tannat. The newly formed compound is well known as organometallic surface complex. It is the one that clings onto the steel surface and prevents the further corrosion process as this complex is absorbed onto the steel surface and protects it [33].

3. Research method

The equipment for this study is as follows: measuring cups, analytical balance, digital caliper, grindstone, oven, Heidolph WB 2000 *rotary evaporator, hot plate,* cutting machine, sand paper with the roughness of 80, 120, 400, 600, 1000, 1500 mesh, 3 mm drilling machine, electron microscope, *Foundry-Master Xpert Spectrometer,* GCMS-QP2010S SHIMADZU, LC–MS-QMicro QAA 842, *Potensiostat* eDAQ, Potentiostat/Galvanostate AUTOLAB PGSTAT 320 N *Electrochemical Impedance Spectroscopy (EIS), Perkin Elmer System 2000 Fourier Transform Infrared Spectroscopy (FTIR), Hitachi S3400 N Scanning Electron Microscopy (SEM)* and *SEM–EDX* by using *EMAX software, Universal Testing Machine type RAT-30P CAP,* and *Rockwell Hardness Tester TH 550.*

4. Results and discussion

4.1. Analysis of chemical composition and microstructure of mild steel

The chemical composition of used mild steel in this study is 0.32% C, 97.8% Fe, 0.22% Si, 0.10% Cr, 0.20% Mo, 0.90% Mn, 0.06% S, 0.30% Cu, and 0.07% P. It is referred to AISI 1035 (ASM) steel standard and supported by microstructure of mild steel testing as seen in **Figure 1**. This microstructure consists of the white color ferrite and the black color pearlite with a ratio of 1:2. This condition shows that the mild steel used in this study contains lesser Fe_3C than pure Fe; therefore, it is resilient [39].

4.2. Analysis of methanol extract of cacao peels

The testing result of phytochemical shows that cacao peel extract contains secondary metabolite compound such as alkaloid, terpenoid, phenolic, flavonoid, and steroid. Further identification is analyzed by using GC–MS. GC–MS spectra result before fractionation indicates that cacao peel extract contains of 42 dominant compounds (spectra peak). To identify the inhabitance of polar compound in cacao peel extract that acts as inhibitor, further analysis using LC–MS is carried out. The LC–MS identification results 10 dominant peaks as seen in **Figure 2** and **Table 2**. **Table 2** displays the compounds which are identified by LC–MS containing compounds with higher molecular weight referred to fragmented ion and molecule ion from the testing data [40]. These compounds have heteroatom group which functions as the corrosion inhibitor as well [26].

4.3. Analysis of Fourier transform infrared

FTIR analysis is carried out on cacao peels, free extract corrosion product, and corrosion product with cacao peel extract in HCl 1.5 M solution. Several peaks on those three spectra undergo friction; however, amount of peaks are occurred on almost similar frequency or adjacent. Identified functional group from cacao peel extract is phenolic, aromatic ring, and

Figure 1. Microstructure of mild steel sample displayed: (a) 20×, (b) 50×, and (c) 100× magnification.

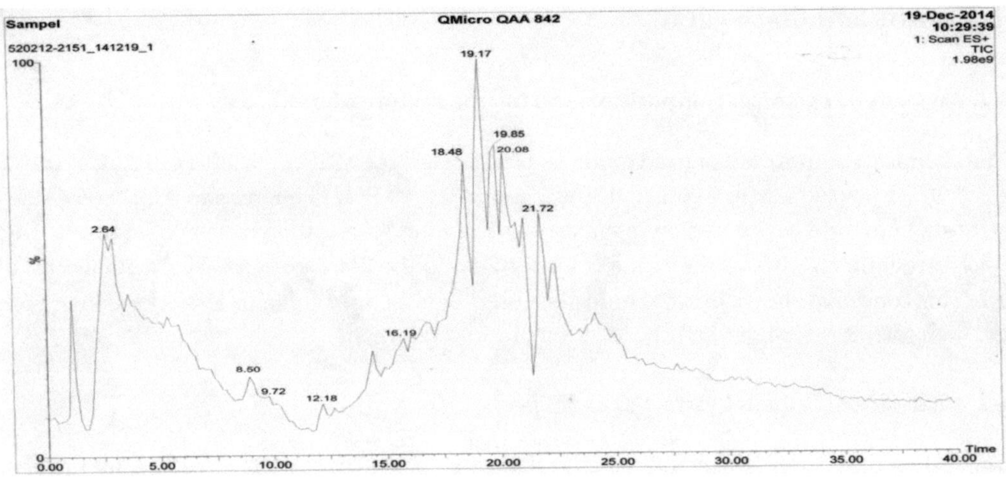

Figure 2. The chromatogram of cacao peel extract by using LC–MS.

Compound name	Formula	Retention time (min)
Quercetin	$C_{15}H_{10}O_7$	2.64
Catechin	$C_{15}H_{14}O_6$	8.50
Gallic acid	$C_7H_6O_5$	9.72
Epicatechin	$C_{15}H_{14}O_7$	12.18
Catechin	$C_{15}H_{14}O_6$	16.19
Caffeic acid derivate	$C_8H_{10}N_4O$	18.48
Salvianolic acid	$C_9H_{10}O_3$	19.17
Kaempferol 3-O-rutinoside	$C_{15}H_{10}O_6$	19.85
Procyanidin B_2	$C_{31}H_{28}O_{12}$	20.8
Kaempferol 3-O-(sinapoyl)-sophoside	$C_{53}H_{56}O_{28}$	21.72

Table 2. Analysis of cacao peel extract compound which has been fragmented by LC–MS.

ether [41]. Some of these functional groups are appeared in corrosion product but with a slight shift of frequency friction. The occurrence of friction and the disappearance of several peaks on particular frequency indicate that there is interaction between cacao peel extract and mild steel in HCl 1.5 M solution. The result shows the interaction and chemical bound between extract compound and metal on the surface area [4]. Identified functional groups from peaks are shown in **Table 3**.

4.4. Analysis of mild steel corrosion rate

The measuring result of corrosion rate and inhibition efficiency by using the weight loss method shows that adding extract cacao peels into corrosion medium causes the corrosion rate becoming

FTIR spectrum peak, ν (cm^{-1})			Functional group identification
Cacao peel extract	Corrosion product		
	In HCl 1.5 M without extract	In HCl 1.5 M with 2.5% cacao peel extract	
—	668	—	Fe—O
1051	1020	1019	C—O (str)
1400	—	—	C—C=C (aromatic)
—	1458	—	C—C (aromatic)
1603	1636	1629	C=O (str)
—	2359–2342	2363	C—H (str)
3422	3385	3376	O—H (phenolic)

Table 3. FTIR spectra analysis of cacao peel extract, corrosion product, and its identification.

lesser and conversely an increase in inhibition efficiency as seen in **Figure 3**. It was due to the larger iron surface coated with molecules on the cacao peel extract. Cacao peel extract molecules that coat the metal surface are adsorbed into the steel surface and form passive layer that protects the steel from active ion's corrosion attack in the acid solution. Formed passive layers are derived from oxygen and functional group, which is in cacao peel extract [42–44].

Natural product extract is compounds containing atom with free electron pair and can be used to covalent coordination bond [7]. This atom is electron donor that can form complex compounds with iron existed in the mild steel. It is a stabile complex compound, not easily oxidized, and coats the mild steel surface so that the corrosion can be restrained [19].

Figure 3 means that the increase of extract concentration decreases the corrosion rate and on the other hand increases its inhibition efficiency. However, the increasing time during submersion makes the corrosion rate increase as well, and decreases its inhibition efficiency on

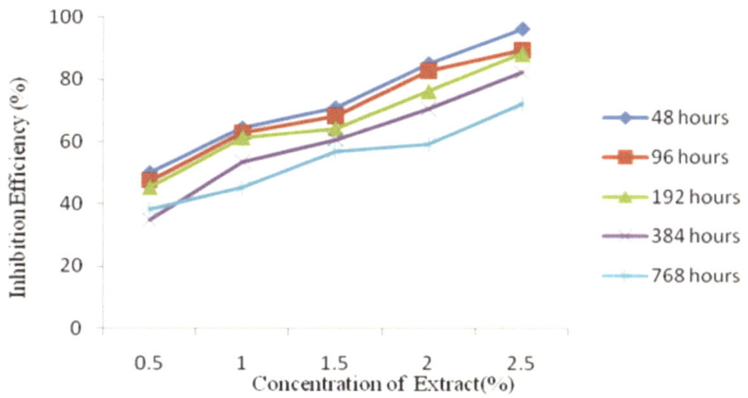

Figure 3. The effect of extract concentration on steel corrosion inhibition efficiency in HCl 1.5 M with different submersion time.

similar concentrate. It is due to the increasing time during the submersion that causes more weight loss and triggers desorption [45], so that the smallest corrosion rate and the biggest efficiency are obtained for 2-day (48-h) submersion duration by adding 2.5% extract.

In order to see the interaction in mild steel in the absence and the presence of cacao peel extract within HCl 1.5 M, temperature variation is applied starting from 303 K to 323 K. In **Figure 4**, it can be seen that the rate corrosion decreases where the inhibition efficiency increases by the rise of added cacao peel extract concentration. However, the corrosion rate is increasing, and inhibition efficiency is decreasing by the rising temperature. The increase of corrosion rate and the decrease of efficiency by the increasing temperature are caused by the increase of adsorption strength on high temperature or inhibitor molecule desorption occurred [46].

Constantly, inhibitor molecule adsorption and desorption have taken place on the steel surface until the balance of the two processes reached the certain temperature. The highest efficiency value is 83.91% for 2.5% extract concentration on 303 K temperature [47].

4.5. Potentiodynamic polarization method

Mild steel polarization curve on varied extract concentration in HCl 1.5 M has been extrapolated into Tafel plot as shown in **Figure 5**. Corrosion current value (I_{corr}), corrosion potential (E_{corr}),

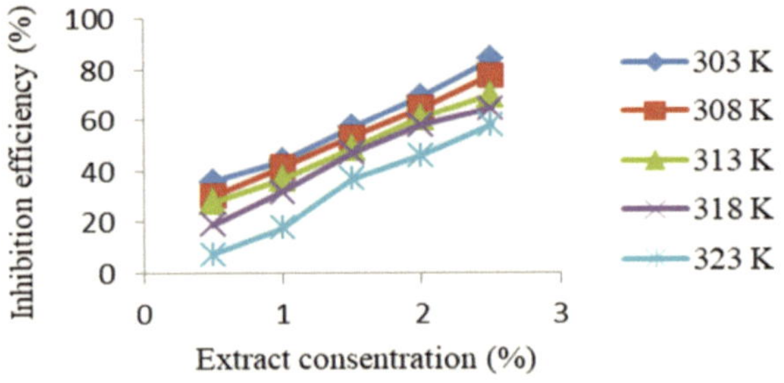

Figure 4. Extract concentration effect on mild steel corrosion inhibition efficiency in HCl 1.5 M of temperature variation.

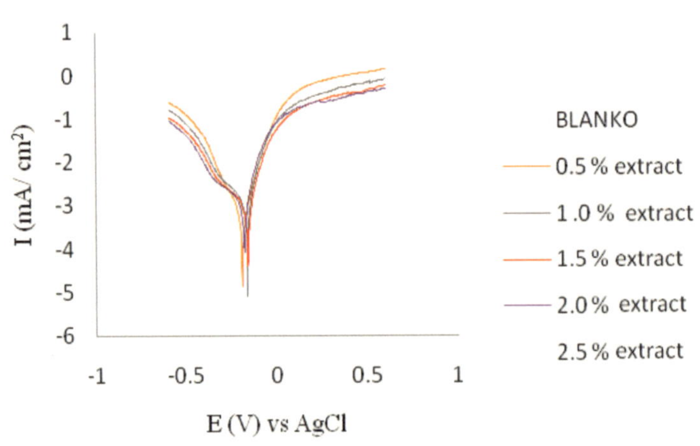

Figure 5. Potential Tafel plot steel polarization in HCl 1.5 M with variation of cacao peel extract concentration.

cathodic and anodic constant (b_a and b_c), and inhibition efficiency on each treatment obtained from Tafel line extrapolarization are shown in **Table 4** [48]. Measuring the corrosion rate by using Tafel plot began with determining the corrosion current value and corrosion potential.

Table 4 shows that the increasing extract concentration in the medium will decrease the I_{corr} value. The highest I_{corr} value is seen on the immersed sample in the medium without 2.5% additional extract, whereas the lowest I_{corr} value belongs to the sample with 2.5% additional extract. Comparing inhibition efficiency testing result weight loss and potentiodynamic polarization results relatively similar value.

The table shows the E_{corr} friction value where the lowest E_{corr} value belongs to medium without additional extract which is −0.28 V, and the highest with additional extract medium which is −0.20 V. This corrosion potential value friction toward more positive and negative ways indicates that the additional extract is anodic and cathodic [49, 50]. Undoubtedly, it demonstrates that cacao peel extract influences the anodic dissolution on mild steels and at the same time slows down hydrogen evolution reaction on cathode [2, 51]. This result is clearly shown that this extract is a mixed-type inhibitor [48, 52].

4.6. Analysis of polarization resistance

The change in value of b_a and b_c displayed in **Table 4** indicates that cacao peel extract adsorption is modification of anodic dissolution and hydrogen evolution on cathodic. In **Figure 5**, it is obviously seen that these two cathodic and anodic reactions are inhibited, and to some extent, its inhibition process is in line with the increase of cacao peel extract concentration in acidic medium; conversely, cathodic is more polarized [53–55]. The polarization value linear resistance is measured using Stern-Geary Eq. (1) which its result can be seen in **Table 4**.

$$\mathrm{Rp} = \frac{b_a \times b_c}{I_{corr} \times 2.303(b_a + b_c)} \tag{1}$$

b_a and b_c are anodic and cathodic constants.

The resistance polarization (Rp) measuring result undergoes an increase with the rise of extract concentration. This increasing polarization resistance shows that the use of cacao peel

Cons. inhibitor (%v/v)	I_{corr} (mA cm^{-2})	E_{corr} (Vdec^{-1})	b_a (Vdec^{-1})	b_c (Vdec^{-1})	R_p (m^2)	IE (I_{corr})	IE (R_p)
Blank	0.06	−0.28	2.4	1.71	6.87	—	—
0.5	0.016	−0.27	2.8	2.3	34.48	74.81	80.07
1.0	0.013	−0.25	3.67	2.57	52.05	80.03	86.6
1.5	0.0078	−0.2	2.00	1.60	57.36	87.48	88.02
2.0	0.007	−0.22	5.25	2.63	115.26	89.54	94.04
2.5	0.005	−0.28	5.71	2.67	157.99	92.08	95.64

Table 4. Parameter of mild steel corrosion in HCl 1.5 M in the absence and the presence of cacao peel extract obtained from polarization measurement.

extract slows down electron transfer on steel surface or is capable to inhibit its oxidation reaction, so that corrosion process can be protected. It makes the steel corrosion rate decrease [53].

Positive relation between polarization resistance with extract concentration and polarization resistance with inhibition efficiency is shown in **Figure 6**. This figure points out the increasing of extract concentration can slow down mild steel corrosion rate. Due to the fact that the additional extract concentration, increases the polarization resistance, the transfer electron from mild steel surface to solution can be inhibited, and it makes the corrosion inhibition efficiency on the mild steel surface increase as well [55].

4.7. Electrochemical impedance spectroscopy

The result of EIS measurement for mild steel corrosion in HCl 1.5 M is recorded by Nyquist diagram as seen in **Figure 7**. This behavior can be related to frequency dispersion of electrode surface roughness [55]. The difference on Nyquist diagram between blanks and the presence of inhibitor is clearly seen. In the medium with additional inhibitor, there is impedance value increase on electrode interface solution, particularly Rct value [56]. The EIS measuring result in **Figure 7** shows that the additional inhibitor impedes the electron transfer from the steel surface into the solution.

The increase of Rct value results in the decrease of iron atoms' oxidation process and ions' H^+ reduction process [8]. Electrochemical parameter on the inhibitor concentration variation can be seen in **Table 5**. This table clearly shows connection between Rct with inhibition efficiency and capacitance (Cdl). The increase of Rct is also followed by the increase of inhibition efficiency value [4] and in similar with the decrease of capacitance with the increase of extract concentration [48]. The capacitance value is obtained by using the following formula (2):

$$Cdl = \frac{1}{2\pi f_{max}\, Rct} \tag{2}$$

where Cdl = double layer capacitance, f_{max} = maximum imaginer frequency from obtained impedance, and Rct = charge transfer resistance.

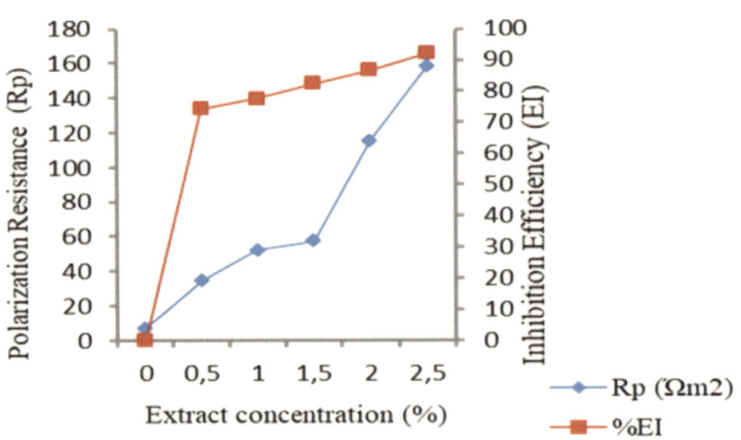

Figure 6. The effect of cacao peel extract concentration on polarization resistance and mild steel inhibition efficiency in HCl 1.5 M.

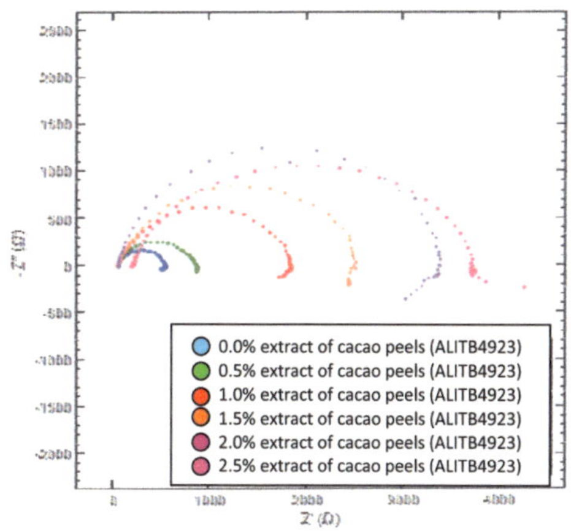

Figure 7. Nyquist plot steel in HCl 1.5 M solution with variation of cacao peel extract concentration.

Inhibitor concentration (%)	Rs (Ω)	Rct (Ωcm²)	Cdl (μFcm²)	EI (%)
0.0	34.0	505	0.63	0.0
0.5	23.0	850	0.31	40.9
1.0	23.3	1680	0.16	69.94
1.5	22.2	2435	0.093	79.26
2.0	32.4	3259	0.081	84.50
2.5	181	3552	0.075	85.78

Table 5. Relationship of extract concentration with electrochemical parameter in HCl 1.5 M medium.

The decrease of Cdl value in **Table 5** can be connected to the decrease of local dielectrical constant or the increase of layers thickness on mild steel surface due to the presence of adsorption on this surface [57, 58]. Besides that, the decrease of capacitance value might occur due to the hydro-molecule changes which adsorbed on the steel surface with cacao peel extract molecules [58]. It can be also the formation from the film layers with the interaction between inhibitor molecules and atom Fe on the mild surface which involves the chemical and physical reactions. Thus, inhibitor molecules are coating more to the active side and press the corrosion reaction with the increase of extract concentration. It causes that the charge transfer between the steel surface and solution is going to be tough and corrosion reaction is inhibited more effectively [59].

The inhibition efficiency of the cacao peel extract produced depends on the inhibitor concentration and corrosive medium used. In the graph, it can be seen that the inhibition efficiency in HCl solution can reach 96.26% (weight loss), 95.64% (IE), and 92.08 (I_{corr}) with 2.5% concentration of inhibitor as shown in **Table 4**. This happens, because the Fe-tannin complex is formed on the surface, covering the entire surface of the steel. Comparison of three methods for testing the efficiency of inhibition of polar extract of cacao peels on mild steel is shown in **Figure 8**.

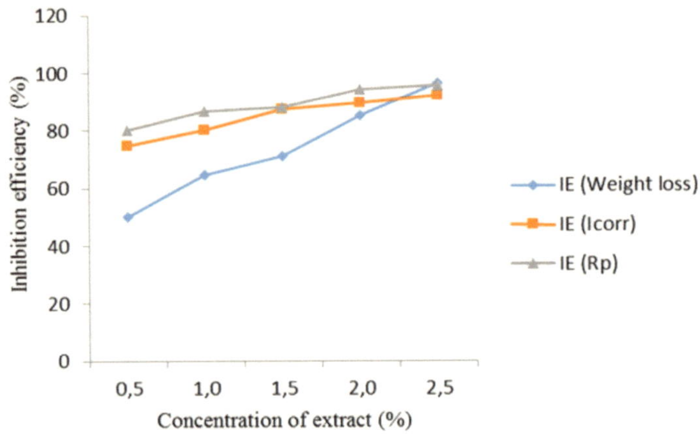

Figure 8. Inhibition efficiency against concentration of extract in HCl 1.5 M.

4.8. Analysis of adsorption isotherm

Cacao peel extract adsorption on mild steel is characterized by using degree data of surface coating from the weight loss data obtaining through experiment with the formula (3) as follow:

$$\theta = \frac{V_0 - V_1}{V_0} \tag{3}$$

In order to obtain clear and accurate data to find out the adsorption on the mild steel surface, adsorption isotherm of Langmuir, Temkin, and Frumkin adsorption is applied. Interaction phenomena between metal surface and inhibitor can be studied from the adsorption isotherm. Those three adsorption isotherms show linear relationship between extract concentration with the adsorption occurred on the mild steel surface. This linear relationship is proven by the correlation coefficient value in **Table 6** in the range of 0.90–0.97, with different time variation.

Nevertheless, the strongest relationship from one out of those three adsorption isotherms is Langmuir with the correlation coefficient reaching up to 0.97. These data indicate that the inhibitor which is used here fulfills the Langmuir adsorption isotherm rule [8, 44]. **Figure 9** shows that the more the immersion time with the increase of extract concentration, the more

Time (h)	Correlation coefficient (R^2)		
	Langmuir	Temkin	Frumkin
48	0.9587	0.9415	0.6081
96	0.9726	0.9595	0.5573
192	0.9532	0.9327	0.3340
384	0.9648	0.9447	0.3454
768	0.9531	0.9082	0.9171

Table 6. Correlation coefficient (R) obtained from various isotherm adsorption on varied immersion time.

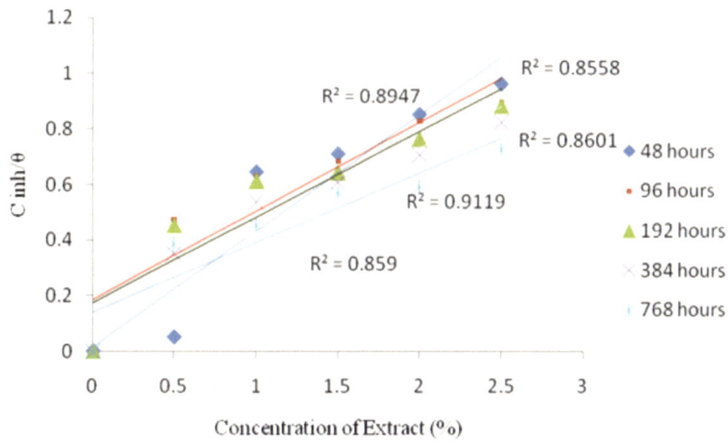

Figure 9. Langmuir adsorption isotherm of cacao peel extract on mild steel surface in HCl 1.5 M with different temperature.

the adsorption occurred on the mild steel surface. It demonstrates that the strong molecule interaction has occurred between adsorption particles and mild steel surface, where the interaction increases with the raise of added inhibitor concentration [60].

4.9. Analysis of thermodynamics

The thermodynamic data are obtained from experiment on temperature effect and adsorption isotherm. These data can complete the basic information on the interaction between cacao peel extract and steel surface. The temperature effect on steel corrosion rate is studied through HCl 1.5 M in the absence and presence of cacao peel extract at the temperature of 303–323 K. The resulted straight line represents good correlation between corrosion rate and temperature, where the increase of the temperature is followed by the increase of corrosion rate; however, there is a decrease in its inhibition efficiency. This sign explains that the film layer is formed during physical adsorption [58].

The increase on entropy value by the addition of cacao peel extract shows that the regularity degree is lower. The friction of entropy value to positive way indicates that complex is formed on certain level, and causes the regularity degree becoming higher. This formed complex will impede the corrosion rate on metal surface [52]. The thermodynamic data in **Table 7** are strengthening the conclusion made that cacao peel extract can act as the corrosion inhibitor. The regularity degree or entropy change (ΔS) is obtained by applying the following formula (4):

$$\Delta G = \Delta H - T\Delta S \qquad (4)$$

4.10. Analysis of surface morphology

Analysis of mild steel surface and the forming of passive layer on its surface in HCl 1.5 M in the absence and the presence of cacao peel inhibitor is studied by using SEM scaled up 2000× as seen in **Figure 10**.

Indicator	Ea (kJ/mol)	ΔH (kJ/mol)	ΔG (kJ/mol)	ΔS (kJ/mol)
Blank	142.3782	217.6073	—	0.32
Blank + inhibitor	198.8048	196.1433	−19.89	0.58

Table 7. Kinetic and thermodynamic parameters of mild steel with and without cacao peel extract in HCl 1.5 M.

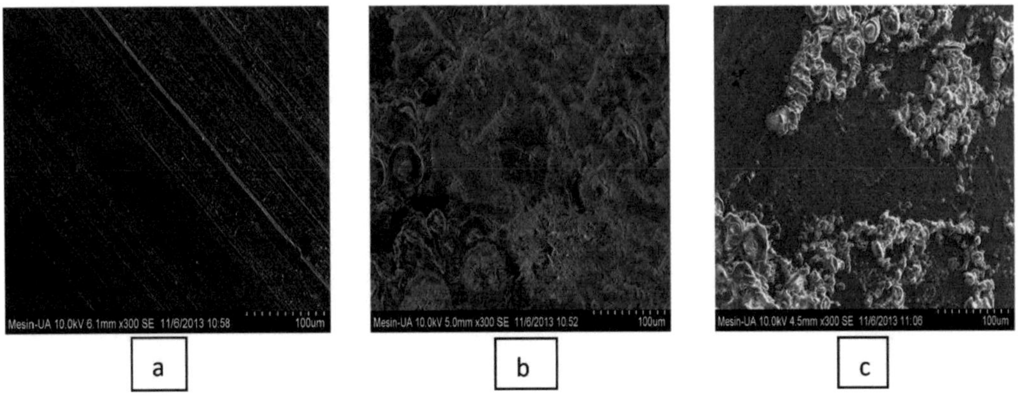

Figure 10. SEM photograph of mild steel (a) before being immersed and has been polished, (b) immersed for 32 days in room temperature by HCl 1.5M in the absence of inhibitor and (c) immersed for 32 days in room temperature by HCl 1.5M in the presence of 2.5% inhibitor.

Figure 10a shows the surface morphology from steel sample where fine lines in white color can be seen and it is relatively thin due to the effect of grindstoning and sanding on the mild steel surface. It is also seen that the surface is still flat, clean, and impermeable and yet has holes. **Figure 10b** is where corrosion product formed and the holes obviously seen and marks that there has been damage occurred on the surface. However, **Figure 10c**, in the presence of cacao peel extract, points out that it minimizes corrosion product and holes on the steel surface by forming passive layers on the surface. This layer becomes barrier on the corrosive ion attack on mild steel surface so that the electrochemical reaction is slowing down, and eventually the corrosion rate is decreasing as well [43, 56].

Analysis of C and Fe elements on mild steel surface in HCl 1.5 M which have been immersed for 32 days in the absence and presence of cacao peel extract is studied by applying SE-EDX. The element recapitulation is represented in **Table 8**. Referred to this table, the percentage of C element decreases by the presence of cacao peel extract, from 0.3% becoming 16.90%. This percentage of C element points out that this atom C of cacao peel extract molecules is adsorbed on the steel surface forming passive layers on the mild steel surface. Meanwhile, the percentage of atom on Fe element decreases by the presence of cacao peel polar extract, from 98.79% becoming 37.43%.

The data presented explain that Fe forms complex compound with cacao peel polar extract, so the percentage of detected Fe atom becomes smaller. Whereas, the Oxygen (O) element which in the beginning has not been detected yet, then by the adding of cacao peel extract can be

Treatment	Element content (% weight)		
	C	Fe	O
The absence of treatment	0.32	98.79	–
Immersion in 2.5% extract	6.19	92.66	4.33
Immersion in HCl 1.5 M	1.50	29.39	63.54
Immersion in HCl 1.5 M with 2.5% extract	16.90	37.43	44.89

Table 8. Identified element contents of the EDX testing on mild steel surface in HCl 1.5 M.

detected with low percentage. Then, there is an increase in percentage of oxygen up to 63.54% due to being immersed in corrosive medium HCl 1.5 M, quickly formed the oxide by the attack of corrosive ions of HCl. Nevertheless, by the help of polar inhibitor cacao peel extract, this corrosive ion attack can be blocked by forming passive layers in the form of metal organocomplex on the mild steel surface, because of the decrease of corrosion rate and lesser oxide been formed which is represented by O percentage becoming 44.89% [2, 50, 54].

4.11. Analysis of mechanical properties

The testing result of hardness and tensile on mild steel is represented in **Figure 11**. This hardness and tensile states an increase by the raise of added extract concentration. It is supported by the testing data of hardness, which shows that the increase of extract concentration adds the hardness on mild steel surface. Therefore, it causes the raise of the steel resistance and decreases corrosion rate due to the higher amount of extract adsorbed on its surface.

These data exhibit that the higher the extract concentration, the wider the mild steel surface which is coated by cacao peel extract. This extract is chemically adsorbed onto the mild steel surface and forms thin film layer that is difficult to be damaged [22]. The amount of extract adsorbed onto the surface is clearly determined by the SEM–EDX testing result as seen in **Table 7**. On contrary, the increase of surface hardness decreases corrosion rate and increases efficiency, and this connection is shown in **Figure 3** [61]. The increase of carbon level on mild steel surface does not only raise the hardness but also improve surface coating degree [62, 63]. The relationship of these four mechanical properties is expressed in **Figure 11**.

4.12. Corrosion inhibition mechanisms

Inhibitor molecules on mild steel surface occurred because of the presence of adsorption. This adsorption is due to the adhesion force between inhibitor and mild steel surface. Inhibitor molecule adsorption on mild steel surface produces thin film layer which can impede corrosion rate. In this case, polar extract inhibitor of cacao peels formed thin layer on the surface which can function as the control of the corrosion rate by making separator between the steel and corrosion media [64]. The process of cacao peel extract adsorption on mild steel surface will occur on functional group [22], where the mechanism can be seen in **Figure 12**.

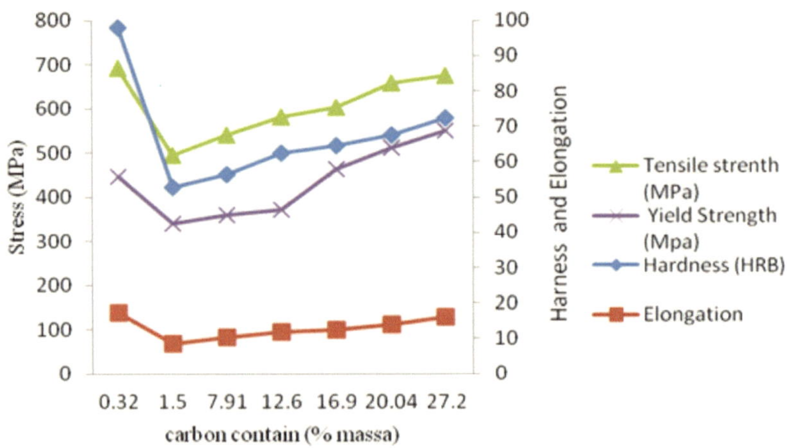

Figure 11. Yield strength, tensile strength, hardness, and elongation against % carbon in surface.

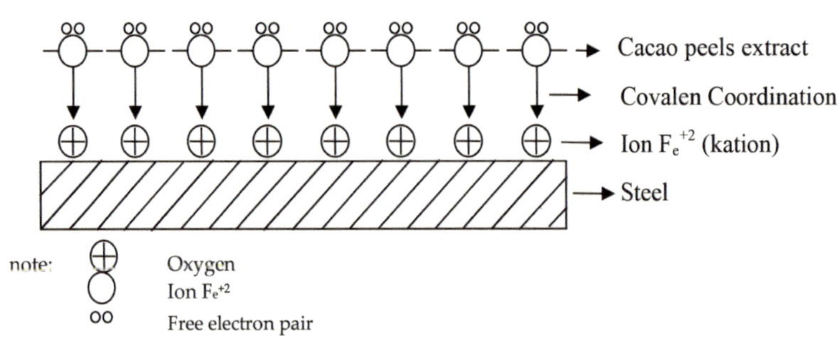

Figure 12. Adsorption mechanism of cacao peel extract on mild steel surface.

The adsorption result expresses that the higher the added inhibitor concentration, the more the metal part which is coated by corrosion inhibitor molecules. The bond occurred during the inhibitor adsorption process on the mild steel surface indicates as coordination covalent bond which involves chemical adsorption. It can be seen that from the struggle to omit the layer and in coherence with perceived enthalpy change value (ΔH) is higher than that with the absence of inhibitor [65].

The mechanism occurred on mild steel surface can be expressed on the basis of adsorption process and constituent structure that is present in the extract. The inhibit process might appear due to the presence of phytochemical constituent adsorption through oxygen atom and/or nitrogen atom on metal surface [22]. Possibly, this complex is absorbed into steel surface through Van der Waals force to form the protection layer in order to prevent the steel from corrosion [66]. Several major constituents of its extract are catechin, kaempferol, gallic acid, procyanidin, and tannin, which are shown in **Figure 13**. Inhibitor effective performance depends on phytochemical constituent measurement of cacao peel extract where its area is wider than the metal surface; so it can minimize the corrosion [2, 22].

Adding cacao peel extract into the solution is highly effective in order to reduce the chloride acid attack on mild steel surface, as compared to other plant extracts as can be seen in **Table 1**.

http://dx.doi.org/10.5772/intechopen.73263

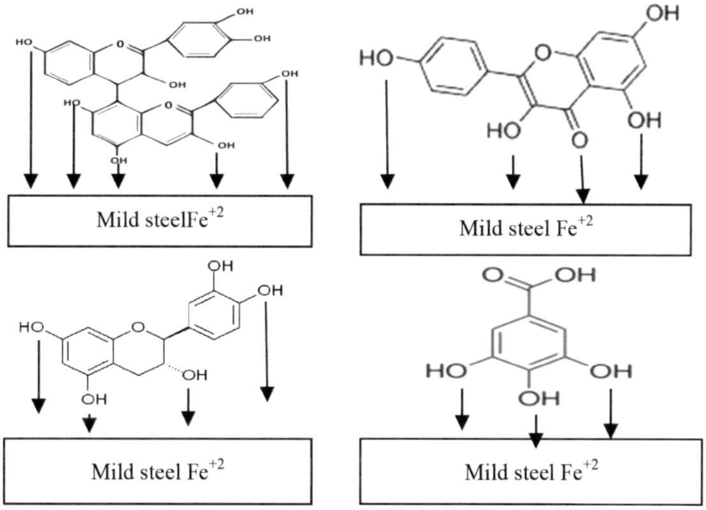

Figure 13. Compound adsorption process existed in cacao peel extract on mild steel surface.

This can be seen from its ability to inhibit on the surface of mild steel greater than other inhibitors at concentrations that are not too different and the same media corrosive hydrochloric acid.

5. Conclusions

By and large, of all researches conducted, it can be concluded that:

1. Cacao peel extract can function as appropriate corrosion inhibitor for mild steel in HCl 1.5 M.
2. GC–MS and LC–MS testing results shows that cacao peel extract contains of many active compounds. These functional group compounds have been confirmed by FTIR test containing heteroatom group which takes a part in corrosion inhibition through coordination covalent bond on mild steel surface.
3. Significantly mild steel corrosion rate decreases by the addition of cacao peel extract. On the contrary, inhibition efficiency increases by the increase of extract concentration up to 2.5%. Even though the efficiency decreases with the raise of work temperature, it is considered high enough reaching the temperature of 323 K. Adding cacao peel extract into the solution is highly effective in order to reduce the chloride acid attack on mild steel surface.
4. Through potentiodynamic measurement and impedance measurements that adding the inhibitor indicate inhibits electron transfer from mild steel surface into solution, so that both oxidation process of iron atoms and reduction process of H^+ ion decrease. Has been determined that the inhibitor extract of cacao peel in HCl are mixed-type with dominant as cathodic inhibitor.
5. Inhibition mechanism between extract and mild steel surface is studied through the interaction between isolated electron pair functioning as donor ion and mild steel surface functioning as acceptor, so that coordination covalent bind formed.

6. Cacao peel extract is chemically adsorbed into mild steel surface by forming Fe organometallic layer on the surface referring to Langmuir isotherm adsorption rules. The adsorption occurred on this mild steel surface influences mechanical properties of hardness and tensile, so it can restore the mechanical properties which has been attacked by corrosion.

Acknowledgements

The first author thanks to Ministry of Education, Republic of Indonesia, for Hibah Fundamental Research Grant II No. DIPA 041/PL.1.4/LT/2016.

Author details

Yuli Yetri[1]*, Gunawarman[2], Emriadi[3] and Novesar Jamarun[3]

*Address all correspondence to: yuliyetriyetri@gmail.com

1 Mechanic Department of Padang State Polytechnic, Padang, West Sumatera, Indonesia

2 Mechanic Department, Faculty of Engineering, Andalas University, Padang, West Sumatera, Indonesia

3 Chemistry Department, Faculty of Mathematic and Natural Science, Andalas University, Padang, Indonesia

References

[1] Vinod KP, Narayanan Pillai MS, Rexin Thusnavis G. Pericarp of the fruit of Garcinia Mangostana as corrosion inhibitor for mild steel in hydrochloric acid medium. Portugaliae Electrochimica Act. 2010;**28**:373-383

[2] Ostovari A, Hoseinieh SM, Peikari M, Shadizadeh SR, Hashemi SJ. Corrosion inhibitor of mild steel in 1 M HCl solution by henna extract: A comparative study of inhibition by henna and its contituents (Lawsone gallic acid, glucose and tannic acid). Corrosion Science. 2009;**51**:1935-1949

[3] Kesavan D, Gopiraman M, Sulochana N. Green inhibitor for corrosion of metals. Chemical Science Review and Letters. 2012;**1**(1):1-8

[4] Okafor PC, Ikpi ME, Uwah IE, Ebenso EE, Ekpe UJ, Umoren SA. Inhibitory action of Phyllanthus Amarus extracts on the corrosion of mild steel in acidic media. Corrosion Science. 2008;**50**(8):2310-2317

[5] Umoren SA, Eduok UM, Solomon MM, Udoh AP. Corrosion inhibitor by leaves and stem extracts of *Sida acuta* for mild steel in 1M H_2SO_4 solutions investigated by chemical and spectroscopic techniques. Arabian Journal of Chemistry. 2011

[6] Rani ABE, Basu BBJ. Green inhibitors for corrosion protection of metals and alloys—An overview. International Journal of Corrosion. 2012;**10**(1):1-15

[7] Okafor PC, Ebenso EE, Ekbe UJ. Azadirachta indica extract as corrosion inhibitor for mild steel in acidic medium. International Journal of Electrochemical Science. 2010;**5**:978-993

[8] Lebrini M, Robert F, Ross C. Alkoloids extract from Palicourea Guianensis plant as corrosion of C38 steel in 1 M hydrochloric acid medium. International Journal of Electrochemical Science. 2011;**6**:5357-5371

[9] Quraishi MA, Singh A, Singh VK, Yadav DK. Green approach to corrosion inhibitor of mild steel in hydrochloric acid and sulphuric acid solutions by the extract of Murayya Koeniggi leaves. Materials Chemistry and Physics. 2010;**122**(1):114-122

[10] Obi-Egbedi NO, Obot IB, Umoren SA. Spondias Mombin L. as a green corrosion inhibitor for aluminium in sulphuric acid: Correlation between inhibitive effect and electronic properties of extracts major constituents using density fuctional theory. Arabian Journal of Chemistry. 2010;**5**(3):361-373

[11] Inzunza RG, Salas B'ı V, Kharshan R, Furman A, SchorrWienner M. Interesting behavior of Pachycormus discolor leaves ethanol extract as a corrosion inhibitor of carbon steel in 1M HCl: A preliminary study. International Journal of Corrosion. 2012;**2012**:1-8

[12] Saratha R, Priya SV, Thilagavathy P. Investigation of Citrus Aurantiifolia leaves extract as corrosion inhibitor for mild steel in 1 M HCl. E-Journal of Chemistry. 2009;**6**(3):785-795

[13] Kasthuri PK, Arulantham A. Eco-friendly extract of Euphorbia Hirta as corrosion inhibitor on mild steel in sulphuric acid. Asian Journal of Chemistry. 2010;**22**(1):430-434

[14] Ekanem UF, Umoren SA, Udousoro SA, Udoh AP. Inhibition of mild steel corrosion in HCl using pineapple leaves (Ananas Comosus L.) extract. Journal of Materials Science. 2010;**45**:5558-5566

[15] Sharma MK, Arora P, Kumar S, Mathur SP, Ratnani R. Inhibitive effect of Prosopis Cineraria on mild steel in acidic media. Corrosion Engineering Science and Technology. 2008;**43**(3):213-218

[16] Hussin MH, Kassim MJ. Electrochemical, thermodynamic and adsorption studies of (+)-catechin hydrate as natural mild steel corrosion inhibitor in 1 M HCl. International Journal of Electrochemical Science. 2011;**6**:1396-1414

[17] Garg U, Tak RK. Inhibition of the corrosion of mild steel in acid media by naturally occuring Acacia senegal. E-Journal of Chemistry. 2010;**7**(4):1220-1229

[18] Afia L, Salghi R, Bazzi EH. Argan hulls extract: Green inhibitor of mild steel corrosion in 1 M HCl solution. Research on Chemical Intermediates. 2012;**10**(7):1708-1717

[19] Shyamala M, Arulanantham A. Corrosion inhibition of Centella asitica (Vallarai) on mild steel in hydrochloric acid. Asian Journal of Chemistry. 2009;**21**(8):6102-6110

[20] Gulsen A. Corrosion inhibition of mild steel by *Laurus nobilis* leaves extract as green inhibitor. Research Chemical Intermediates. 2011;**10**(7):451

[21] Gunavathy N, Murugavel SC. 2012, Corrosion inhibition studies of mild steel in acid medium using Musa Acuminata fruit peel extract, E- Journal of Chemistry, vo. 9, no. 1, pp. 487-495

[22] Leelavathi S, Rajalakshmi R. Dodonaea Viscosa (L.) leaves extract as` acid corrosion inhibitor for mild steel as green approach. Journal of Material and Environment Science. 2013;**4**(5):625-638

[23] Yetri Y, Sukatik. Green inhibitor for mild steel in acidic solution by using crude extract and polar extract of cacao peels, Oriental Journal of Chemistry. 2017;**33**(3):1550-15541

[24] Julia Witjaksono, Asmin. Cocoa farming system in Indonesia and its sustainability under climate change. Agriculture, Forestry and Fisheries. 2016;**5**(5):170-180

[25] Sustainable Cocoa Production Program Indonesia. Annual Report Swisscontact; 2013

[26] Sastri VS. Adsorption in Corrosion Inhibition, Green Corrosion Inhibitor. Winston Revie, Canada; 2011. pp. 103-105

[27] Ahmad Z, Patel F. Development of novel corrosion techniques for a green environment. International Journal of Corrosion. 2011;**10**:1-8

[28] Noor EA. Temperature effects on the corrosion inhibition of mild steel in acidic solutions by aqueous extract of fenugreek leaves. International Journal of Electrochemical Science. 2007;**2**:996-1017

[29] Rosales BM, Vera R, de Rincon OT, DiSarli A, Valenzuela JAR, Tidblad J. Atmospheric corrosion. International Journal of Corrosion. 2012;**10**:1-3

[30] Obot IB, Ebenso EE, Obi-Egbedi NO, Afolabi AS, Gasem ZM. Experimental and theoretical investigations of adsorption characteristics of itraconazole as green corrosion inhibitor at a mild steel in hydrochloric acid interface. Research Chemical Intermediates. 2012;**38**:1761-1779

[31] Hussin MH, Kassim MJ. Electrochemical studies of mild steel corrosion inhibition in aqueous solution by Uncaria Gambir extract. Journal of Physical Science. 2010;**21**(1):1-13

[32] Schweitzer PA. Fundamental of Corrosion Mechanisms, Cause, and Preventive Methods. America: CRC Press; 2010. pp. 116-140

[33] Okuda T, Ito H. Tannin of constant structure in medical and food plant, hydrolyzable tannins and polyphenol related to tannins. Journal of Moleculer Structure. 2011;**16**:2191-2217

[34] Hii CL, Law CL, Suzannah S, Misnawi, Cloke M. Polyphenols in cocoa (Theobroma Cacao L). Asian Journal of Food Agriculture Industry. 2009;**2**(4):702-722

[35] Alemanno L, Ramos T, Gargadenec A, Andary C, Ferriere N. Lacalization and identification of phenolic compounds in *Theobroma cacao* L somatic embryogenesis. Annals of Botany. 2003;**92**:613-623

[36] Osman H, Nasarudin R, Lee SL. Extracts of cocoa (*Theobroma cacao* L.) leaves and their antioxidation potential. Food Chemistry. 2004;**8**(6):41-46

[37] Helmestein AM. Theobromine Chemistry: Theobromine is Chocolates. 2010. Taken from: http://chemistry.about.com/old/factsstructures/a/theobrominechemistry.htm

[38] Chung BY, Ilyama K, Han K-W. Compositional characterization of cacao (Theobroma Cacao L) hull. Agriculture Chemical Bioethanol. 2003;**46**(1):12-16

[39] Callister WD, Rethwisch DG. Material Science Engineering an Introduction. 9th ed. New York: Wiley & Son; 2013

[40] Andujar I, Recio MC, Giner RM, Rios JL. Cocoa polyphenol and their potential benefits for human health, oxidative medicine and cellular longevity. Hindawi Publishing. 2012;**2012**:23. Article ID 90625. DOI: 10.1155/2012/906252

[41] Kim H, Keeney PG. Method of analysis for (−) epicatechin in cocoa beans by HPLC. Journal of Food Science. 1983;**48**(2):548-551

[42] Singh MR. A green approach: A corrosion inhibition of mild steel by *Adhatoda vasica* plant extract in 0.5M H_2SO_4. Journal of Material Environment Science. 2013;**4**(10):119-126

[43] Vatshala, Pawethra MK, Nayana KO, Vermetesha TV. Synergetic effect of halide ions on improving corrosion inhibitor behavior of benzi so thiozole-3-piperizim hydrochloride on mild steel in 0,5M H_2SO_4 medium. Corrosion Science. 2010;**32**(11):3811-3819

[44] Znini M, Cristofari G, Majidi L, Ansari A, Bauyanzer A, Paolini J, Costa J, Hammouti B. Green approach to corrosion inhibition of mild steel by essential oil leaves of *Astericus graveolus* (Forsk.) in sulphuric acid medium. Corrosion Science. 2012;**7**:3959-3981

[45] Satapathy AK, Gunasekaran G, Sahoo SC, Amit K, Rodrigues PW. Corrosion inhibition by *Justicia gendarussa* plant extract in hydrochloric acid solution. Corrosion Science. 2009;**51**(12):2848-2856

[46] Phongying S, Aiba A, Chirachanchai. Direct chitisan nanoscaffold formation via chitin whiskers. Polymer. 2007;**48**:393-400

[47] Yetri Y, Emriadi NJ, Gunawarman. Corrosion inhibitor of mild steel by polar extract of Theobroma Cacao peels in hydrochloric acid solution. Asian Journal of Chemistry. 2015;**27**(3):875-881

[48] Mohanty US, Lin KL. Potentiodynamic polarization measurement of Sn–8.5Zn–XAl–0.5Ga alloy in 3.5% NaCl solution. Journal of Electrochemical Science. 2006;**153**(8):B319-B324

[49] Amin MA, Sayed S, Abd E-R, El-Sherbini EEF, Bayoumi RS. The inhibition of low carbon steel corrosion in hydrochloric acid solution by succinic acid (part I, weight loss, polarization, EIS, PZC, EDX and SEM studies). Electrochemical Acta. 2007;**52**:3588-3600

[50] Vishnudevan M. Migitation of mild steel corrosion in acidic solution using inhibitor. E-Journal of Chemistry. 2011;**8**(1):53-60

[51] Abo EOM. Novel corrosion inhibition for carbon steel alloy: Synthesis and characterization. Tikrit Journal of Pure Science. 2006;**11**(1):1-6

[52] Cang H, Fei Z, Shao J, Shi W, Xu Q. Corrosion inhibition of mild steel by aloes extract in HCl solution medium. International Journal Electrochemical Science. 2013;**8**:720-743

[53] Afolabi AS. Corrosion and stress corrosion behaviors of low and medium carbon steels in agro-fluid media. Leonardo Electronic Journal of Practices and Technologies. 2007;**10**:55-66

[54] Yetri Y, Emriadi NJ, Gunawarman. Corrosion behavior of environmental friendly inhibitor of *Theobroma cacao* peels extract for mild steel in NaCl 1.5M. Environment Asia. 2016;**9**(1):45-59

[55] Chauhan LR, Gunasekaran G. Corrosion inhibition of mild steel by plant extract in dilute HCl medium. Corrosion Science. 2007;**49**:1143-1161

[56] Dang Nam N, Mathesh M, Hinton B, Mike JYT. Rare earth 4-hydrocinnamate compounds as carbon dioxide corrosion inhibitors for steel in sodium chloride solution. Journal of Electrochemical Science. 2014;**161**(12):C527-C534

[57] Hui X, Min G, Xuan ZY, Qin W, Rong HB. Corrosion inhibition and mechanism of mild steel in hydrochloric acid by centriaxone and amoxicill. SCIENCE CHINA Chemistry. 2011;**54**(9):1529-1536

[58] De Souza FS, Goncalves RS, Spinelli A. Assessment of caffeine adsorption onto mild steel surfaces as an eco-friendly corrosion inhibitor. Journal Brazil Chemistry Society. 2014;**25**(1):81-90

[59] Cruz J, Martinez R, Genesca J, Garc J. Experimental and theoretical study of 1-(2-ethylamino)-2-methylimidazoline as an inhibitor of carbon steel in acid media. Journal of Electroanalytical Chemistry. 2011;**566**:111-121

[60] Nahle A, Abu-Abdoun I, Abdel-Rahman I, Al-Khayat M. UAE neem extract as corrosion inhibitor for carbon steel in HCl solution. International Journal of Corrosion. 2010;**10**:1-9

[61] Gunawarman, Yetri Y, Emriadi NJ, Ken-Cho MN, Niinomi M. Effect of polar extract of cocoa peels inhibitor on mechanical properties and microstructure of mild steel exposed in hydrochloric acid. Applied Mechanics and Materials. 2015;**776**:193-200

[62] Singh MR. A green approach: A corrosion inhibition of mild steel by Adhatoda Vasica plant extract in 0.5M H_2SO_4. Journal of Material Environment Science. 2013;**4**(10):119-126

[63] Afolaby AS. Corrosion and stress corrosion behaviors of low and medium carbon steel in agro-fluid media. Leonardo Electronic Journal of Practices and Technologies. 2007;**10**:55-66

[64] Gerengi H, Sahin HI. Shinopsis lorentzii extract as a green corrosion inhibitor for low carbon steel in 1 M HCl solution. Industrial Engineering Chemical Research. 2011;**51**:780-787

[65] Yetri Y, Emriadi, Novesar J and Gunawarman, 2016, Inhibitory action of Theobroma Cacao peels extract on corrosion of mild steel in different media, Rasayan Journal of Chemistry, vol. 9, no. 4, pp. 716-727

[66] Li X, Deng S, Hui F. Inhibition of the corrosion of steel in H_2SO_4 solution by bamboom leaf extract. Corrosion Science. 2012;**62**:163-175

Chapter 11

Screening and Evaluating Environmentally-Friendly Corrosion Inhibitors for Amine-Based CO_2 Absorption Process

Sureshkumar Srinivasan, Amornvadee Veawab and Adisorn Aroonwilas

Additional information is available at the end of the chapter

http://dx.doi.org/10.5772/intechopen.72752

Abstract

This chapter evaluated the performance of environmentally friendly organic corrosion inhibitors on carbon steel in the amine-based carbon dioxide (CO_2) absorption process. The evaluation was experimentally conducted using electrochemical techniques in 5.0 kmol/m^3 monoethanolamine (MEA) solutions in the absence and presence of process contaminants, namely formate and chloride, at 80°C and 0.55 mol/mol CO_2 loading. The results show, in the absence of process contaminants, that 2-aminobenzene sulfonic acid, 3-aminobenzene sulfonic acid, 4-aminobenzene sulfonic acid, sulfapyridine, and sulfolane yielded 85–92% corrosion inhibition efficiencies, while sulfanilamide yielded the lowest efficiency of 20–42%. Sulfolane was the only tested inhibitor whose performance could be maintained in chloride- and formate-containing MEA solutions. On the contrary, the performance of 3-aminobenzene sulfonic acid and sulfapyridine was decreased by chloride. The performance of all the tested aminobenzene sulfonic acids was compromised by formate.

Keywords: corrosion inhibitor, CO_2 capture, CO_2 absorption, gas treating, alkanolamine, electrochemical tests, carbon steel

1. Introduction

Corrosion is one of the most severe operational problems in a typical amine-based carbon dioxide (CO_2) absorption process [1]. To mitigate corrosion, a number of organic and inorganic corrosion inhibitors have long been applied to the process. The corrosion inhibitors that were found effective are those containing heavy metals such as vanadium-, antimony-, cobalt-, and nickel-based compounds. For example, 0.05–0.10% sodium metavanadate and antimony tartrate were reported to yield 90–95% inhibition efficiency (IE) when used in 15–30% monoethanolamine (MEA) solution for mild steel in the ammonia plant [2]. A mixture of sodium metavanadate and cobalt nitrate and a mixture of ammonium metavanadate and amines performed well on mild steel in 50% MEA solution in acid gas treating process. They reduced corrosion rates (CRs) of mild steel down to less than 1 mile per year [3]. A combination of copper carbonate, dihydroxyethyl glycine, alkali metal thiocyanate, ammonium permanganate, and nickel or bismuth oxide provided 99% of inhibition efficiency for the MEA-based acid gas treating plants using mild steel [4]. A mixture of 50 ppm thiocyanate and 100 ppm bismuth citrate was reported to significantly reduce the corrosion of mild steel, stainless steel (304 and 316), and monel with approximately 94% in the natural gas treating plants using 30% MEA solution [5]. Dodecylbenzyl chloride with alkyl pyridine and nickel acetate was successfully applied with 93% efficiency in natural gas treating plants using 20-60% diethanolamine (DEA) solutions [6]. A mixture of amino ethyl piperazine, formaldehyde-thiourea, polymer and nickel sulfate yielded close to 100% efficiency on carbon steel in 30% MEA solution in the refinery gas conditioning unit [7]. More reviews on the corrosion inhibitors used for the amine-based CO_2 absorption process can be found in [8].

Despite their inhibition effectiveness, the usage of some heavy metal inhibitors is now restricted while others would be banned in the near future due to incoming stringent environmental regulations on the use of toxic chemicals. The objective of this chapter is thus to search for an environmentally friendly corrosion inhibitor with comparable inhibition performance that can replace conventional highly toxic corrosion inhibitors for the CO_2 absorption process. To achieve such an objective, a number of organic compounds were selected as potential corrosion inhibitors based on the principles of Hard and Soft Acids and Bases (HSAB), toxicity properties, and quantum chemical analysis. The inhibition performance of the selected inhibitors on carbon steel was experimentally tested using electrochemical techniques under a test condition simulating the process solution environment, that is, 5.0 kmol/m^3 monoethanolamine (MEA) solution at 80°C and 0.55 mol/mol CO_2 loading. The effects of the corrosion inhibitor concentration and process contaminants (namely, formate and chloride) on the inhibition performance were also studied.

2. Selection of tested corrosion inhibitors

2.1. Compound selection

The principle of Hard and Soft Acids and Bases (HSAB) was employed as the first screening criteria to select the corrosion inhibitors to be tested. Since ferrous iron (Fe^{+}) which is a borderline acid is typically found in the amine-based CO_2 absorption process [1, 9], eight aniline- and two pyridine-based compounds which are the borderline basic inhibitors were the preferred choices. As shown in **Table 1**, the aniline compounds, including 2-, 3-, 4-aminobenzene sulfonic acids, sulfanilamide, and sulfapyridine, contain nitrogen (N), sulfur (S), and oxygen (O) as the reaction centers for adsorption, while 2-, 3-, 4-bromoaniline and bromopyridine contain N and bromine (Br).

http://dx.doi.org/10.5772/intechopen.72752

Compound	Formula	Structure[a]
Aniline-based compounds		
2-aminobenzene sulfonic acid	$C_6H_7NO_3S$	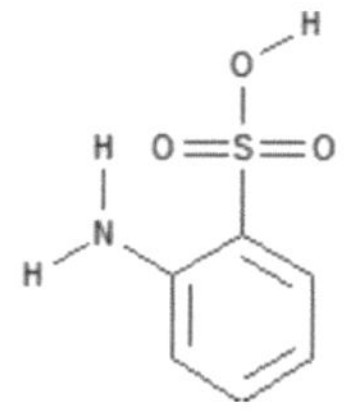
3-aminobenzene sulfonic acid	$C_6H_7NO_3S$	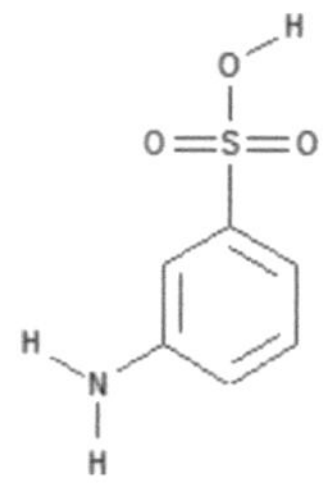
4-aminobenzene sulfonic acid	$C_6H_7NO_3S$	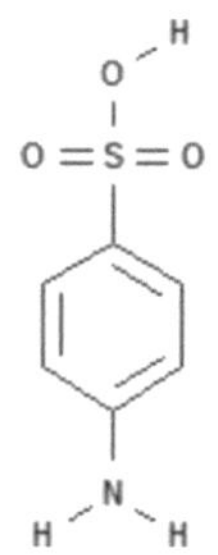
2-bromoaniline	C_6H_6BrN	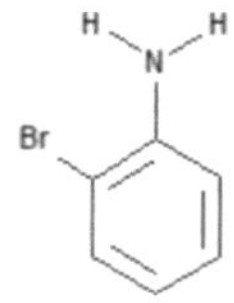
3-bromoaniline	C_6H_6BrN	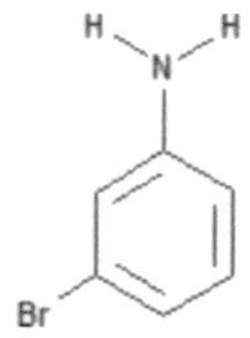
4-bromoaniline	C_6H_6BrN	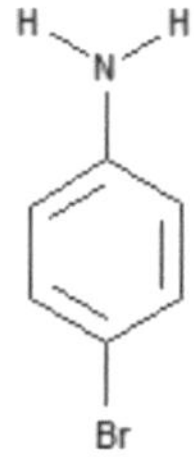

Compound	Formula	Structure[a]
Sulfapyridine	$C_{11}H_{11}N_3O_2S$	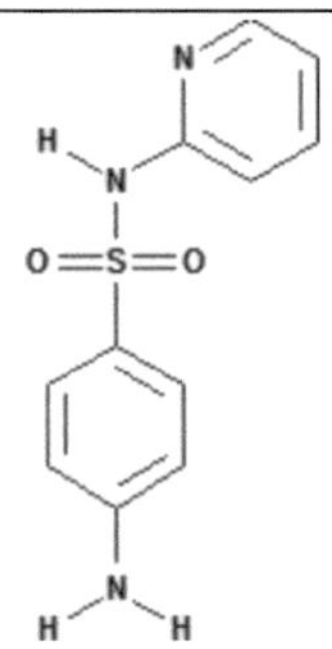
Sulfanilamide	$C_6H_8N_2O_2S$	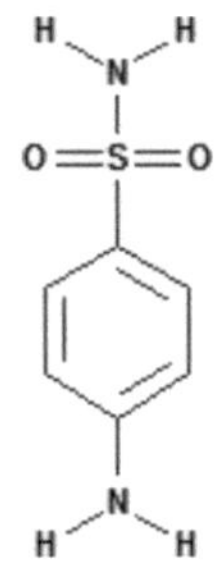
Pyridine-based compounds		
2-bromopyridine	C_5H_4BrN	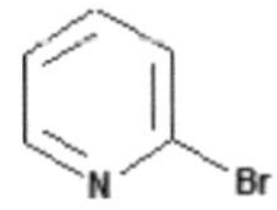
3-bromopyridine	C_5H_4BrN	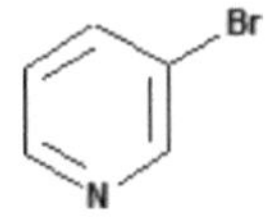
Sulfur-based compounds		
Sulfolane	$C_4H_8O_2S$	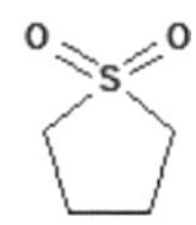
Thiosalicylic acid	$C_7H_6O_2S$	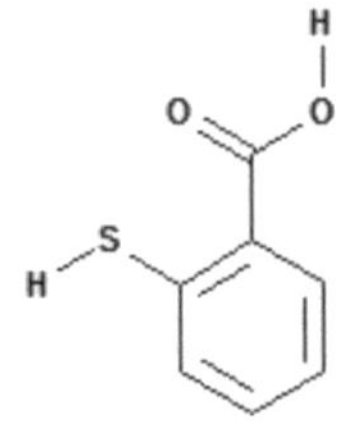

[a]Molecular structures redrawn from PubChem database [PubChem[1]].

Table 1. List of selected compounds.

http://dx.doi.org/10.5772/intechopen.72752

In addition to the abovementioned compounds, two sulfur-containing compounds, namely sulfolane and thiosalicylic acid, were also chosen since they were reported to be effective due to their superior electron-donating ability compared to nitrogen and oxygen [10–12]. Sulfolane is a heterocyclic compound with a sulfonyl reaction center. Thiosalicylic acid has one S- and two O-reaction centers.

2.2. Toxicity evaluation

The selected compounds were first screened based on compound toxicity using the criteria in that the corrosion inhibitors must not be more toxic than the absorbents used in the CO_2 absorption process. That is, their lethal dosage (LD_{50}) values must be equal or greater than those of the absorbents. Since MEA is the benchmark absorbent for the amine-based CO_2 absorption process, its toxicity value was used as the basis for comparison. Thus, any compound having LD_{50} (oral rat) less than LD_{50} of MEA (1720 mg/kg) was removed from the list of selected compounds to be tested. Note that, in addition to LD_{50} values, carcinogenicity and hazard rating for toxicity were also used for compound screening when toxicity information was not available.

Table 2 shows that 4-bromoaniline and 2-bromopyridine have LD_{50} lower than MEA, and all bromine (Br^- substituted) compounds are suspected carcinogens with hazard ratings of 3 (moderate hazard). This indicates greater toxicity and health risk of these compounds compared to other selected compounds. As a result, these bromine compounds (i.e., 2-bromoaniline, 3-bromoaniline, 4-bromoaniline, 2-bromopyridine, and 3-bromopyridine) were removed from the list of tested inhibitors. Seven remaining compounds were selected for further

Inhibitor	**LD_{50} (oral rat) mg/kg**	**Carcinogenicity**	**Hazard rating for toxicity[b]**
2-aminobenzene sulfonic acid	—	—	0
3-aminobenzene sulfonic acid	—	—	2
4-aminobenzene sulfonic acid	12,300	—	0
Sulfapyridine	15,800	—	2
Sulfanilamide	—	—	2
2-bromoaniline	—	Suspected carcinogens	3
3-bromoaniline	—	Suspected carcinogens	3
4-bromoaniline	456	Suspected carcinogens	3
2-bromopyridine	92	Suspected carcinogens	3
3-bromopyridine	—	Suspected carcinogens	3
Sulfolane	1941	—	2
Thiosalicylic acid	—	—	0

[a]All details were extracted from MSDS of each compound [chemwatch, 2012].
[b]Hazard ratings in a scale of 0–4 (0, min/nil; 1, low; 2, moderate; 3, high and 4, extreme).

Table 2. Toxicity information of the selected corrosion inhibitors[a].

Compound	E_{HOMO} (eV)	ΔN	Z
2-aminobenzene sulfonic acid	−9.17	0.25	2.54
3-aminobenzene sulfonic acid	−9.10	0.25	2.45
4-aminobenzene sulfonic acid	−9.45	0.22	2.48
Sulfanilamide	−9.18	0.25	2.52
Sulfapyridine	−9.01	0.27	2.56
Thiosalicylic acid	−9.27	0.21	−0.18
Sulfolane	−10.82	0.18	2.44

Table 3. Quantum chemical parameters of selected compounds.

analysis, that is, 2-aminobenzene sulfonic acid, 3-aminobenzene sulfonic acid, 4-aminobenzene sulfonic acid, sulfapyridine, sulfanilamide, sulfolane, and thiosalicylic acid.

2.3. Quantum chemical analysis

Quantum chemical analysis was carried out to predict the inhibition performance of selected compounds prior to corrosion tests. This was done to eliminate the selected compounds with the predicted inhibition performance that was much poorer than others. The analysis was carried out using the semiempirical quantum chemistry calculations package MOPAC2007 to perform the parameterization (PM-6) calculations [13]. The quantum parameters of interest included the highest occupied molecular orbital energy (E_{HOMO}), fraction of electron transferred (ΔN), and charge of the donor atom (Z). These parameters relate directly to the inhibition efficiency [10, 12, 14–16]. The results in **Table 3** show that E_{HOMO}, ΔN, and Z values of all selected compounds are in comparable ranges, suggesting comparable inhibition performance. Sulfapyridine has the highest E_{HOMO}, ΔN, and Z, implying that sulfapyridine might be the most efficient inhibitor compared to others. Thus, no selected compounds should be removed from the list.

3. Experiments

3.1. Experimental setup

Figure 1 illustrates the experimental setup used for electrochemical corrosion tests. It comprises a microcell, a water bath, a gas supply set, a water-cooled condenser, and a potentiostat. The microcell was a 100-ml three-electrode corrosion cell consisting of a cylindrical working electrode, a silver/silver chloride (Ag/AgCl) reference electrode, and a platinum counter electrode. The water bath with a temperature controller was used to control the temperature of the corrosion cell at a required temperature. The gas supply set consisted of CO_2 and nitrogen (N_2) cylinders with gas regulators and flow meters. The water-cooled condenser was used for minimizing evaporation losses from the corrosion cell. The potentiostat was PAR 263A (Princeton Applied Research, USA) interfaced with an impedance system (Model 5210 Lock-in amplifier). The Powercorr (Version 2.53) software was used to record and analyze the

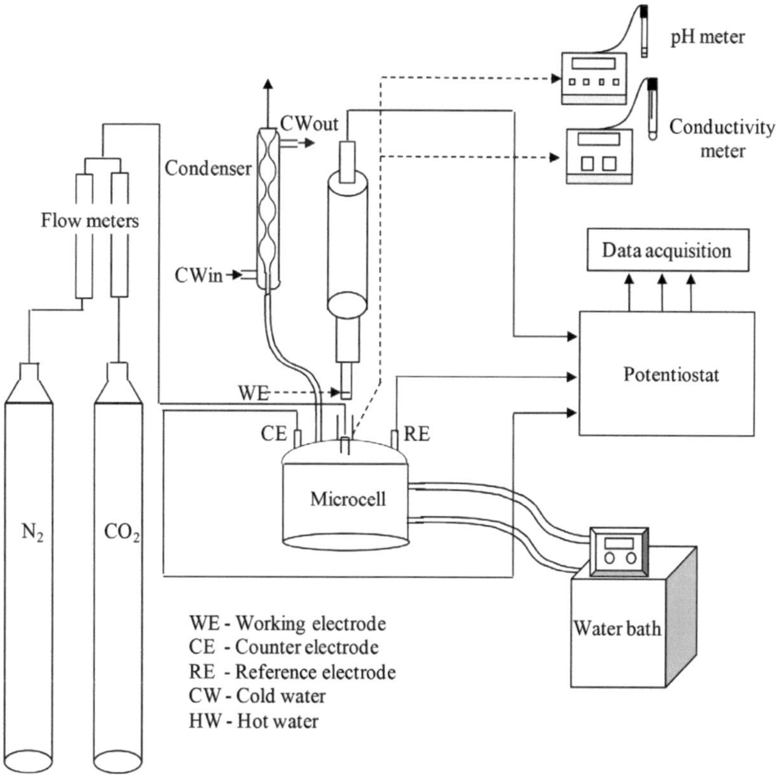

Figure 1. A schematic diagram of the electrochemical corrosion testing system.

results. A pH meter (Oakton pH 510 series) and a conductivity meter (YSI 3200 conductivity instrument) were connected to the setup for sample measurements.

3.2. Specimen preparation

Carbon steel (CS 1018) with the composition of 0.175% carbon, 0.75% manganese, and balance iron was chosen as the working electrode due to its common use as the construction material in amine-based CO_2 absorption plants [17]. The specimens were cylindrical with a dimension of 0.8 cm in height, 1.2 cm in diameter, and 0.6 cm central hole. Prior to each experiment, the specimens were surface finished with up to 600 grit silicon, degreased with methanol, rinsed with deionized water, and dried with air.

3.3. Solution preparation

A 5.0 kmol/m^3 aqueous solution of monoethanolamine (MEA) was used as the CO_2 absorption solvent to simulate the service process solution [1]. The MEA solution was prepared from 99% reagent grade MEA and deionized water, purged with CO_2 to achieve a saturation loading of 0.55 ± 0.05 mol CO_2/mol MEA and added with a tested corrosion inhibitor. The inhibitor concentrations were in the range of 250–10,000 ppm. To evaluate the performance of the tested inhibitors in the MEA solution containing process contaminants, 10,000 ppm of the representative process contaminants, namely sodium chloride (NaCl) and formic acid

(CH_2O_2), were introduced to the MEA solution. A titration-based Chittick apparatus together with hydrochloric acid (HCL) and methyl orange indicators was used for analyzing MEA concentration and CO_2 loading.

3.4. Validation of experimental setup and procedure

Prior to tests, the anodic polarization and impedance scans of stainless steel (SS430) specimens were carried out in accordance with the ASTM G5-94 (reapproved in 2004) [18] and the ASTM G106-89 (reapproved in 2010) [19], respectively. The anodic polarization scan was done in a 1.0-N sulfuric acid (H_2SO_4) solution that was deaerated with nitrogen (N_2) at 30°C. The impedance scan was conducted using a deaerated 0.495-M sodium sulfate (Na_2SO_4) solution containing 0.005 M·H_2SO_4 at 30°C. The results of both scans were comparable to those of the ASTM standards, thus validating the experimental setup and the polarization and impedance procedures.

3.5. Experimental procedure

Prior to each test, the prepared MEA solution was transferred to a microcell where the solution temperature was controlled at 80°C, the CO_2 loading was maintained at saturation by CO_2 purging, and the solution concentration was kept at 5.0 kmol/m^3. The open circuit potentials (OCP) were monitored with time until they reached equilibrium. The impedance scan was performed using 10 mV AC amplitude over a frequency range of 10 kHz–10 MHz. The corresponding impedance results were recorded in respect of frequency. After the impedance scan, the cyclic potentiodynamic polarization scan was then performed at 10 mV per minute. The solution samples were taken for the analysis of MEA concentration, CO_2 loading, pH, and conductivity before and after the electrochemical scans. The tested specimen was taken for surface analysis using the scanning electron microscopy (SEM).

3.6. Data analysis

The corrosion current density (i_{corr}) in the unit of μA/cm^2 was determined from the obtained potentiodynamic polarization data using the Tafel extrapolation. It was subsequently translated to corrosion rate (CR) in the unit of mmpy using the following equation:

$$CR = \frac{3.27 \times 10^{-3}\, a i_{corr}}{nD} \quad (1)$$

where a, n, and D represent atomic weight, number of electrons, and density of specimen in g/cm^3, respectively. The obtained corrosion rates of the uninhibited and inhibited specimens were then used for determining the inhibition efficiency (IE).

$$IE = \left(\frac{CR_{uninhibited} - CR_{inhibited}}{CR_{uninhibited}}\right) \quad (2)$$

4. Results and discussion

The corrosion inhibition performance of seven selected compounds in aqueous solutions of MEA was evaluated under test conditions listed in **Table 4**. The evaluation was done in the

http://dx.doi.org/10.5772/intechopen.72752

Parameter	Test condition
Amine type	Monoethanolamine (MEA)
Amine concentration (kmol/m^3)	5.0 ± 0.1
Temperature (°C)	80.0 ± 1.0
CO_2 loading (mol CO_2/mol amine)	Saturation (0.55 ± 0.05)
Process contaminants	Chloride formate
Tested corrosion inhibitors	2-aminobenzene sulfonic acid
	3-aminobenzene sulfonic acid
	4-aminobenzene sulfonic acid
	Sulfapyridine
	Sulfanilamide
	Sulfolane
	Thiosalicylic acid
Inhibitor concentration (ppm)	
2-aminobenzene sulfonic acid	250, 500, 1000, 2000, 3000, 10,000
3-aminobenzene sulfonic acid	250, 500, 1000, 2000, 3000, 10,000
4-aminobenzene sulfonic acid	250, 500, 1000, 2000, 3000, 10,000
Sulfapyridine	500, 1000, 2000, 3000, 5000, 7000, 10,000
Sulfanilamide	3000, 10,000
Sulfolane	1000, 2000, 3000, 5000, 10,000
Thiosalicylic acid	3000, 10,000

Table 4. Summary of experimental parameters and conditions.

absence and presence of process contaminants (i.e., chloride and formate). Chloride was chosen because it is in the water used for the amine solution preparation and also found in the coal-fired flue gas fed to the carbon capture process. Formate was chosen because it is one of the predominant heat-stable salts produced from oxidative degradation of the amine solutions.

4.1. 2-aminobenzene sulfonic acid

4.1.1. Absence of process contaminants

The corrosion inhibition performance of 2-aminobenzene sulfonic acid was evaluated as a function of the inhibitor concentration. It is seen in **Figure 2a** that the corrosion rate of carbon steel in the 2-aminobenzene sulfonic acid-inhibited MEA solution significantly reduced from 4.27 mmpy (uninhibited) to 0.46–0.57 mmpy with inhibition efficiencies of 87–89%, when the inhibitor concentrations were in the range of 250–3000 ppm. However, at 10,000 ppm, the corrosion rate increased to 1.62 mmpy and the inhibition efficiency reduced to 62%.

From the potentiodynamic polarization curves in **Figure 2b**, it is apparent that, in the inhibited solutions, the carbon steel specimens were in an active state similar to the specimen in the uninhibited solution. The cathodic polarization curves shifted to lower current densities

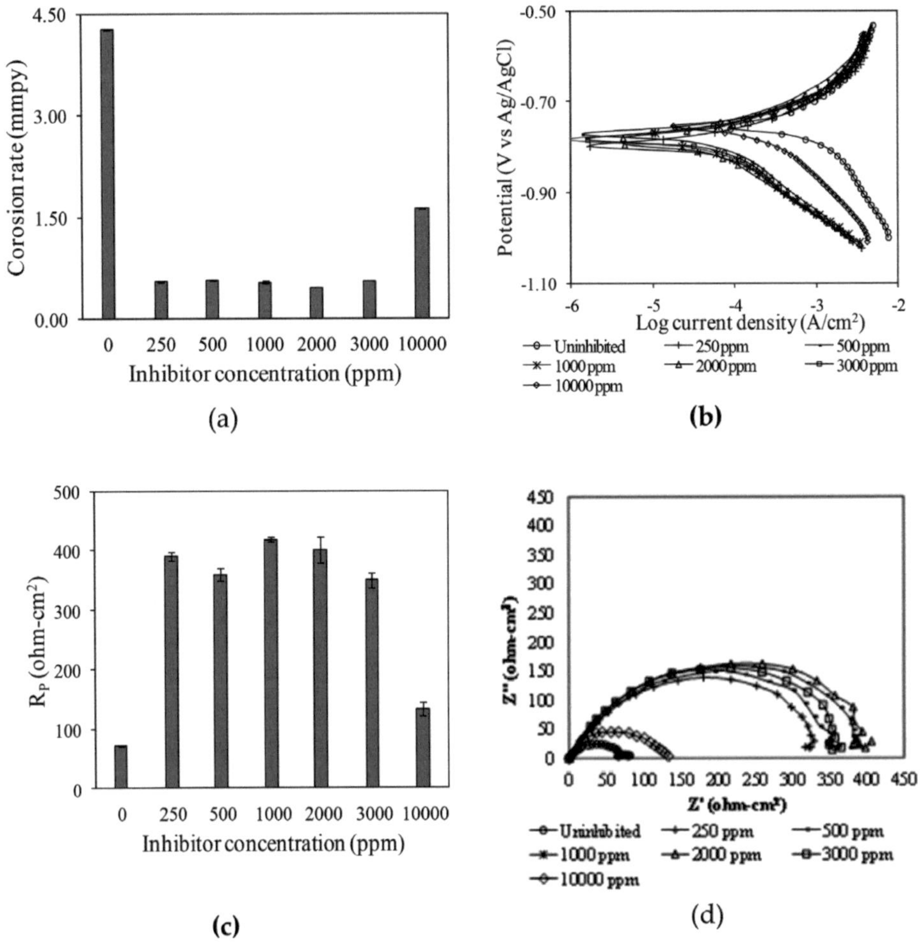

Figure 2. Corrosion inhibition performance of 2-aminobenzene sulfonic acid in 5.0 kmol/m^3 MEA solutions containing 0.55 mol/mol CO_2 loading in the absence of process contaminants at 80°C. (a) corrosion rate, (b) polarization curve, (c) polarization resistance, and (d) impedance behavior.

compared to the uninhibited condition while the anodic polarization curves remained unchanged. This suggests that the corrosion inhibition was cathodic in nature and was due to the preferential adsorption of inhibitor molecules onto the cathodic sites of the metal surface. The change in Tafel slopes of the inhibited potentiodynamic polarization curves compared to the uninhibited condition suggests a change in the corrosion mechanism. From the cyclic polarization curve obtained, no pitting was induced by the presence of 2-aminobenzene sulfonic acid at all tested concentrations.

The polarization resistance (R_p) values were analyzed from the obtained impedance data of the uninhibited and inhibited MEA solutions to reveal the effect of inhibitor concentration on inhibition performance. **Figure 2c** shows that the values of R_p increased from 72 (uninhibited) to 418 ohm-cm^2 when the concentration of 2-aminobenzene sulfonic acid was raised to 3000 ppm. However, the R_p value decreased to 134 ohm-cm^2 when the concentration of 2-aminobenzene sulfonic acid further increased to 10,000 ppm.

From **Figure 2d**, the Nyquist plots of inhibited solutions were in a semicircle shape representing a capacitive loop due to charge transfer kinetics. This suggests that the inhibitor protected the carbon steel surface by adsorption of inhibitor molecules, not by forming a passive barrier. The reduction in the inhibition performance at 10,000 ppm may have been caused by the lateral interactions of the adsorbed inhibitor on the metal surface [20].

4.1.2. Presence of process contaminants

Results in **Figure 3** show that chloride has a negligible effect on the inhibition performance of 2-aminobenzene sulfonic acid. The inhibited corrosion rate in the presence of chloride (i.e., 0.62 mmpy) was slightly higher than that with no process contaminants (i.e., 0.54 mmpy). This was confirmed by a small reduction in the R_p value from 418 to 380 ohm-cm^2 due to the presence of chloride. The inhibition efficiency was slightly reduced from 87 to 86%, which is considered negligible.

In contrast, formate appears to deteriorate the performance of 2-aminobenzene sulfonic acid. When 10,000 ppm formate was present in the solution, the inhibited corrosion rate increased from 0.54 to

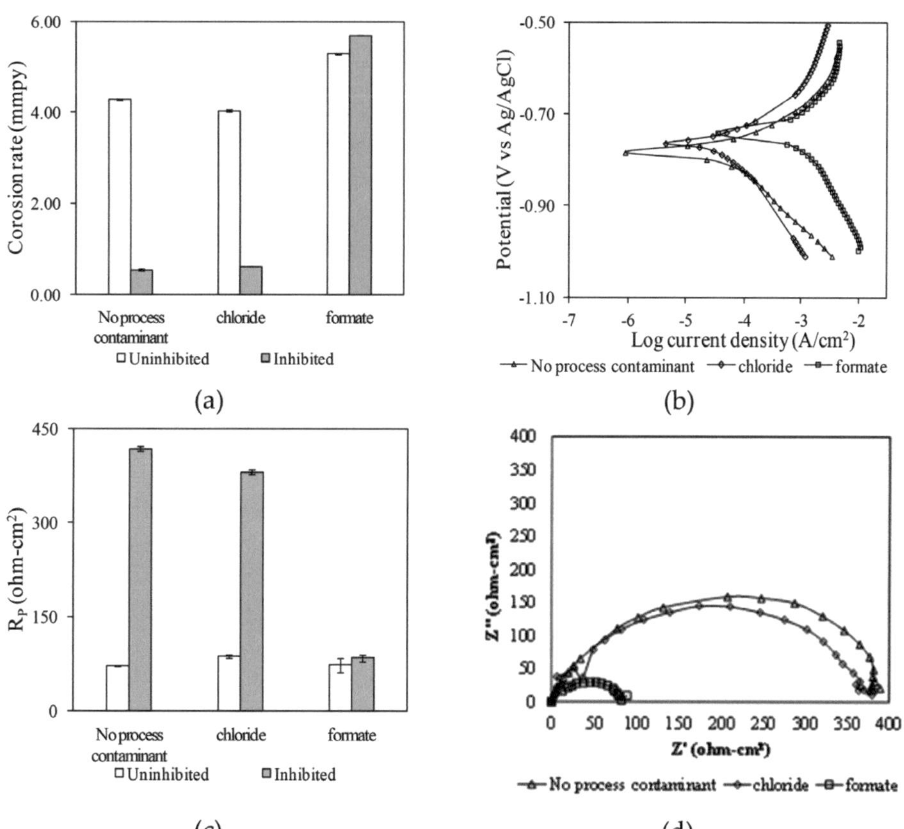

Figure 3. Corrosion inhibition performance of 1000 ppm of 2-aminobenzene sulfonic acid in 5.0 kmol/m^3 MEA solutions containing 0.55 mol/mol CO_2 loading in the presence of process contaminants at 80°C. (a) Corrosion rate, (b) polarization curve, (c) polarization resistance, and (d) impedance behavior.

5.70 mmpy with a significant reduction in the R_p value from 418 to 85 ohm-cm^2. Such an increase in the corrosion rate can also be observed by the significant shift of cathodic polarization curves to greater current densities. This suggests that the adsorption of inhibitor molecules on the metal surface may have been disrupted, thus allowing higher transport rate of corroding agents between the metal surface and the solution. No pitting was induced by the presence of chloride and formate.

4.2. 3-aminobenzene sulfonic acid

4.2.1. Absence of process contaminants

The 3-aminobenzene sulfonic acid performed well in the 5.0 kmol/m^3 MEA containing 0.55 mol/mol CO_2 loading and no process contaminants at 80°C. From **Figure 4**, the corrosion rates of carbon steel specimens were 0.48–0.49 mmpy with 89% inhibition efficiency when the concentrations of 3-aminobenzene sulfonic acid were in the range of 250–3000 ppm. This was confirmed by the R_p values that increased from 72 to 369–428 ohm-cm^2. Such a reduction in corrosion rates was a result of cathodic inhibition by 3-aminobenzene sulfonic acid. The cathodic polarization curves of the inhibited solutions shifted to lower current densities compared to that of the uninhibited. This suggests that this inhibitor acted as the cathodic inhibitor that impeded cathodic reactions by the adsorption of inhibitor molecules onto the metal surface. No pitting was observed from the cyclic polarization curves and surface analysis.

Results in **Figure 4** also show that applying this inhibitor at a high concentration of 10,000 ppm reduced the inhibition effectiveness. That is, the corrosion rate increased to 3.81 mmpy with the inhibition efficiency of 11% and the R_p of 88 ohm-cm^2. Such reduction in the inhibition performance was observed from the cathodic polarization curves of the inhibited solutions that shifted toward the cathodic curve of the uninhibited solution. Their current densities were slightly lower than the uninhibited curve. The shift of the cathodic curve in this direction illustrates deterioration of adsorption performance of the inhibitor which might be due to the increased lateral attractions in the adsorbed layer of inhibitor molecules on the metal surface.

4.2.2. Presence of process contaminants

The presence of chloride and formate caused a negative effect on corrosion inhibition performance of 3-aminobenzene sulfonic acid. In **Figure 5** the corrosion rate in the inhibited solution containing chloride increased from 0.48 mmpy (in the absence of process contaminants) to 1.65 mmpy. As a result, the inhibition efficiency was reduced from 89 to 61%. In case of formate, the corrosion rate in the inhibited solution increased to 5.31 mmpy, thus reducing the inhibition efficiency to −24%. This indicates that the inhibitor did not retard the corrosion but in fact aggravated corrosion. The impedance results also support such behavior. That is, the R_p value decreased from 369 ohm-cm^2 (in the absence of process contaminants) to 88 ohm-cm^2 in the presence of chloride and to 105 ohm-cm^2 in the presence of formate.

The deterioration of inhibition performance in the presence of chloride and formate can be observed from the potentiodynamic polarization curves. The tested carbon steel underwent active corrosion, and the cathodic current densities were greater in the presence of contaminants than those in the absence of contaminants. This might be due to the disruption of the adsorption of 3-aminobenzene sulfonic acid molecules onto the metal surface in the presence of process contaminants. The effect was more pronounced in the presence of formate than chloride.

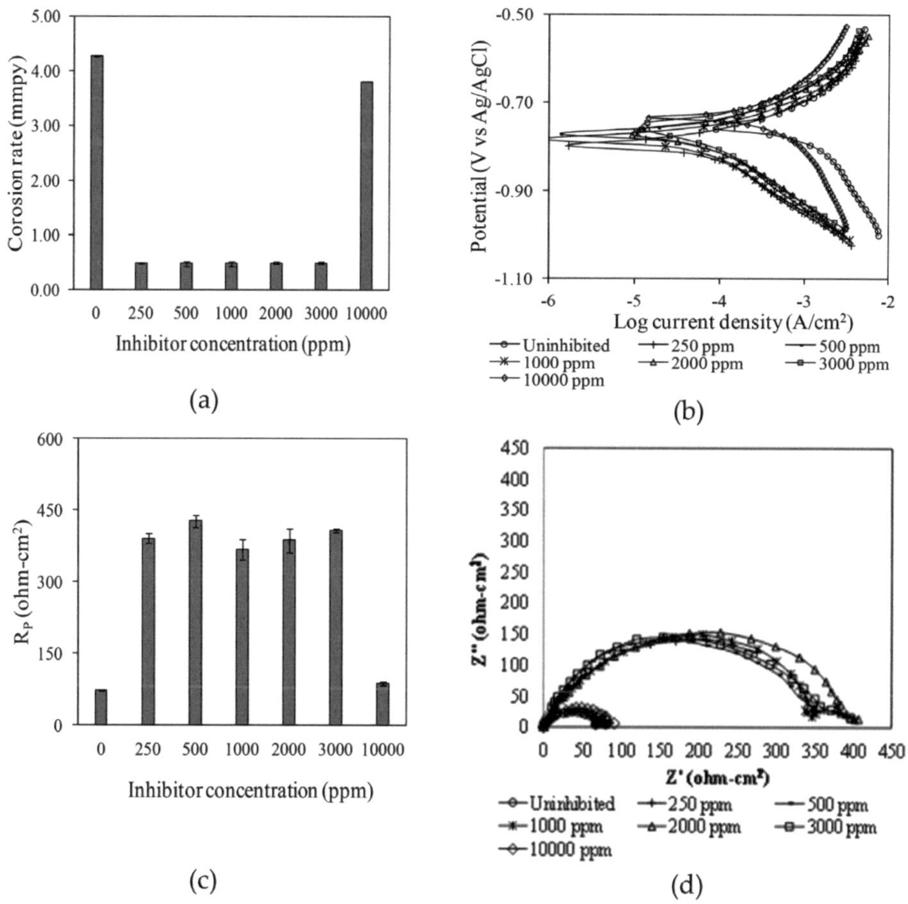

Figure 4. Corrosion inhibition performance of 3-aminobenzene sulfonic acid in 5.0 kmol/m³ MEA solutions containing 0.55 mol/mol CO_2 loading in the absence of process contaminants at 80°C. (a) Corrosion rate, (b) polarization curve, (c) polarization resistance, and (d) impedance behavior.

It should be noted that the presence of chloride caused pitting corrosion on carbon steel in the MEA solution inhibited by 3-aminobenzene sulfonic acid. Pits were detected from the cyclic polarization curve in **Figure 6a** in the form of positive hysteresis, that is, the reverse curve lies to the right of the forward curve. These pits were also seen in an SEM image of the specimen after test (**Figure 6b**).

4.3. 4-aminobenzene sulfonic acid

4.3.1. Absence of process contaminants

4-Aminobenzene sulfonic acid significantly reduced the corrosion rates of carbon steel in the 5.0 kmol/m³ MEA containing 0.55 mol/mol CO_2 loading and no process contaminant at 80°C. As shown in **Figure 7a**, the corrosion rate decreased from 4.27 to 0.38–0.56 mmpy with inhibition efficiencies of 87–91% when the inhibitor concentrations were in the range of 250–3000 ppm. However, the corrosion rate increased to 3.77 mmpy and the inhibition efficiency dropped to 12% when 10,000 ppm was used.

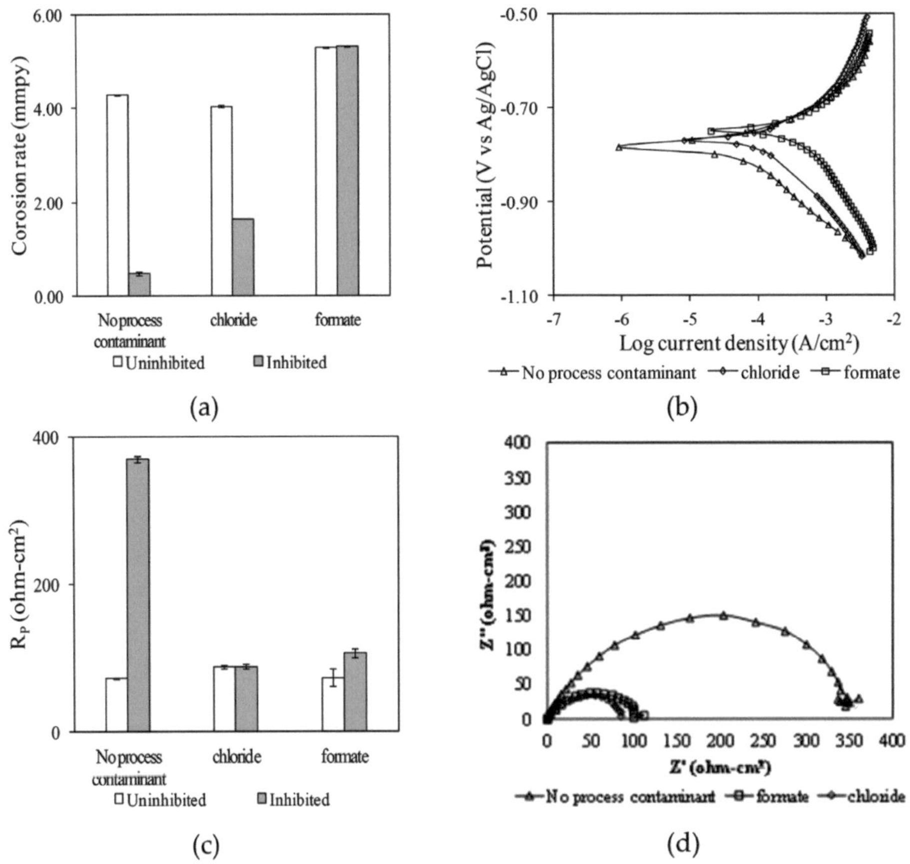

Figure 5. The corrosion inhibition performance of 1000 ppm of 3-aminobenzene sulfonic acid in 5.0 kmol/m^3 MEA solutions containing 0.55 mol/mol CO_2 loading in the presence of process contaminants at 80°C. (a) Corrosion rate, (b) polarization curve, (c) polarization resistance, and (d) impedance behavior.

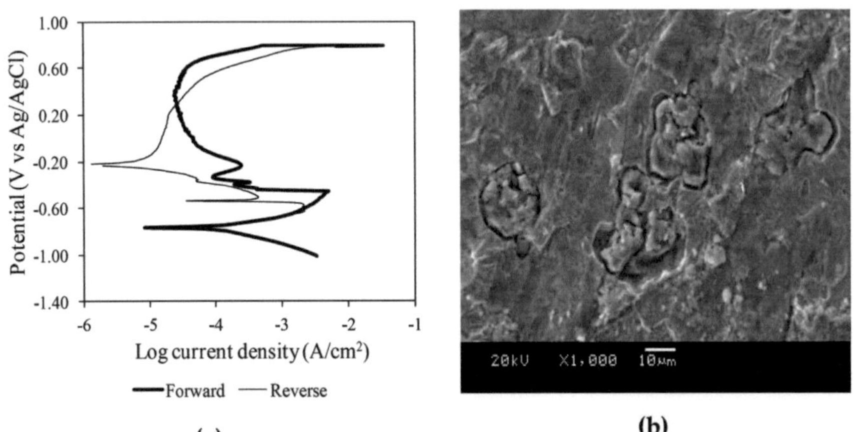

Figure 6. Pitting corrosion in 5.0 kmol/m^3 MEA solutions containing 0.55 mol/mol CO_2 loading, inhibited by 1000 ppm of 3-aminobenzene sulfonic acid in the presence of 10,000 ppm chloride at 80°C (a) cyclic polarization curve showing positive hysteresis (b) SEM image showing pits with 1000× magnification.

Such inhibition performance of 4-aminobenzene sulfonic acid was observed by the potentiodynamic polarization curves in **Figure 7b**. The tested specimens were in an active state. The 4-aminobenzene sulfonic acid functioned as a cathodic corrosion inhibitor. At 250–3000 ppm, the cathodic current densities were lower than the uninhibited cathodic current densities suggesting that the cathodic reactions were impeded by the adsorption of the tested inhibitor. No pitting tendency was induced by the presence of 4-aminobenzene sulfonic acid at any concentrations.

The values of R_p obtained from impedance analysis (**Figure 7c**) showed similar results to the polarization analysis. The R_p was in the range of 355–460 ohm-cm^2 at the inhibitor concentrations of 250–3000 ppm and then dropped to 99 ohm-cm^2 at the inhibitor concentration of 10,000 ppm. The impedance analysis (**Figure 7d**) yielded a semicircle characteristic of charge transfer kinetics at the metal-solution interface, suggesting that no passive layer was present.

4.3.2. Presence of process contaminants

The corrosion inhibition performance of 4-aminobenzene sulfonic acid at 1000 ppm was unaffected by the presence of chloride but deteriorated in the presence of formate. No pitting tendency was found in either case. From **Figure 8a**, the corrosion rate of a carbon steel specimen remained close to the uncontaminated corrosion rate (0.45 mmpy) when 10,000 ppm chloride was added to the inhibited MEA solution. The presence of chloride did not alter anodic and cathodic polarization behavior of the inhibited MEA solution. On the contrary, the addition of 10,000 ppm formate into the inhibited MEA solution aggravated corrosion and caused the corrosion rate of the inhibited MEA solution to increase to 5.45 mmpy (−28% inhibition efficiency) which exceeded the corrosion rate of the uninhibited solution. This effect was evidenced by the shift of the cathodic polarization curve to higher current densities in **Figure 8b**. Such a shift of the cathodic curve implied a higher flux of corroding agents on the metal surface, which resulted from the deficiency of adsorption of inhibitor molecules onto the metal surface in the presence of formate.

The impedance results (**Figure 8c**) also confirm the abovementioned findings. The R_p in the presence of chloride (385 ohm-cm^2) was slightly lower than the no contaminant condition (441 ohm-cm^2) whereas the R_p in the presence of formate (87 ohm-cm^2) was drastically lower. A semicircular loop in **Figure 8d** exhibited charge transfer kinetics.

4.4. Sulfapyridine

4.4.1. Absence of process contaminants

From **Figure 9a**, it was apparent that sulfapyridine was not effective at 500 and 1000 ppm in the 5.0 kmol/m^3 MEA containing 0.55 mol/mol CO_2 loading and no process contaminant at 80°C. At 500 ppm, the corrosion rate of carbon steel remained at 4.27 mmpy which was the corrosion rate of the uninhibited solution. At 1000 ppm, the corrosion rate was reduced to 3.07 mmpy with an inhibition efficiency of 28%. The inhibition efficiency was significantly increased to 90–92% with a reduction in corrosion rates to 0.33–0.44 mmpy when sulfapyridine concentration increased to the range of 2000–10,000 ppm.

The effectiveness of sulfapyridine can be observed from potentiodynamic polarization curves in **Figure 9b**. The addition of sulfapyridine caused the cathodic polarization curves to shift to

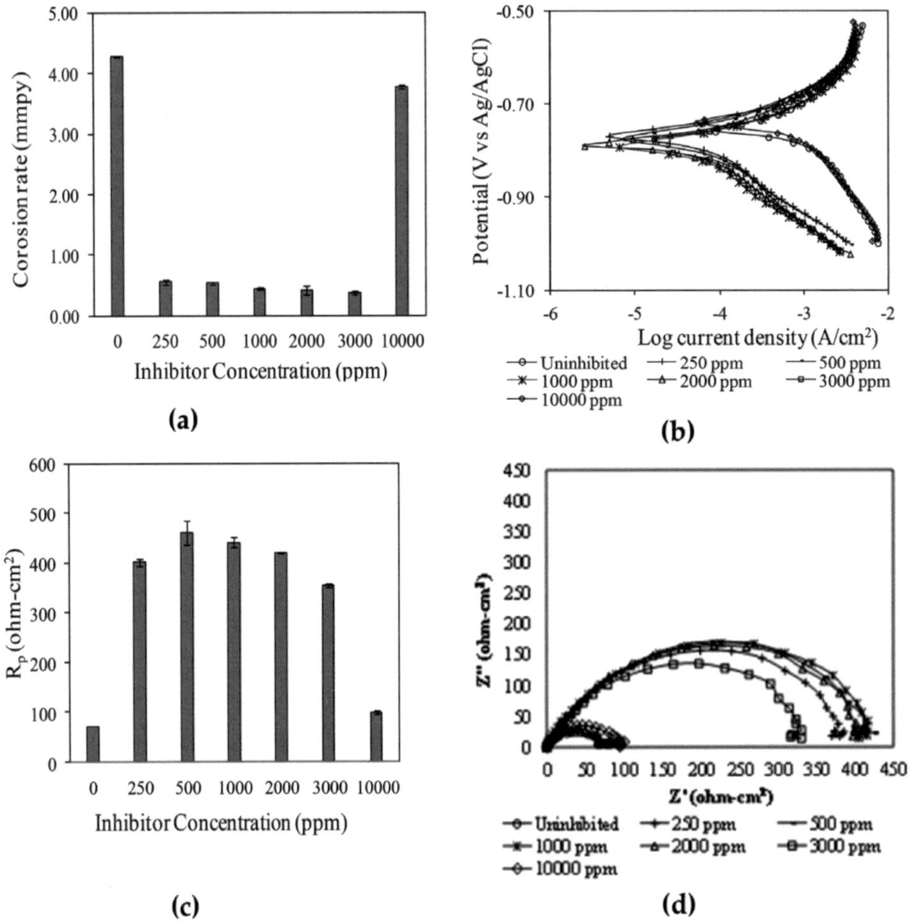

Figure 7. The corrosion inhibition performance of 4-aminobenzene sulfonic acid in 5.0 kmol/m^3 MEA solutions containing 0.55 mol/mol CO_2 loading in the absence of process contaminants at 80°C. (a) Corrosion rate, (b) polarization curve, (c) polarization resistance, and (d) impedance behavior.

lower current densities but did not alter the anodic curves. This suggested that sulfapyridine acted as a cathodic inhibitor which formed an adsorption layer onto the metal-solution interface and thus retarded the mass transfer rate of corroding agents. The cyclic polarization curves and the surface analysis revealed no pitting corrosion at any concentrations of sulfapyridine.

The R_p obtained from the impedance analysis (**Figure 9c**) reinforced the above findings in that sulfapyridine did not function effectively below 2000 ppm. The R_p value of the MEA solution containing 500 ppm sulfapyridine was 86 ohm-cm^2, which was similar to the R_p of the uninhibited solution. However, once the sulfapyridine concentration increased to 1000 ppm, the R_p value began to rise to 105 ohm-cm^2. The R_p was further increased to 397–528 ohm-cm^2 when the sulfapyridine concentrations were 2000–10,000 ppm. The impedance analysis (**Figure 9d**) also showed the semicircular loop characteristic of charge transfer kinetics at the interface, suggesting the absence of passive layer. Hence, the higher R_p values obtained were due to the adsorption of sulfapyridine onto the metal surface.

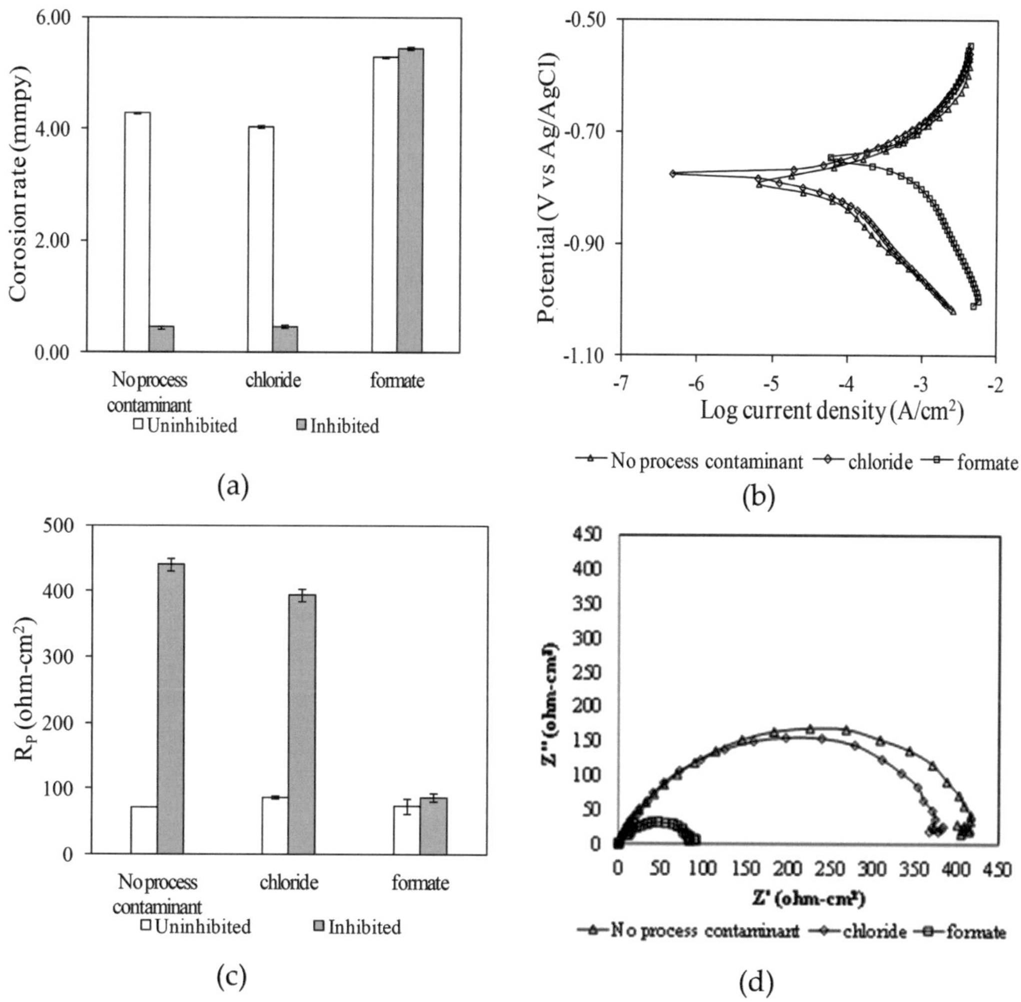

Figure 8. Corrosion inhibition performance of 1000 ppm of 4-aminobenzene sulfonic acid in 5.0 kmol/m³ MEA solutions containing 0.55 mol/mol CO_2 loading in the presence of process contaminants at 80°C. (a) Corrosion rate, (b) polarization curve, (c) polarization resistance, and (d) impedance behavior.

4.4.2. Presence of process contaminants

Figure 10 shows that the presence of formate in the MEA solution did not have an apparent effect on the inhibition performance of sulfapyridine. When 10,000 ppm formate was present in the MEA solution, the inhibition efficiency of sulfapyridine slightly reduced from 90% (in the absence of formate) to 88%, and the R_p value also reduced from 468 to 358 ohm-cm². This result can also be observed from a slight shift of the cathodic polarization curve to lower current densities.

Unlike formate, chloride had a detrimental effect on the inhibition performance of sulfapyridine. The presence of 10,000 ppm chloride in the MEA solution caused the corrosion rate of carbon steel to increase from 0.44 mmpy (in the sulfapyridine-inhibited MEA solution) to 4.34 mmpy, reflecting a reduction of the inhibition performance from 90% to −2%. The R_p value was significantly

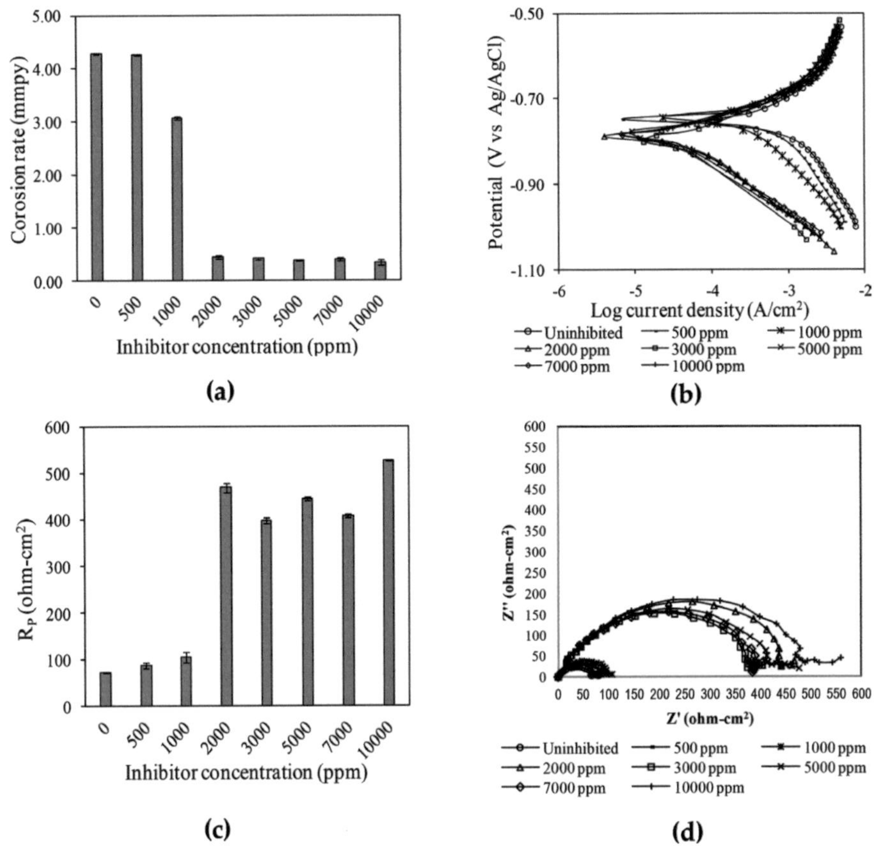

Figure 9. Corrosion inhibition performance of sulfapyridine in 5.0 kmol/m³ MEA solutions containing 0.55 mol/mol CO_2 loading in the absence of process contaminants at 80°C. (a) Corrosion rate, (b) polarization curve, (c) polarization resistance, and (d) impedance behavior.

reduced from 468 to 91 ohm-cm², which is close to the R_p of the uninhibited MEA solution. The cathodic polarization curve of the sulfapyridine MEA solution shifted to much higher current densities in the presence of chloride. This suggested that chloride could destroy the inhibition adsorption ability of sulfapyridine. No pitting was induced in the presence of both formate and chloride.

4.5. Sulfanilamide

Sulfanilamide did not perform well in the 5.0 kmol/m³ MEA solution containing 0.55 mol/mol CO_2 loading at 80°C. When 3000 and 10,000 ppm sulfanilamide were introduced to the MEA solutions, the corrosion rate of carbon steel reduced from 4.27 mmpy to 2.50 and 3.41 mmpy, respectively, with corresponding inhibition efficiencies of 42 and 20% (**Figure 11a**). Such poor inhibition performance of sulfanilamide was apparent from small shifts of cathodic curves from the uninhibited curve (**Figure 11b**) and small increments of R_p in the presence of sulfanilamide (**Figure 11c**). The R_p increased from 72 ohm-cm² in the uninhibited MEA solution to only 104 and 84 ohm-cm² at 3000 and 10,000 ppm sulfanilamide, respectively. This reflected poor resistance of the adsorption layer on metal surface to the transport of corroding agents.

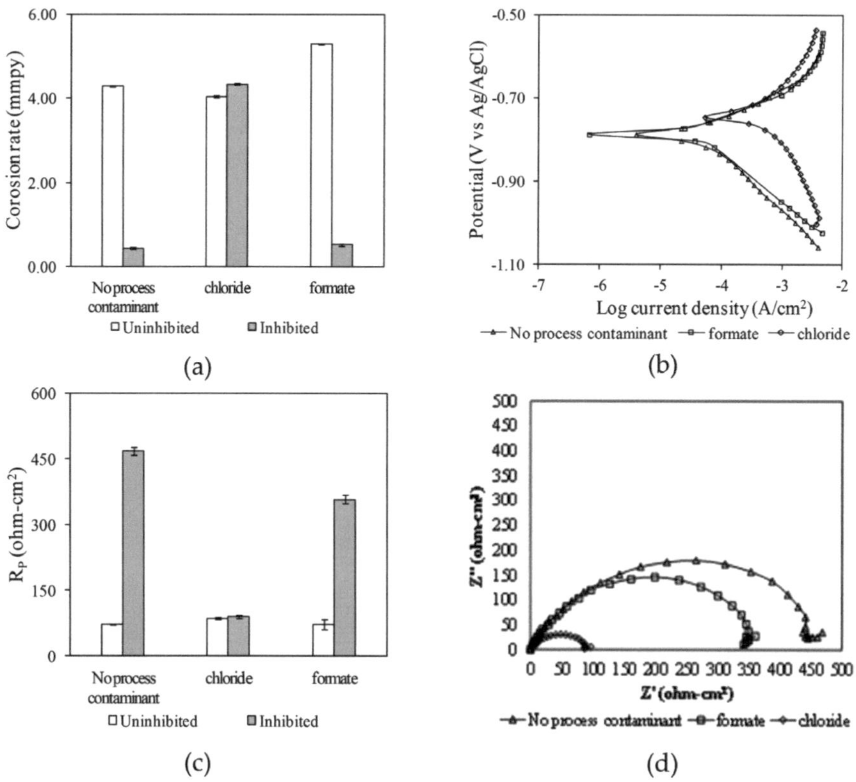

Figure 10. The corrosion inhibition performance of 2000 ppm of sulfapyridine in 5.0 kmol/m³ MEA solutions containing 0.55 mol/mol CO_2 loading in the presence of process contaminants at 80°C. (a) Corrosion rate, (b) polarization curve, (c) polarization resistance, and (d) impedance behavior.

The cyclic polarization curve and surface analysis showed no pitting in the presence of sulfanilamide at any concentrations. The impedance analysis (**Figure 11d**) traced a semicircle or a capacitive loop due to charge transfer kinetics similar to that of the uninhibited system, suggesting that no passive film developed on the surface. Since its corrosion inhibition efficiencies were much lower than other tested compounds, sulfanilamide was not further tested for the effect of process contaminants.

4.6. Sulfolane

4.6.1. Absence of process contaminants

Sulfolane was an effective corrosion inhibitor when an adequate concentration was introduced to the solution. As seen in **Figure 12a**, an addition of 1000 ppm sulfolane reduced the corrosion rate of carbon steel from 4.27 mmpy in the uninhibited solution to 3.63 mmpy with 15% inhibition efficiency. Once the concentration of sulfolane increased to 2000–10,000 ppm, the corrosion rate substantially reduced to 0.43–0.62 mmpy, with an inhibition efficiency of 85–90%. This finding was supported by the impedance analysis results (**Figure 12c** and **d**). The R_p increased from 116 to the range of 331–445 ohm-cm² when the concentration of sulfolane increased from 1000 to 2000–10,000 ppm. No pitting was induced by the presence of sulfolane at any concentrations.

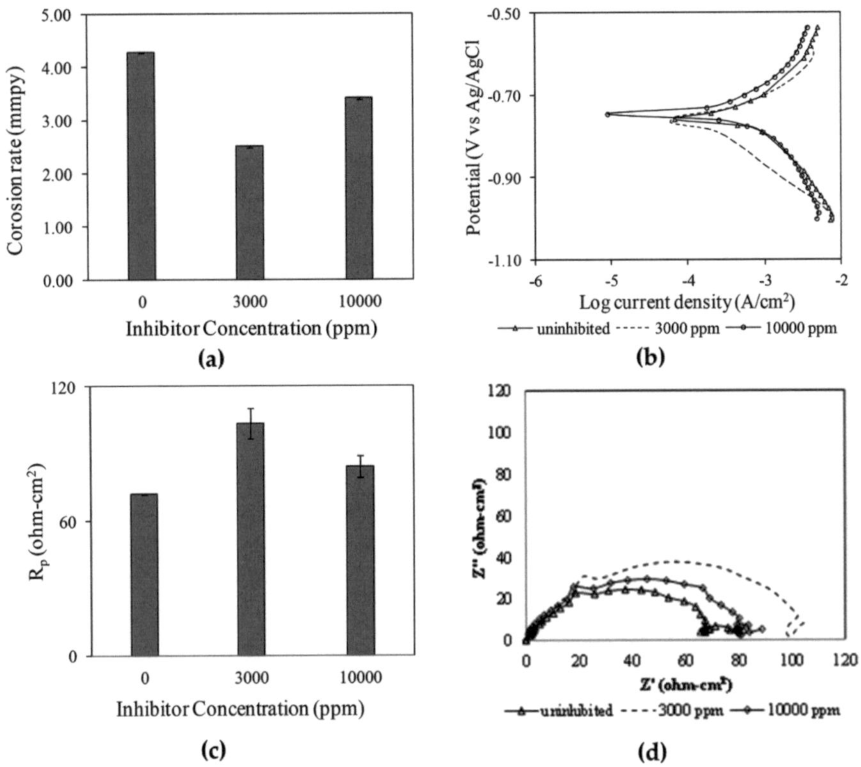

Figure 11. The corrosion inhibition performance of sulfanilamide in 5.0 kmol/m³ MEA solutions containing 0.55 mol/mol CO_2 loading in the absence of process contaminants at 80°C. (a) Corrosion rate, (b) polarization curve, (c) polarization resistance, and (d) impedance behavior.

The effectiveness of sulfolane can be explained by considering its polarization behavior in **Figure 12b**. The presence of sulfolane in the solution caused the cathodic polarization curve to shift to the direction of lower current densities but did not alter the anodic curve. This suggested that sulfolane acted as a cathodic inhibitor which was adsorbed onto the metal surface to inhibit the transport of corroding species between the metal surface and MEA solution. As a result, the rate of the cathodic reaction was retarded which was exhibited by the shift of the cathodic curve.

Due to the nature of inhibitor adsorption onto the metal surface, it is important to have sufficient inhibitor coverage on the metal surface. This was why the increasing sulfolane concentration from 1000 to 2000 ppm resulted in a much higher inhibition efficiency. However, an addition of a higher sulfolane concentration than 2000 ppm yields very small increases in efficiency. This suggested that the minimum concentration required for effective corrosion inhibition was 2000 ppm.

4.6.2. Presence of process contaminants

The inhibition performance of sulfolane was not considerably affected by the presence of chloride or formate. As illustrated in **Figure 13a**, the corrosion rate of carbon steel in the

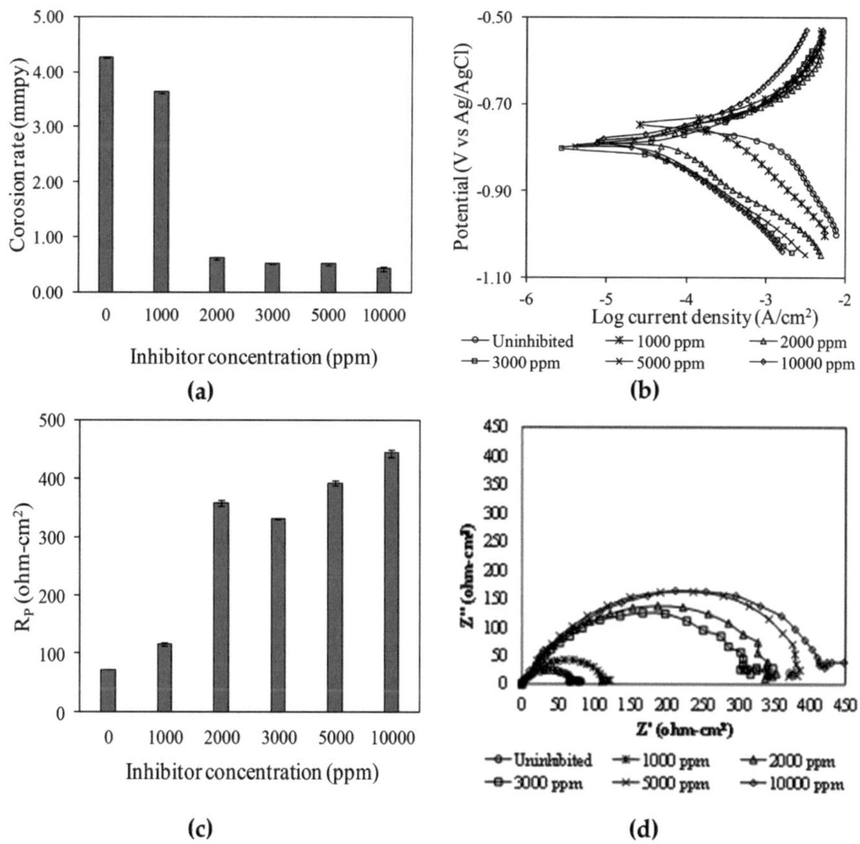

Figure 12. The corrosion inhibition performance of sulfolane in 5.0 kmol/m^3 MEA solutions containing 0.55 mol/mol CO_2 loading in the absence of process contaminants at 80°C. (a) Corrosion rate, (b) polarization curve, (c) polarization resistance, and (d) impedance behavior.

sulfolane-inhibited solution did not change much when chloride or formate was added to the solution. The corrosion rates in the presence of chloride and formate were 0.48 and 0.44 mmpy, respectively, which were slightly lower than the corrosion rate in the absence of these two contaminants (0.62 mmpy). This reflects slight increases in inhibition efficiencies from 85% (the absence of contaminants) to 89–90% (the presence of formate and chloride). The insignificant effects of chloride and formate were also observed from the cathodic polarization curves (**Figure 13b**) that were not shifted from the no contaminant condition in the presence of chloride and formate. No pitting was induced by these two contaminants.

From the impedance analysis (**Figure 13c** and **d**), the R_p of formate-MEA-sulfolane solution (431 ohm-cm^2) was higher than that of the MEA-sulfolane solution (358 ohm-cm^2). This showed a small increase in the resistance developed on the metal surface in the presence of formate, which supported the above findings from the potentiodynamic polarization tests. However, the R_p of the chloride-containing solution decreased to 314 ohm-cm^2, which indicated a slight reduction in inhibition efficiency in the presence of chloride. Thus, this implied that chloride may slightly deteriorate the inhibition performance of sulfolane.

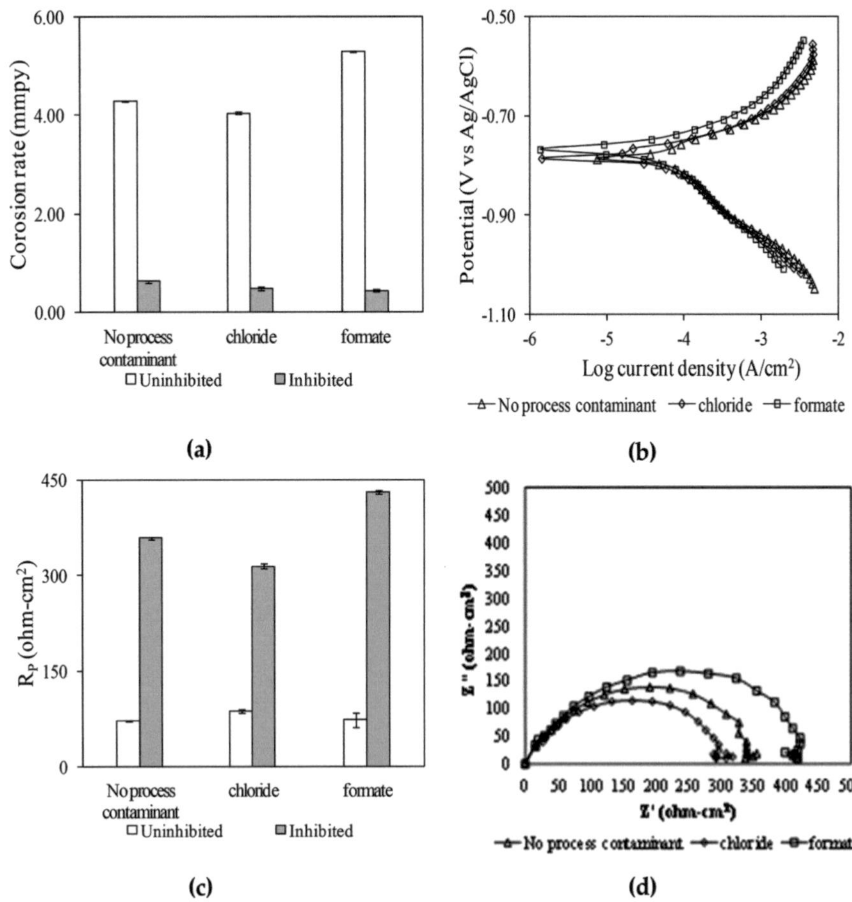

Figure 13. The corrosion inhibition performance of 2000 ppm of sulfolane in 5.0 kmol/m³ MEA solutions containing 0.55 mol/mol CO_2 loading in the presence of process contaminants at 80°C. (a) Corrosion rate, (b) polarization curve, (c) polarization resistance, and (d) impedance behavior.

4.7. Thiosalicylic acid

Thiosalicylic acid was found to reduce the corrosion rate of carbon steel from 4.27 mmpy in the uninhibited MEA solution to 0.84 and 2.24 mmpy with the inhibition efficiencies of 80 and 47% when its concentrations were 3000 and 10,000 ppm, respectively (**Figure 14a**). This finding was confirmed by the values of R_p from the impedance analysis (**Figure 14c**). That is, the R_p increased from 72 in the uninhibited MEA solution to 190 and 101 ohm-cm² for 3000 and 10,000 ppm thiosalicylic acid, respectively.

Similar to other tested inhibitors, thiosalicylic acid acted as a cathodic inhibitor. As shown in **Figure 14b**, an addition of thiosalicylic acid caused the cathodic curve to shift to lower current densities but did not shift the anodic curve. This indicated a resistance to the transportation of corroding species developed onto the metal surface. Such findings were also supported by

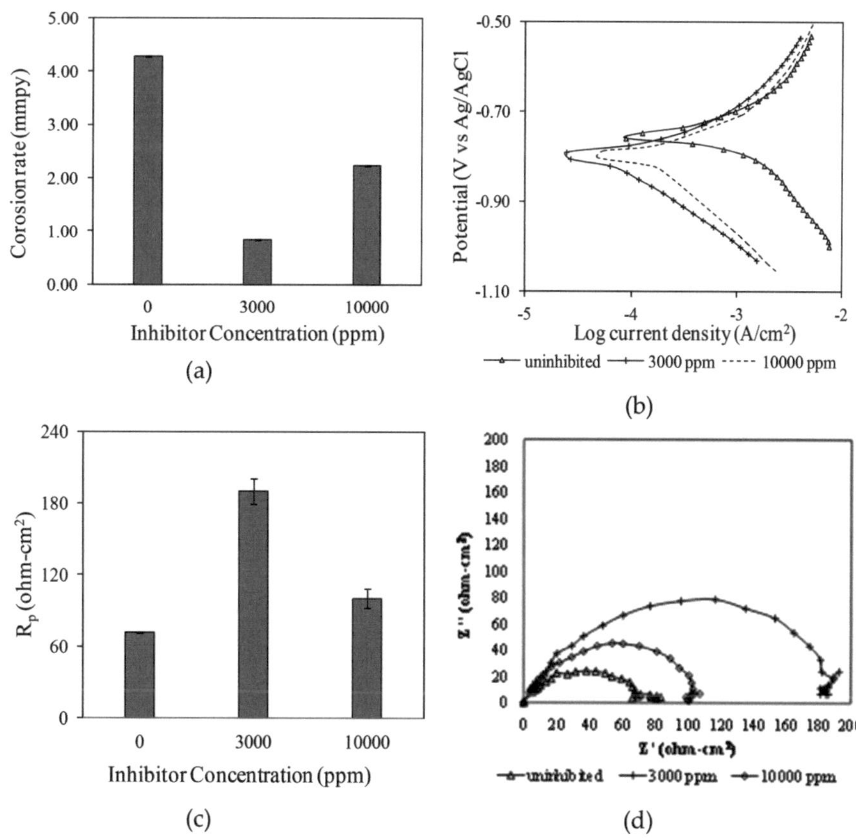

Figure 14. The corrosion inhibition performance of thiosalicylic acid in 5.0 kmol/m^3 MEA solutions containing 0.55 mol/mol CO_2 loading in the absence of process contaminants at 80°C. (a) Corrosion rate, (b) polarization curve, (c) polarization resistance, and (d) impedance behavior.

the semicircle characteristic of a capacitive loop due to charge transfer kinetics at the interface in **Figure 14d**.

It should be noted that, in the course of the experiment, the test solutions containing thiosalicylic acid turned black at both 3000 and 10,000 ppm. This might be due to the incompatibility of the inhibitor with the MEA solution. Thus, thiosalicylic acid was not tested further at other concentrations and not tested for the effect of process contaminant.

4.8. Surface analysis

For the uninhibited MEA solution, a uniform corrosion product was formed over the metal surface as seen from the scanning electron microscopy (SEM) images in **Figure 15a**. A significant increase in the amounts of oxygen and carbon in the tested specimens was found in the energy-dispersive X-ray spectroscopy (EDS) analysis (**Figure 15b**) compared to the uncorroded specimens (**Figure 15d**). The peaks obtained from X-ray diffraction (XRD) analysis

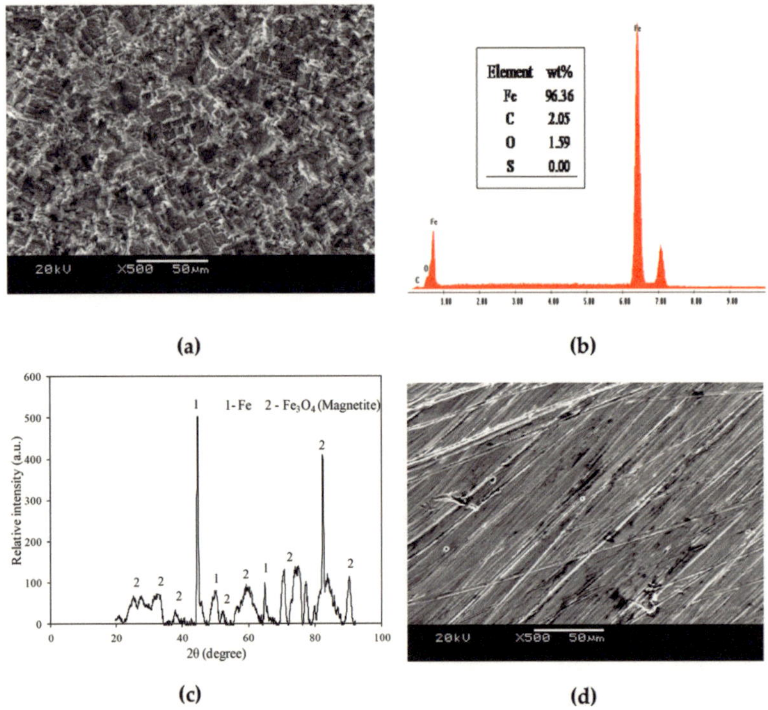

Figure 15. Surface analysis of CS 1018 in uninhibited 5.0 kmol/m^3 MEA solutions at 80°C and 0.55 mol/mol CO_2 loading. (a) An SEM image of corroded specimen (500× magnification) after 28 days, (b) EDS spectra of corroded specimen after 28 days, (c) XRD spectra of corroded specimen after 28 days, and (d) an SEM image of uncorroded specimen (before test).

(**Figure 15c**) revealed poorly crystalline corrosion products identified as magnetite (Fe_3O_4). In addition to Fe_3O_4, peaks corresponding to iron (Fe) were present suggesting a poor protective layer on the surface. The increase in carbon amount suggests that the metal surface may contain amorphous iron carbonate ($FeCO_3$).

For the MEA solution containing 4-aminobenzene sulfonic acid, the corrosion product was uniform over the metal surface similar to that found on the uninhibited surface (**Figure 16a**). Larger quantities of oxygen and carbon than in the uninhibited condition were observed from EDS spectra, indicating a significant amount of poorly protective corrosion product that was formed. This was confirmed by the XRD results, where peaks corresponding to Fe were characterized in addition to the primary corrosion product Fe_3O_4.

The presence of sulfapyridine and sulfolane led to tenacious surface layers on the metal surface. Fe_3O_4 was the primary corrosion product. The sulfapyridine induced a nonuniformly distributed corrosion product (**Figure 16b**) with large quantities of both oxygen and carbon observed from EDS and XRD spectra. The sulfolane led to a uniform intact surface layer (**Figure 16c**) with the least amounts of carbon and oxygen (**Figure 16d**) compared to the rest of inhibitors.

4.9. Performance comparison of tested corrosion inhibitors

Table 5 summarizes the performance of tested organic corrosion inhibitors. Five inhibitors, that is, 2-, 3-, and 4-aminobenzene sulfonic acids, sulfapyridine, and sulfolane showed promise for corrosion control in the MEA-based CO_2 absorption process. They could achieve up to 85–92%

http://dx.doi.org/10.5772/intechopen.72752

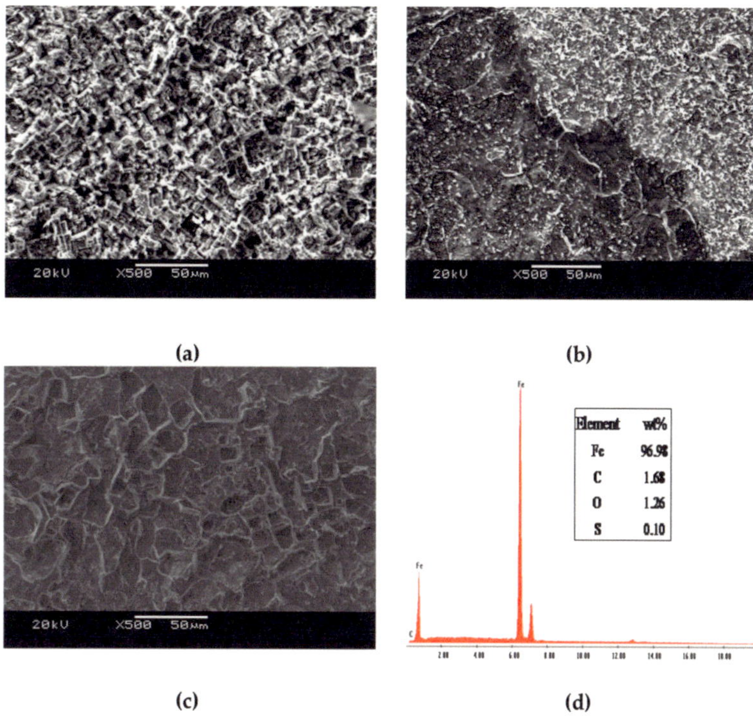

Figure 16. Surface analysis of tested specimen in inhibited solutions after 28 days (5.0 kmol/m^3 MEA, 80°C, and 0.55 mol/mol CO_2 loading): (a) SEM images (500× magnification) of 1000 ppm 4-aminobenzene sulfonic acid, (b) SEM images (500× magnification) of 2000 ppm sulfapyridine, (c) SEM images (500× magnification) of 2000 ppm sulfolane, and (d) EDS spectra of 2000 ppm sulfolane.

inhibition efficiency when applied at effective concentrations. The effective concentrations of 2-, 3-, and 4-aminobenzene sulfonic acid were in the range of 250–3000 ppm, while those of sulfapyridine and sulfolane were in higher ranges, that is, 3000–10,000 ppm and 2000–10,000 ppm, respectively. Reducing or increasing concentrations of these inhibitors from the effective concentration ranges may lessen their effectiveness.

Corrosion inhibitor	Effective concentration (ppm)	Inhibition efficiency (%)	Effective in chloride environment	Effective in formate environment
2-aminobenzene sulfonic acid	250–3000	87–89	Yes	No
3-aminobenzene sulfonic acid	250–3000	89	No (pitting)	No
4-aminobenzene sulfonic acid	250–3000	87–91	Yes	No
Sulfapyridine	3000–10,000	90–92	No	Yes
Sulfanilamide	(Not effective)	—	—	—
Sulfolane	2000–10,000	85–90	Yes	Yes
Thiosalicylic acid (incompatibility with MEA)	3000	80	—	—

Table 5. Performance summary of the tested organic corrosion inhibitors.

The effectiveness of most tested inhibitors was reduced when the MEA solution contained chloride or formate. 2- and 4-aminobenzene sulfonic acids can maintain their effectiveness in the presence of chloride but not in the presence of formate. On the contrary, sulfapyridine could function well in the presence of formate but not in the presence of chloride. The effectiveness of 3-aminobenzene sulfonic acid was lessened by both chloride (with possibility of pitting corrosion) and formate. Sulfolane was the only tested inhibitor that was not affected by chloride and formate.

Apart from the abovementioned inhibitors, two tested inhibitors were not promising. Sulfanilamide was not effective as it could only achieve up to 42% inhibition efficiency. Although thiosalicylic acid at 3000 ppm could achieve 80% inhibition efficiency, it was not compatible with the MEA solution.

5. Conclusions

2-, 3-, and 4-aminobenzene sulfonic acids, sulfapyridine, and sulfolane are effective corrosion inhibitors with inhibition efficiencies of up to 85–92%. Sulfolane is the most promising inhibitor compared to the other six tested compounds since it can yield up to 92% inhibition efficiency and maintain its effectiveness in the presence of both chloride and formate. 2- and 4-aminobenzene sulfonic acids do not work well in the presence of formate while sulfapyridine is not effective in the presence of chloride. 3-aminobenzene sulfonic acid can achieve up to 89%, but its effectiveness is deteriorated by both chloride and formate.

Acknowledgements

The authors would like to thank the Natural Sciences and Engineering Research Council of Canada (NSERC) for financial support.

Author details

Sureshkumar Srinivasan, Amornvadee Veawab* and Adisorn Aroonwilas

*Address all correspondence to: veawab@uregina.ca

Energy Technology Laboratory, Faculty of Engineering and Applied Science, University of Regina, Saskatchewan, Canada

References

[1] Kohl AL, Nielsen RB. Gas Purification. 5th ed. Houston: Gulf Publishing Co.; 1997

[2] Mago BF, West CW. Antimony-vanadium corrosion inhibitors for alkanolamine gas treating system. US Patent 3808140; 1974

[3] Nieh ECY. Vanadium-amine corrosion inhibitor system for sour gas conditioning solutions. US Patent 4372873; 1983

[4] Pearce RL. Process for removal of carbon dioxide from industrial gases. US Patent 4440731; 1984

[5] Dupart MS, Oakes BD, Cringle DC. Method and composition for reducing corrosion in the removal of acidic gases from gaseous mixtures. US Patent 4446119; 1984

[6] Jones LW, Alkire JD. Corrosion inhibitors for amine gas sweetening systems. US Patent 4541946; 1985

[7] Hensen ER, Tipton TM, Courtwright JG. Corrosion inhibitors for alkanolamines. US Patent 4595723; 1986

[8] Srinivasan S. Environmentally friendly corrosion inhibitors for amine-based CO_2 absorption process [M.Sc. thesis]. Saskatchewan: University of Regina; 2012

[9] Hamah-Ali B, Ali BS, Yusoff R, Aroua MK. Corrosion of carbon steel in aqueous carbonated solution of MEA/[bmim] [DCA]. International Journal of Electrochemical Science. 2011;**6**:181-198

[10] Sastri VS. Corrosion Inhibitors: Principles and Applications. West Sussex: John Wiley & Sons Ltd.; 2001

[11] Hackerman N, Makrides AC. Action of polar organic inhibitors in acid dissolution of metals. Industrial and Engineering Chemistry. 1954;**46**(3):523-527

[12] Khaled KF, Hackerman N. Investigation of the inhibitive effect of *ortho*-substituted anilines on corrosion of iron in 1 M HCl solutions. Electrochimica Acta. 2003;**48**(19):2715-2723

[13] PubChem. 2011. Available from: http://pubchem.ncbi.nlm.nih.gov/ [Accessed: January 22, 2011]

[14] Chakrabarti A. Quantum chemical study of the corrosion inhibition of mild steel in 6 percent (wt/wt) HCl by means of cyanoguanidine derivatives. British Corrosion Journal. 1984;**19**:124-126

[15] Sastri VS, Perumareddi JR. Molecular orbital theoretical studies of some organic corrosion inhibitors. Corrosion. 1997;**53**(8):617-622

[16] Khalil N. Quantum chemical approach of corrosion inhibition. Electrochimica Acta. 2003; **48**(18):2635-2640

[17] Dupart MS, Bacon TR, Edwards DJ. Understanding corrosion in alkanolamine gas treating plants, part 1 & 2. Hydrocarbon Processing. 1993;**93**:75-80

[18] ASTM Standard G5-94 (Reapproved 2004). Standard reference test method for making potentiostatic and potentiodynamic anodic polarization measurements. In: Annual Book of ASTM Standards. West Conshohocken: American Society of Testing and Materials

[19] ASTM Standard G106-89 (Reapproved 2010). Standard practice for verification of algorithm and equipment for electrochemical impedance measurements. In: Annual Book of ASTM Standards. West Conshohocken: American Society of Testing and Materials

[20] Vračar LM, Dražić DM. Adsorption and corrosion inhibitive properties of some organic molecules on iron electrode in sulfuric acid. Corrosion Science. 2002;**44**(8):1669-1680